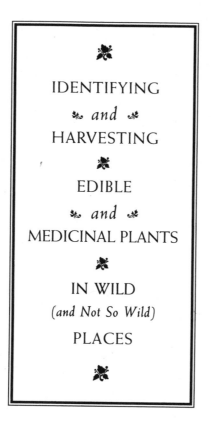

IDENTIFYING

and

HARVESTING

EDIBLE

and

MEDICINAL PLANTS

IN WILD

(and Not So Wild)

PLACES

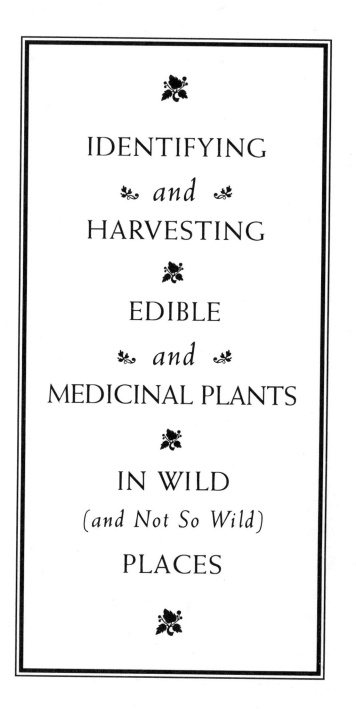

IDENTIFYING

❧ *and* ❧

HARVESTING

EDIBLE

❧ *and* ❧

MEDICINAL PLANTS

IN WILD

(and Not So Wild)

PLACES

"Wildman" STEVE BRILL

WITH

EVELYN DEAN

HEARST BOOKS   *New York*

Wild poisonous plants sometimes resemble edible plants, and they often grow side by side. It is the responsibility of the reader to identify and use correctly the edible plants described in this book.

The information about the medicinal uses of plants is for educational purposes only. It is not the intention of the authors for the readers to use these plants as a substitute for consultation with a licensed physician for treating illness. The authors and publisher assume no responsibility for problems arising from the reader's misidentification or use of wild plants.

Text copyright © 1994 by Steve Brill
Text and Illustrations copyright © 1994 by Evelyn Dean

It is the policy of William Morrow and Company, Inc., and its imprints and affiliates, recognizing the importance of preserving what has been written, to print the books we publish on acid-free paper, and we exert our best efforts to that end.

Library of Congress Cataloging-in-Publication Data

Brill, Steve.
    Identifying and harvesting edible and medicinal plants in wild
(and not so wild) places / "Wildman" Steve Brill with Evelyn Dean.—
1st ed.
      p.   cm.
    Includes bibliographical references and index.
    ISBN 0-688-11425-3
    1. Wild plants, Edible—United States—Identification.
2. Medicinal plants—United States—Identification. 3. Cookery
(Wild foods)   I. Dean, Evelyn. II. Title.
QK98.5.U6B75 1994
581.6'3'0973—dc20
                              93-31796
                              CIP

Printed in the United States of America

First Edition

1 2 3 4 5 6 7 8 9 10

581.63

BOOK DESIGN BY PATRICE FODERO

This book is dedicated to the memory of David Drazen, my nephew, who passed away at the age of seventeen in 1989 after a heroic five-year struggle with leukemia. Dave actively participated in his treatment since the age of twelve. Instead of complaining, he made the most out of every minute of his life, despite years of suffering, even as his chances of survival dwindled to nothing.

If he could make so much out of so little, we can certainly work to achieve our potential as individuals, act as positive influences in the lives we touch, and contribute in some way to preserving or restoring the integrity of the environment, no matter what lesser problems we must deal with.

It was an honor for me to be part of Dave's life, and he will always be an inspiration to me.

ACKNOWLEDGMENTS

"WILDMAN" STEVE BRILL'S ACKNOWLEDGMENTS

Many people helped bring this book into existence, directly or indirectly: My mother, Helga Brill, got me interested in foraging by showing me how to pick wild berries when I was a child. My friend Mindy Sperling encouraged me to explore natural areas as an adult, rekindling my interest in foraging. Euell Gibbons's wonderful foraging books provided further inspiration. Evelyn Brooks gave me my first public-speaking opportunity and instruction when she had me teach cooking at her Yoga studio. Lewis Harrison gave me my first job as a foraging teacher and initiated my interests in herbs and publicity. The late "Dr." John Moore provided a first hand look at folk medicine. Rosemarie Liuzzi gave me lots of practice with home remedies, by coming down with all sorts of ailments. Former New York City parks commissioner Henry Stern made me famous by having me arrested for eating a dandelion in Central Park, then dropping the charges and hiring me as a Parks Department naturalist. "Botany" Bill Greiner gen-erously shared a lifetime of botanical and scientific wisdom with me for over a decade, and he edited the botanical content of this book. Sharon Vollmuth helped with botanical research. Carol Lebrecht encouraged me to write the book and assisted me with endless patience. Linda Runyon sent specimens from afar, while Sol Goldberg chased down and brought in many local specimens. Dena Falconi, Lorna and Eric Saltzman, and Rose Starr contributed recipes. Bill Whitley contributed a great story by almost making and drinking a poisonous wild wine. J. D. Martignon and Claire Baksa lent me my first computer. Computer consultant George Struk contributed superhuman patience by volunteering his services and teaching me how to use Macintosh computers, and Paul Martin also added his computer expertise.

EVELYN DEAN'S ACKNOWLEDGMENTS

I'd like to thank my father, Lionel Dean, a talented writer, for his help and encouragement, for patiently lending his ear to our endless ideas, and for his feedback. My mother, Marjorie Dean, always encouraged my creativity, and provided limitless positive energy. My sister, Andrea Dean, was a willing subject for all my games and art projects since childhood. Caroline Lewit was a guiding light. My friend Gabriele Woisin lent enthusiastic support and was a source of original cooking experiments. My friend, Sharon Ross, a talented artist, was generous with her good advice.

CONTENTS

❧

Chapter 1: An Introduction to Foraging
for Wild Plants   1
Preparing to Forage: Conservation and   2
    Safety
Equipment   4
Wild Foods, Nutrition, and Herbal   5
    Medicine
    Herbal Actions   7
    Herbal Preparations   8
Identifying Wild Plants   10
    Wild Plants and Natural History   10
    Leaves   11
    Flowers   13
    Fruit   15
    Roots   16
    Thorns and Natural History   17
    Shoots   17
Life Cycles   18
Plant Types: Herbaceous Plants, Trees,   18
    Shrubs, and Vines
Native and Exotic Species   18
Natural Habitats   19
    Lawns, Meadows, and Humans   19
    Overgrown Fields   20
    Thickets   20
    Woodlands   20
    Disturbed Areas   21
    Deserts   21

Mountains   21
The Seashore   21
Other Wetlands   22
Seasonal Cycles   22

Chapter 2: Edible Wild Plants of Mid-   23
Spring
Plants of Lawns and Meadows in Mid-   24
    to Late Spring
    Clovers • Asiatic Dayflower •   24
        Lady's Thumb • Purslane •
        Wood Sorrel
    Other Plants of Lawns and   30
        Meadows in Mid- to Late Spring
    Plants of Cultivated Areas in Mid-   30
        to Late Spring
    Plants of Disturbed Areas in Mid- to   31
        Late Spring
    Bracken • Burdock • Epazote •   31
        Pineapple Weed • Poison Ivy
        and Its Relatives • Pokeweed
    Other Plants of Disturbed Areas   41
        in Mid- to Late Spring
    Plants of Fields in Mid- to Late Spring   42
    Black Locust • Cow Parsnip •   42
        Lamb's-Quarters • Common
        Milkweed • Mints

Other Plants of Fields in Mid- to   56
Late Spring
Plants of Thickets in Mid- to Late   56
Spring
    Greenbrier • Hercules'-Club •   56
    Wisteria
Other Plants of Thickets in Mid- to   60
Late Spring
Plants of Woodlands in Mid- to Late   60
Spring
    Honewort • Kentucky Coffee Tree   60
    • Linden • Ostrich Fern •
    Redbud
Other Plants of Woodlands in Mid-   66
to Late Spring
Plants of Freshwater Wetlands in Mid-   66
to Late Spring
    Angelica • Cattail • Coltsfoot •   66
    Jewelweed • Marshmallow and
    Relatives
Other Plants of Freshwater Wetlands   77
in Mid- to Late Spring
Plants of the Seashore in Mid- to Late   78
Spring
    Asparagus • Bayberries •   78
    Glasswort • Seaweeds •
    Rockweed • Sea Lettuce • Irish
    Moss • Dulse • Laver • Edible
    Kelp • Horsetail Kelp
Other Plants of the Seashore in Mid-   87
to Late Spring
Plants of Mountains in Mid- to Late   87
Spring

Chapter 3: Edible Wild Plants of   89
Summer
Plants of Lawns and Meadows in   89
Summer
Plants of Cultivated Areas in   89
Summer
Plants of Disturbed Areas in   90
Summer
Plants of Fields in Summer   90
    Wild Bean • Ground Cherry •   90
    Common Strawberry •
    Common Sunflower • Yarrow
Other Plants of Fields in Summer   96
Plants of Thickets in Summer   96
    Blackberries • Blueberries •   96
    Cornelian Cherry • Currants
    and Gooseberries • Common
    Elderberry • Passionflower •
    Peach and Nectarine • Pear •

Wild Plums • Raspberries •
Sumac
Other Plants of Thickets in Summer   119
Plants of Forests in Summer   119
    Cherries • Juneberry • Mayapple •   119
    Mulberry
Other Plants of Forests in Summer   128
Plants of Freshwater Wetlands in   128
Summer
    Water Hemlock • Wild Rice   128
Other Plants of Freshwater Wetlands   130
in Summer
Plants of the Seashore in Summer   131
    Beach Pea   131
Other Plants of the Seashore in   132
Summer
Plants of the Mountains in Summer   132
Plants of Deserts in Summer   132
    Glandular Mesquite, Mesquite,   132
    Honey Mesquite, and
    Screwbean • Yucca
Other Plants of Deserts in Summer   136

Chapter 4: Edible Wild Plants of   137
Autumn
Plants of Lawns and Meadows in   137
Autumn
    Chickweed   137
Other Plants of Lawns and   139
Meadows in Autumn
Plants of Cultivated Areas in Autumn   139
    Gingko • Kousa Dogwood • Yew   139
Other Plants of Cultivated Areas in   145
Autumn
Plants of Disturbed Areas in   145
Autumn
Plants of Fields in Autumn   145
    Amaranth • Foxtail Grass •   145
    Hawthorns
Other Plants of Fields in Autumn   150
Plants of Thickets in Autumn   151
    Apple • Autumn Olive • Hazelnuts   151
    • Roses
Other Plants of Thickets in Autumn   160
Plants of Woodlands in Autumn   160
    Black Walnut • Butternut • Grapes   160
    • Chestnut • Hackberry •
    Hickories • Oak • Pawpaw •
    Persimmon • Common
    Spicebush • Viburnums
Other Plants of Woodlands in   185
Autumn
Plants of Freshwater Wetlands in   185

Autumn
Pickerelweed　185
Other Plants of Freshwater Wetlands　186
in Autumn
Plants of the Seashore in Autumn　186
Plants of Mountains in Autumn　186
Plants of Deserts in Autumn　187
Prickly Pear　102　187
Other Plants of Deserts in Autumn　188

Chapter 5: Edible Wild Plants of Late　189
Fall Through Early Spring
Plants of Lawns and Meadows in Late　190
Fall Through Early Spring
Common Dandelion　190　190
Other Plants of Lawns and　192
Meadows in Late Fall Through
Early Spring
Plants of Cultivated Areas in Late　192
Fall Through Early Spring
Plants of Disturbed Areas in Late Fall　193
Through Early Spring
Common Evening Primrose •　193
Jerusalem Artichoke • Common
Parsnip • Goatsbeard • Sheep
213　Sorrel • Sow Thistle • Wild
Potato Vine　204
Other Plants of Disturbed Areas in　203
Late Fall Through Early Spring
Plants of Fields in Late Fall Through　203
Early Spring
Wild Carrot • Wild Onions　69　203
Other Plants of Fields in Late Fall　209
Through Early Spring
Plants of Thickets in Late Fall　210
Through Early Spring
Groundnut • Juniper　120　210
Other Plants of Thickets in Late Fall　212
Through Early Spring
Plants of Woodlands in Late Fall　212
Through Early Spring
Aniseroot • Black Birch • Hog　212
Peanut • Pine • Sassafras • Wild　76,130
Ginger • Wintergreen
Other Plants of Woodlands in Late　224
Fall Through Early Spring
Plants of Freshwater Wetlands in Late　224
Fall Through Early Spring
Cranberries　224
Other Plants of Freshwater Wetlands　225
in Late Fall Through Early Spring
Plants of the Seashore in Late Fall　226
Through Early Spring

Plants of Mountains in Late Fall　226
Through Early Spring
Plants of Deserts in Late Fall　226
Through Early Spring

Chapter 6: Edible Wild Plants of Early　227
Spring
Plants of Lawns and Meadows in Early　227
Spring
206　Plantain • Common Blue Violet　93　227
Other Plants of Lawns and　231
Meadows in Early Spring
Plants of Cultivated Areas in Early　232
Spring
Daylily　241　232
Other Plants of Cultivated Areas in　234
Early Spring
Plants of Disturbed Areas in Early　234
Spring　108
190　Chicory • Curly Dock • Japanese　234
Knotweed • Mugwort • Nettles　160
194 • Storksbill • Wild Lettuce
Other Plants of Disturbed Areas in　247
Early Spring
Plants of Fields in Early Spring　247
Mullein • Mustards　247
Other Plants of Fields in Early　259
Spring
Plants of Thickets in Early Spring　259
Plants of Woodlands in Early Spring　259
Goutweed • Solomon's Seal • Wild　259
Leek
Other Plants of Woodlands in Early　266
Spring
Plants of Freshwater Wetlands in Early　266
Spring
Waterleaf　266
Other Plants of Freshwater Wetlands　267
in Early Spring
Plants of the Seashore in Early　267
Spring
Plants of Mountains in Early Spring　268
59　Miner's Lettuce • Orpine ? 61　268
Other Plants of Mountains in Early　270
Spring
Plants of Deserts in Early Spring　270

Chapter 7: Cooking with Edible Wild　271
Plants
Wild Foods Are Different　271
Wild Greens　271
Edible Flowers　272
Wild Fruits and Berries　272

Wild Roots 272
Nuts and Seeds 273
All-Natural Cuisine 273
Special Equipment 274
Cleaning Wild Foods 274
Preserving Wild Foods 275
Freezing Wild Foods 275
Drying Wild Foods 275
Unrefined and Wild Flours 276

Chapter 8: Wild Food Recipes 277
Appetizers and Side Dishes 277
Black Locust Flower Oatmeal 277
Wild Carrot Seed Millet 277
Dandelion Sauté 277
Dena and Gabi's Cattail Flower 278
  Pickles
Herbed Jerusalem Artichokes 278
Lamb's-Quarters Stuffing 278
Lentil Delight 278
Nettle and Carrot Casserole 279
Nutty Amaranth 279
Park Nuts 279
Pokeweed and Milkweed: Basic 279
  Preparation
Rockweed Crisps 280
Salsify Roots in Nut Sauce 280
Stuffed Wild Grape Leaves 280
Waterless-Steamed Greens 281
Salads 281
Autumn Green Salad 281
Early Spring Seasonal Salad 281
Late Fall Green Salad 281
Mid-Spring Wild Blossom Salad 282
Milkweed Flower Salad 282
Sea Lettuce Salad 282
Seashore Salad 282
Summer Green Salad 282
Winter Green Salad 282
Soups 283
Early Spring Chick-pea and Violet 283
  Soup
Scandinavian Fruit Soup 283
Breads 284
Blueberry Bread 284
Clover Corn Bread 284
Corn and Daylily Flatbread 284
Plantain Seed–Quinoa Pilaf 284
Dulse Scones 285
Irish Soda Elderberry Bread 285
Pawpaw Bread 285

Entrées 286
Cattail Shoots and Carrots in Peanut 286
  Sauce
Coconut–Curly Dock Curry 286
Common Evening Primrose Leaf 286
  Burgers
Common Evening Primrose Root 286
  Chili
Curly Dock–Cheese Rolls in 287
  Tomato Sauce
Curried Tofu and Chickweed 287
Potato-Purslane Patties 288
Savory Burdock Patties 288
Sunflower Patties 288
Jams, Dressings, Spreads, and Sauces 288
Autumn Olive Jam 288
Black Cherry Jam 289
Creamy Chickweed Dressing 289
Curry Oil 289
Dracula's Delight—Prickly Pear 289
  Sauce
Garlic Mustard Horseradish 289
Garlic Mustard Tofu Cream Cheese 290
Poor Man's Mustard 290
Sassafras Root Jelly 290
Sumac Hollandaise Sauce 290
Tofu-Glasswort Spread 290
Wild Apple Chutney 291
Wild Onion Sauce 291
Desserts 291
Autumn Olive–Banana Tofu 291
  Creme Pie
Blackberry Peach Crunch 292
Carob-Chip Black Walnut Cookies 292
Carob-Pecan Fudge 292
Creamy Wild Raspberry Pudding 293
Crepes with Wild Berries 293
Japanese Knotweed Surprise 293
Marillen Knödel 293
Mulberry Crumble 294
Mulberry Pudding 294
American Persimmon Pudding 294
Tofu-Plum Pie 294
Wild Apple Cake 295
Wild Strawberry Parfait 295

Bibliography and References for Further 297
  Reading

Index 299

CHAPTER 1

AN INTRODUCTION TO FORAGING

FOR WILD PLANTS

There are hundreds of fascinating, delicious wild vegetables, fruits, nuts and seeds, and herbs growing in our neighborhoods, backyards, parks, and forests that we overlook and disregard. Many are easy-to-recognize renewable resources you can easily collect and enjoy, with no harm to the environment. Many are the same prolific "weeds" we unsuccessfully try to destroy. When you know what they are and begin to use them, you'll discover that they are tastier, as well as more nutritious, than anything you can buy, and they're completely free.

This book will introduce you to the world of foraging and nature. We cover the most common useful wild plants of the continental United States (with the exception of subtropical Florida, which has a different, specialized flora) and southern Canada, with the basics of what you need to know to identify, collect, use, and appreciate them. My years as a naturalist draw on pertinent science, nutrition, folklore, and personal experiences to put the plants in context.

Since our focus is on the best, as well as the most practical, common, and widespread wild edible plants, many species you may find in other field guides are omitted. Wildflower, tree, and mush-

room field guides in the Bibliography will help you identify a wide range of nonedible and marginally useful species. This book covers the choice edible species in full detail.

Many marginally edible plants just aren't worth your effort: I've waded through frigid, muddy ponds to collect the tender young unfolding leaves of water lilies (*Nymphaea* species). After washing off the mud, cooking them, and serving them for dinner, I discovered that this "vegetable" tastes just like the mud it grows in. Later on, I found that some of my field-walk participants, inspired by other sources, had similar experiences.

I once dried the leaves of skunk cabbage (*Symplocarpus foetidus*) in my food dehydrator for a week, again following recommendations elsewhere, then added them to a pot of chili. Chili should be hot, but this brew was something else. After I tasted one spoonful, the calcium oxalate crystals that should have been dissipated by drying began piercing my tongue, like thousands of microscopic needles. After half an hour of drinking and spitting out water, jumping up and down, and cursing the author, I flushed the entire concoction down the toilet. Subsequently, a friend rendered this plant harmless by drying it for six months, but then it tasted like paper. This plant, and edibles that require hours of preparation for a few bites, are omitted.

I've been using many plants for food and home remedies since the late 1970s, and I'd like to share my favorites with you. However, it's beyond the scope of this book to cover the thousands of purely medicinal, nonedible plants that also grow in our midst. We concentrate only on the delicious wild foods you can sink your teeth into. Also, we don't cover mushrooms, because you'd need a whole book to do them justice. Fortunately, there are many wonderful mushroom books on the market.

Most works on edible wild plants include recipes in which healthful wild plants are adulterated with refined and processed foods containing artificial chemicals. It's easy to make fantastic-tasting, nutritious meals without any of these. A large portion of this book is devoted to recipes and the ideas behind them, so you'll learn to create all-natural wild meals yourself.

This book includes many detailed pencil drawings that clarify the plants' identifying characteristics, and all their parts, as they appear throughout the seasons. Beginners seek out color photos, but these have disadvantages: Because leaves are green and flower colors can be described, color mainly raises a book's price. This limits you to one view per plant, usually when it's in flower, even when different stages are edible. Photos also contain extraneous details absent in good illustrations. You'll learn more from the greater number of excellent pencil illustrations we provide.

PREPARING TO FORAGE: CONSERVATION AND SAFETY

Poisonous plants, some of which are deadly, sometimes grow alongside edible plants, so you must identify every plant with 100 percent certainty before you eat it. There is no other foolproof method to determine whether something is edible. Look up *all* of a plant's identifying characteristics, and make sure they match *all* your observations. Then, check the accompanying descriptions of any possible look-alikes (especially toxic species), to make sure their key identifying characteristics *don't* match.

Cross-check any reference books you use, especially regarding medicinal plants. Other sources may simply repeat folklore from earlier books, without regard for accuracy or safety. When you look up a plant in more than one book, make sure it has the same scientific name, or you may poison yourself with sloppy semantics. Edible and poisonous plants sometimes have the same common name, and names vary from region to region and country to country. Scientific names are universal, and longer-lasting.

Start by learning a few easy-to-identify plants well. There are some people who've accompanied me on my well-attended nature tours in and around New York City who want to learn the whole country's flora in one afternoon. I am forced to announce that learning about too many plants at once causes not only confusion but permanent brain damage. The malady is called Dementia Botanica, and its first symptom is total destruction of good judgment: Victims laugh at my jokes.

Don't work with plants that have poisonous parts, or become poisonous when they mature, until you have years of experience with safer plants. There are plenty of completely safe species to start off with. Wild foods and instant gratification don't always mix. You must sometimes follow a plant through an entire year before you know it well enough to eat. Many plants are the easiest to identify in seasons when they're not edible. If you return to the same location throughout the year, you'll come to recognize them at their edible stages.

Learn the common poisonous plants in your region, especially those that resemble edible ones. It will make foraging safer, and it will provide a last resort if the in-laws get out of hand.

Get permission to forage on private property. In New York City, I'd occasionally snatch a few handfuls of chickweed from someone's lawn with no problem. But when I was in New Orleans and was tempted to examine a plant in an overgrown area on posted property, my local friend told me never, ever to do that there. In some regions, people literally do shoot first and ask questions later.

Find out if your foraging grounds have been sprayed before you pick. Many parks prohibit spraying, but some of the best plants grow in the partially shaded, disturbed habitats alongside railroad tracks. However, railroads often spray their right-of-way with very dangerous herbicides.

Wash your plants thoroughly under running water before eating them. Any undesirable natural deposits on their surfaces, such as traces of animal droppings, will wash away. (Nobody in their right mind would pick anything with visible animal droppings.)

Never collect rare or legally protected plants. There aren't usually enough of them to be worth eating anyway, and we want to encourage environmental recovery. Don't even pick common plants where they're rare. You'll collect them more rapidly where they're common, and they'll probably be of superior quality where they're thriving. Collect only the parts of the plants you're going to use. Don't uproot plants when you're going to use only the leaves. This is a common error small children make. You can gently explain the correct procedure.

Use common sense to assure that you don't harm the plants you're collecting. If you trample down bramble bushes to get every last berry, you're destroying future crops. Leave more than enough mature plants to reproduce and ensure that you have a future supply. Take only what you're going to use, and collect no more than 10 percent of any plant—less if you're sharing the foraging ground with other people. If you're collecting roots, be especially careful to collect only a tiny fraction of the most common root vegetables, since you're destroying them.

Don't collect near heavy traffic. Lead, the worst pollutant in exhaust fumes, usually settles within fifty feet of the road. Leave a much wider margin just in case. Most gasoline today is unleaded, but the soil may be contaminated from earlier times.

Leaves, roots, and stems contain the most lead. Fruits, berries, and nuts accumulate the least. Some plants, like wild onions, have a greater affinity for accumulating heavy metals than others. Environmental geologists are just beginning to experiment with such plants to remove contamination from soil.

Faster-growing species tend to pick up less heavy metal than more slowly growing ones. As a jazz aficionado, I have no affinity for heavy metal. But if you find a tempting plant growing in a contaminated area, chances are that with some persistence, you'll also eventually find it in a cleaner location. Most wild foods are called "weeds" because they're so widespread and prolific. You just may not be spotting them yet.

Collect the right part of the plant in the appropriate season. Be careful not to collect parts of poisonous plants along with your edibles. A foreign family who knew barely any English attended many of my New York City park tours. Once I pointed out a nonpoisonous, flavorless mushroom, then moved on to lady's thumb. The family insisted on picking the mushrooms. I don't know whether they were being stubborn, or they didn't understand me. Everyone else was standing in the hot sun, waiting for me to discuss lady's thumb. I blew my whistle, signaled with my arms, and even turned red in the face, but they wouldn't come over to where the rest of the group was. I finally gave up and went over the new species, cautioning everyone about the poisonous dogbane growing alongside. Suddenly, a few minutes later, while everyone was happily collecting lady's thumb, I noticed half a dogbane plant, oozing toxic white sap. Where was the other half? I inspected the family's bag first (they had finally joined us), and sure enough, there it was.

Never collect water plants such as watercress without having the water tested, especially if you're going to eat it raw. There are dangerous microorganisms that can infect you, as well as other forms of pollution that can contaminate the water. Your local Environmental Protection Agency branch will test your water for pathogenic microorganisms free of charge.

It's much easier to avoid putting debris into your bag than to sort out trash later on. Put as little nonedible material in your bags as possible. This saves immeasurable time at home.

Put each species in a separate bag. Enclose an

index card or label in each bag, with the name of the plant and any notes, especially if you're attending a field walk where the instructor identifies many plants. Seal wild vegetables in plastic bags to keep them from wilting.

If you do bring poisonous plants home for study, be sure they're clearly labeled. Take further precautions if necessary. Other family members may assume anything in the refrigerator is fair game for snacking, and small children may eat anything they can get into their mouths.

Wash all edibles in luke-cool running water just before you use them, but don't store them wet, or they'll spoil more quickly (microorganisms thrive in damp settings). If you think a wild plant or a wild-food dish has gone bad, throw it out. Don't risk food poisoning.

EQUIPMENT

It makes a big difference if you're properly equipped when you depart for your local wild areas. You'll need sealable plastic for your vegetables, and plastic containers for delicate berries. Be sure to bring plenty of water, especially if it's hot, and don't forget your lunch or a snack.

In very warm weather, bring snacks that don't spoil quickly, like nuts or fruit. When it's hot, it helps to sip ice water. Fill a water container three-quarters full the night before, and freeze it. Wrap it in aluminum foil and put it in a heavy sock, then fill to the top just before you leave. You'll have cold water for hours, and this funny-looking device is lighter and less bulky than a thermos bottle. A broad-brimmed hat is very helpful in the heat. Sun shining on your head and face makes you much hotter.

In cold weather, bring a wool cap, even if you think you won't need it. Thirty percent of lost body heat escapes from your uncovered head. Before you go out, always listen to the weather forecast, but don't believe a word of it. Dress in layers and bring an extra sweater, just in case.

It's best to wear long pants and long sleeves unless it's unbearably hot, for greater protection against thorns, poison ivy, and insects. In mosquito season, I spray my clothing with insect repellent containing the chemical D.E.E.T. These insects have had a special affinity for me ever since I was a child. They loved me when I ate junk food, and they adored me when I became health-conscious. They appreciated the extra nutrition when I took vitamin B_1 to repel them, and they feasted on herbal-flavored "Wildman" when I used garlic, citronella, pennyroyal, or other herbal repellents. The latest device I tried was an ultrasonic buzzer insect repellent. The mosquitoes buzzed back, and this increased their appetite. It's only the female mosquitoes that bite. They use your blood to nourish their eggs. If female humans loved me as much as female mosquitoes, I'd live a very short life.

Colors do affect insects. Beekeepers wear white, which bees usually avoid, and wearing white makes ticks show up on your clothing. Beige and green don't attract insects, but blue, red, and black do. When I once put my black backpack on the ground in a mosquito-infested marsh, many mosquitoes landed on the backpack because of its color alone. Never wear a fluffy wool sweater as your outer layer. An entangled bee could easily sting you.

If you live in tick country, wear light-colored clothing and a hat. If you have long hair, tie it. Tuck your pants into your socks, and avoid brushing against vegetation. When you get home, put your clothes in the laundry, shower, wash your hair, and inspect yourself in the mirror.

Ticks may be around most of the foraging season. The deer tick, which carries Lyme disease, is the size of the head of a pin. The dog tick, which rarely carries Rocky Mountain spotted fever, is much larger. Both have eight legs. They're not insects, which have six legs, but arachnids, relatives of spiders.

Ticks usually don't bite right away, so you have

time to find them. Use petroleum jelly and tweezers to remove them. If you are bitten by a deer tick, and your skin looks like a target with a bull's-eye in the center, seek medical attention. However, this rash isn't always present, and the tick is so small, people sometimes don't realize they've been bitten. Antibiotics cure Lyme disease in its early stages; herbs don't. If you wait until you develop arthritis and other symptoms, it's much more difficult to treat.

Don't let ticks scare you into becoming a couch potato. Thanks to following the precautions above, I've never been bitten, with only one close call. After a day in the brush, I attended the dullest lecture ever. The speaker carefully avoided using any public-speaking methods that spark excitement, and he said nothing interesting.

As I was sitting there scratching a tiny bump on my head, wondering why I was there, I realized I had no bump on my head. It was a dog tick attached to my scalp poised to follow the example of the speaker and begin boring. The small room was well lighted, and I felt very conspicuous trying to dislodge the tiny parasite. Oblivious to my plight, the speaker droned on: "Blah, blah, blah! Blah, blah, blah!"

At last I yanked out the parasite and flicked it away. It tumbled through the air. The speaker, paying no heed, continued: "Blah, blah, blah! Blah, blah, blah!" But the tick landed in my boot. After several embarrassing minutes of useless squirming, trying to reach in and remove it, accompanied by the speaker's drone, I excused myself, went to the men's room, took off the boot, and flushed the tick down the toilet. I never returned to the lecture, but went home.

Sneakers are the best footwear for most outdoor situations. Hiking boots tend to be heavy and weigh you down, and most expensive waterproof brands leak. Waterproof Gore-Tex rain pants and jackets, albeit expensive, are good for hot, rainy weather because they let your skin breathe. For a cheap, if unstylish waterproof poncho, cut holes for your head and arms in a heavy-duty Hefty garbage bag.

An inexpensive whistle is good for communication. One whistle lets your friends know where you are. Two whistles means come over, I've found something. Three whistles is for SOS. This system is not infallible. When an aspiring botanist became separated from my group in the woods, she kept walking toward the sound of the whistle but never quite caught up. She finally located the source, emerging on a field of soccer players.

A magnifying lens of 15 to 20 power is very useful for identifying plants. Some important features are too small to see with the naked eye. Edible milkweed and poisonous dogbane shoots look very similar, but milkweed's stem has tiny hairs, visible under magnification. Dogbane's stem is bald. Some of the most beautiful features of small flowers and other plant parts are also revealed under magnification.

People often make the mistake of getting Sherlock Holmes–type reading lenses. They're usually only 5 power—not much help. Jeweler's loupes, consisting of two back-to-back lenses, are best. The small field size is fine, since we're looking at tiny plant parts. A large stationery or camera store can provide one at half the price of one purchased from a jewelry store. Tie a string to the loupe and loop it around your neck.

WILD FOODS, NUTRITION, AND HERBAL MEDICINE

Supplementing your diet with wild foods will help you live a longer, healthier life. Most Americans are overfed and malnourished. We eat a diet high in calories and low in vitamins, minerals, and fiber, and we consume large amounts of artificial chemicals.

We use huge quantities of refined carbohydrates today—white-sugar and white-flour products. In Colonial times, white sugar was so rare that it was kept under lock and key. We consume large quantities of meat, high in saturated fat and pumped full of carcinogenic growth hormones.

Our early ancestors subsisted on irregularly available scavenged or caught lean wild game. They also collected plenty of edible wild plants.

Fiber is vital for a healthy digestive tract. Societies with high-fiber diets have much lower rates of cancer and heart disease, our two leading killers, than cultures on high-fat, high-refined carbohydrate diets. We can accurately call the white-flour baked goods we stuff into ourselves today the staff of death.

Turning to a whole-foods diet centered on complex carbohydrates, with plenty of fiber, makes a big difference. Junk-food propaganda tells you sugar gives you "quick energy." But it's dissolved and brought into the bloodstream so quickly, it overtaxes the pancreas, which has to quickly produce inordinate amounts of insulin to metabolize this unnatural influx. The excess is stored as fat, and the energy is gone. You're fatter—and soon become hungry again. As all dieters know, fat is very hard to break down.

The complex carbohydrates of whole grains and root vegetables, agricultural or wild, take hours to break down, so you get an even flow of energy, and stay full longer. This helps you stay healthier and makes it easier to keep off excess weight.

Wild foods have more vitamins, minerals, and fiber than anything else, so they combat the subclinical deficiencies, as do vitamin supplements. Wild foods also have all the enzymes and cofactors, some still undiscovered, that make the nutrients work better.

Free-radical–induced cellular damage is an important factor in aging, heart disease, and cancer. Cells derive their energy from the transfer of electrons incorporated in carbohydrate and fat fuels. But through a complex set of reactions, electrons can attach themselves to oxygen molecules and create free radicals—charged particles ready to neutralize their charges by taking or giving electrons to other molecules, and damaging them in the process. Free-radical production can cause core genetic material to undergo mutation. This causes DNA to divide and reproduce improperly—a prime cause in aging, and a step toward cancer.

The enzyme xanthine oxidase, present in homogenized milk, produces free radicals that create lesions in the arteries, leading to heart disease. Margarine and low-quality oils produce free radicals that contribute to cancer. So do artificial preservatives, artificial colors (derived from coal tar), and artificial flavors. Foods that come from the earth, not ones that are manufactured, provide antioxidants that repair free-radical damage.

When we attune ourselves to the environment, and understand that we're biological products of this environment, we create a constructive context for understanding and taking charge of our planet and our health. The body heals many illnesses itself with proper nutrition (along with proper rest, exercise, and psychological/spiritual well-being). We use wild plants for health in two overlapping ways—as superior sources of nutrients, and as home remedies for minor health problems. People are often confused over whether a plant is a food or a medicine. Often, it's both. Plants that cured deficiency diseases before we understood these ailments were thought to contain medicinal substances, whether or not they were also used as food. Today, we simply consider them to be highly nutritious foods. People treated scurvy with vitamin C–rich "medicinal" plants because they worked.

But you don't need clinical vitamin-deficiency diseases to benefit from superior nutrition. There's a big difference between the minimal vitamin levels required to stave off deficiency diseases in healthy adults (the basis of the Recommended Daily Allowance, or RDA) and optimal vitamin levels for people of different ages in varying stages of health. Chronic degenerative diseases don't begin the day the doctor identifies the first clinical sign or symptom. They develop over years. Although wild foods won't make you live forever, their extra nutrients often help forestall or prevent degenerative disease.

All foods consist of carbohydrates, fats, proteins, vitamins, minerals, fiber, and water. We break down these passive substances, absorb them, and use them for energy, as structural components, etc. Plants also make biologically active compounds for their own needs, and we exploit them as medicines. They actively affect our body chemistry, affecting specific organs or organ systems—they're more than just building blocks.

Drugs are also biologically active, but they're usually more concentrated, with more marked, immediate effects. Since toxicity varies with dosage (two tablespoons of table salt may be lethal), drugs are potentially more harmful, especially with long-term use. What drug companies call "side effects" are biologically identical to poisoning.

Using the same substances as both food and medicine seems strange in a society where medicines are normally drugs, and drugs are dangerous. In general, most edible herbs' active substances don't approach toxic levels. Still, some medicinal plants are dangerous, and everyone's needs are different, so you shouldn't start taking herbs helter-skelter.

Herbs' "active" ingredients are often concentrated to make drugs. One quarter of all drugs come from plants. However, the "inactive" ingredients seem to work together with the "active" ones in whole plants, making them safer. The entire complex of substances evolved together, so it makes sense that they interact chemically with one another. This is why I prefer herbs over drugs whenever possible.

For example, people used willow (*Salix* species) bark tea for centuries to reduce pain and inflammation. Small overdoses of the concentrated active ingredient, methyl salicylate, are toxic, so chemists transformed this extract to the less toxic aspirin, acetylsalicylic acid. But this convenient drug sometimes causes stomach ulcers, which you never get from the whole herb.

More concerned with speed and power than long-term harm, the medical-pharmaceutical establishment promotes concentrated drugs. Many doctors don't like herbs because you can't measure their dosage as precisely as you can do with drugs: If a leaf is genetically programmed with a million resin dots of active ingredient, and it grows in the shade, the leaf may get larger, to get more sunlight. It still contains the same million units as a smaller leaf growing in the sun, but at a lower concentration. A gram of shade-dwelling leaf will contain less medicine than a gram of the smaller sun-dwelling leaf. In some cases, the doctors are right: I'd rather take a measured dose of digitalis than foxglove (a very dangerous plant that digitalis comes from), because the foxglove might kill me. Unfortunately, American medicine ignores milkweed and hawthorn (pages 47 and 148), two much safer herbs that strengthen the heart. The recom-

mended herbs in this book are safe, and the dangerous ones are clearly noted.

I'd rather you use plants without substances that approach toxic levels, when a little more or less isn't a matter of life and death. This is especially important for avoiding the inevitable side effects of long-term drug use.

Conditioned by irresponsible TV commercials, Americans take drugs to suppress symptoms, and disregard the symptoms' causes until the drugs' toxicity produces more symptoms. Then they take more drugs to suppress the side effects—great business for the drug companies.

Unfortunately, some people carry this thinking over to herbs. They want to know which herb will relieve their symptoms. Ideally, we need to determine what's wrong with us, use herbs, food, and other natural therapies to help the body heal itself, and take measures to stay healthy.

Things are never black-and-white. Conventional medicine, using drugs, is very effective for acute, life-threatening disease and trauma. Also, herbs can't diagnose your problem—a major factor people sometimes disregard. Those diagnostic procedures that are noninvasive (pregnancy tests, for example) provide invaluable information with no danger.

It's safe to treat yourself with safe herbs for those minor problems most people treat with over-the-counter medications, and, unlike drugs, you can interchange herbs that do the same thing. However, don't treat yourself with safe herbs for major health problems. The herb may not do you any harm, but the illness could kill you. After you become experienced with herbs, you can experiment with different combinations.

HERBAL ACTIONS

Here are the major ways herbs affect the body:

Anodynes or *analgesics* are substances in plants that reduce pain. Some of the same plants also contain *antispasmodics*, which reduce muscle tension and cramping. *Nervines*, which calm the nerves, are also related. So are *soporifics*—substances that induce sleep.

Astringents make tissue contract. They're good for diarrhea, and as a gargle for sore throat. Some are also *hemostatic*—stopping bleeding. Externally, they're good for runny sores.

Alteratives are over-all strengtheners and detoxifiers, similar to tonics but directed toward helping people recover from illness.

Antiseptics, which you use externally, and *antibiotics,* which you use internally, interfere with microbial action.

Cardiacs are substances that strengthen the heart muscle. Don't use them without medical supervision. Even if they're safe, heart disease is dangerous without proper treatment and monitoring.

Carminatives are herbs that stimulate digestion and dispel gas. Different carminatives, or combinations, work better for different people. See which ones grow in your area, and determine what's best for you. In the nineteenth century, George Bizet wrote the opera *Carmen.* Since opera is not to everyone's taste, so the story goes, somebody likened the opera singer's voice to other, less pleasant emissions of the human body, and named all herbs that work to dispel gas carminatives, after *Carmen.* The name stuck. However, an opera lover told me the word *carminative* comes from the verb *to card,* a form of purification, as in carding wool. Let your musical preferences determine which version you accept.

Demulcents are soothing substances you take internally. They usually contain mucilage, a form of fiber that absorbs water and protects damaged or inflamed tissues.

Diaphoretics are herbs that cause perspiration—good for fevers. We used to think you could "sweat out" fevers. This is oversimplified. Fever may be a symptom of an activated immune system, as is sleep: The immune system induces normal sleep. When it's activated to fight infection, you become inordinately sleepy. Sleeping when you're sick doesn't make you better any more than sweating does. They're both effects of an active immune system. Perhaps diaphoretics work by stimulating the immune system.

Diuretics stimulate the kidneys to increase the flow of urine. Pharmaceutical diuretics, the so-called water pills, are routinely prescribed to suppress high blood pressure, even though exercise, dietary modification, garlic, and meditation may better address the underlying causes. Diuretics, even natural ones, may leach valuable minerals through the urine, causing serious imbalances, so use them sparingly; find out why you're retaining water and work on the cause.

Emollients are soothing substances you use externally, on the skin. Many contain mucilage.

Expectorants stimulate the lungs to expel mucus—very helpful for respiratory infections.

Emmenogogues promote menstruation. They often help the uterus contract, eliminate cramps, and induce menstruation to occur earlier. Harmful overdoses are still sometimes suggested to cause abortion. This is very dangerous, because you're in effect poisoning yourself. Medical abortions are much safer.

Rubefacients increase blood flow near the skin's surface. They're often used for arthritis.

Stimulants increase metabolism and circulation, often breaking up obstructions and warming the body. Strong stimulants, like caffeine, are harmful in the long run.

Tonics are herbs that strengthen the entire body. Because they're nonspecific, medical science has a difficult time accepting that they're real. This is changing because so much research on ginseng (*Panax* species), mostly overseas, shows that it works. (Ginseng isn't listed [see page 1] because it's much too rare to collect, and this wonderful medicinal plant has no food uses.)

Vulneraries encourage injured tissue to heal.

HERBAL PREPARATIONS

There are a number of ways to prepare and use herbs:

Eating an herb is the simplest way of getting its medicinal benefits. This works best for plants you'd eat anyway, like chickweed or mint. However, the medicinal parts of some plants taste terrible.

Capsules are a good way to ingest bad-tasting herbs. Dry the herb, grind it into a powder, put it into a capsule, and swallow it. Note:

Commercial capsules are made of gelatin, derived from cows' and pigs' hooves—not a good choice for animal lovers or vegetarians.

An *infusion* involves steeping the herb. Use this for relatively delicate herbs, like leaves and flowers, especially if they contain volatile substances that would be driven away by boiling. Add from one teaspoon to a small handful of fresh or dried herbs to one cup of water just off the boil. A glass or enamel container is better than a metal one, which could interact chemically with the herb. Cover the container and let the herb stand, away from the heat, 15 to 20 minutes. (Some people let infusions stand for hours.) Strain out the herb and drink the infusion. You can store an infusion in the refrigerator for several days.

There are other variations: You may chill some infusions, to serve cold. For very delicate herbs, you may want to try a sun infusion: Put the herbs in a glass container with room-temperature water, and let it stand in direct sunlight for a few hours.

In a *decoction,* you simmer the herb over low heat in a covered, nonmetal pot, 15 to 20 minutes. This is good for tougher, thicker plant material, like twigs, bark, and roots. There are exceptions. Twigs of black birch contain a volatile oil you'd lose by boiling, so you must make an infusion.

Douches of warm infusions or decoctions, applied vaginally, are commonly used for vaginal infections, such as candida. Avoid routine use, because they may upset the healthy, natural balance of microorganisms.

A *tincture* is an alcohol extract. Fill a jar with an herb or herbs. Cover with vodka. Seal tightly, and let it stand for a few days or weeks. If you're mystically oriented, begin your tincture the night of a full moon, and strain out the herbs on the night of the new moon.

Use vodka because this is distilled alcohol, without impurities. Vodka is not 100 percent proof, so you're extracting both alcohol-soluble and water-soluble principles. Also, you're not using heat, which may harm some of the constituents. Because alcohol is the end product of microbial action, no further microbial action

is possible, so a tincture doesn't spoil. However, light can cause unwanted chemical reactions, especially over long periods of time, so store tinctures in the dark.

Alcohol is toxic to the liver, and it damages the stomach lining. It's rapidly converted to sugar, creating an undue demand on the pancreas for insulin. However, it has a low boiling point. To get rid of the alcohol, put one teaspoon to one tablespoon of tincture in one cup of water just off the boil. Let it sit, uncovered, 10 to 15 minutes, or until cool enough to drink. By then, virtually all the alcohol will have evaporated. (My apologies to the drinkers.)

A *compress* or *fomentation* is an external application. Wet a towel or cloth with an infusion or decoction, and apply it to the affected area. You can wrap it in plastic to keep the surroundings from getting wet. The compress may be hot or cold. Heat stimulates circulation, and cold reduces swelling and cools fever.

A *poultice* consists of warm, moist ground or powdered herbs applied directly to the skin, usually with a bandage. A *plaster* is similar, except the herbs are placed between layers of linen. These preparations draw out infections or irritants. You may mix the herbs and some warm water with clay or cornmeal to increase absorption.

A *salve* or *ointment* is a thickened, oil-based extract. Cook a handful of the herbs 15 to 20 minutes on low heat, just below simmering, in a quart of light oil (such as sunflower or safflower oil) in a glass or enamel pot. Strain out the herbs, and add two to three tablespoons of beeswax for thickness, plus the contents of an oil-soluble vitamin-E capsule—a natural preservative. Remove one tablespoon and cool to room temperature, to test the thickness. Add more beeswax or oil if you want it thicker or thinner. Cool, and store refrigerated, in jars.

Smoking herbs gets them directly into the bronchial tubes and lungs. You can use a pipe or water pipe. However, you're also inhaling carbon monoxide and ash, so this should be reserved for short-term emergencies.

A *syrup* is a concentrated herbal extract mixed with honey or glycerin—thick enough to remain in direct contact with the affected area: Gently boil down a quart of water with two ounces of herbs to about one pint. Strain out the herbs, and add one to two ounces of honey and/or glycerin. Syrups are used for sore throats and coughs.

Black-cherry bark is the traditional herb of choice. Although honey is as unhealthful as white sugar, it's also a natural antiseptic, so a syrup is good to use for acute infection.

IDENTIFYING WILD PLANTS

WILD PLANTS AND NATURAL HISTORY

To know which plants are good to eat and which ones can kill you, you must make 100 percent certain identifications. A few plants, like dandelions and apples, are easy to recognize, but most require careful examination of identifying characteristics—those features that are stable throughout the species—especially if there are poisonous lookalikes. Some features are not shared by all members of a species. If you're trying to identify *Homo sapiens*, determining whether your specimens have freckles will distract you from what's important.

Before we look at plants' parts, and examine to see how their features vary, let's examine what's behind these features. Since Darwin, we know that living things gradually developed their forms because they worked, and more living things came into being than survived, so over countless generations, the ones that survived and reproduced the best are the ones we find. Anyone who's tried to rid a lawn of "weeds" knows how effective natural selection has been for producing organisms that survive and multiply.

Within that framework, almost anything goes. However, there are some influences and constraints. Everything must have a predecessor. Our middle ear, for example, is a greatly modified fish's gill arch. As our ancestors left the water, what had originally been part of a respiratory structure that could also conduct sound gradually changed into a specialized sound conductor. You can see still these changes paralleled in developing embryos, which start off with gill arches and end with the tiny hammer, anvil, and stirrup bones. If our early ancestors hadn't been fish, we might still be able to detect sound, but the mechanism would have been completely different.

On the other hand, if the anatomical or genetic raw material isn't available, certain lines of evolution are blocked. Woodpeckers evolved in North America because their ancestral forms could be modified by natural selection into a bird that pecks for insects under tree bark. On other continents, insect-eating birds were anatomically incapable of occasionally pecking through wood, so woodpeckerlike forms never evolved, and insects hidden beneath the bark on other continents are safe from birds.

Still, nature's plasticity provides plenty to work with. Flower parts, for example, are highly specialized leaves that are modified for reproduction. Showy petals are still much like their ancestral leaves. They simply exchanged chlorophyll for other pigments, and changed shapes and configuration—adaptations that attract pollinators.

Physical and biological laws also constrain evolution. Human ancestors developed ever larger brains, but this trend has stopped. Our babies' heads became so large that many women and babies died during childbirth, offsetting the advantage of a larger brain. We can't further enlarge the birth canal because our pelvis is already distorted in comparison to other hominids': Our leg bones already angle inefficiently to the side. The leg bones of our smaller-brained immediate ancestor, *Homo erectus*, were placed below the pelvis. These creatures could probably run circles around us and outendure our best athletes. Some scientists think they stalked prey to exhaustion.

But nonbiological physical factors can also provide evolution with raw material: Repetition, or variation on a theme, is a common organic and inorganic motif. Something that works is worth repeating. Physically, if it's the course of least resistance, it probably will be repeated. Crystal molecules repeat the same sequence over and over because that configuration entails the least energy.

Of course, flowers with many petals, and plants with many leaves, represent advantageous repetition. Reproduction of the organism is even one of the criteria for determining whether something is alive. But repetition also occurs on different scales. The branching patterns of the limbs of a tree reiterate on a smaller scale to create the same branching in the twigs. The leaflets of a huge compound leaf work like small simple leaves. A blackberry, raspberry, or mulberry really consists of many tiny fruits, each with its own seed, united into a whole that's greater than the sum of its parts. Animals are attracted to this large fruit from a distance, and if they eat part and drop the rest, they disperse the seeds.

Another factor that determines what we see in the natural world is history. (As a schoolchild, history meant memorizing boring dates and the uninteresting names of long-gone political and military leaders, cities, and countries. I hated it. Now I realize that the unfolding of human events is fascinating if it's presented properly. It's also important: To find solutions to our environmental problems, for example, it helps to understand how they came about.) Knowing the history of a natural phenomenon may be as important as understanding the pertinent physical laws.

In the physical and biological realms, laws are regular occurrences that underlie our observations. We often abstract them from phenomena and express them mathematically. Although these laws set limits to what may or may not happen, they don't dictate which of the "legal" possibilities actually do occur. Quantum theory, and the new science of chaos, have shattered all pretense of determinism. We no longer see the universe as a precise, potentially predictable, mechanism. What does actually happen within the limits of physical law greatly depends on what occurred earlier.

Along with evolution and natural physical laws, happenstance affects what's going on now. Before we sent probes to explore the outer planets (except Pluto), we thought they would be monotonous ice giants—huge globes of frozen gases, all similar, according to the physical laws as we understood them. But the individual histories of these planets made them all unique, and their many satellites proved even more individualistic. How they formed, past collisions, and gravitational influences made them richly varied. Living things, and the history of life, are infinitely more complex.

We all realize humans wouldn't have evolved if the dinosaurs hadn't become extinct and opened the way for mammals to take over the earth and diversify. But if the dinosaurs hadn't preceded us, we might not have had the flowering plants that provide our most important foods: Weeds thrive in open, sunny, disturbed habitats, such as lawns, because they grow and reproduce rapidly. When the earth was dominated by earlier, nonflowering plants, there were no habitats where the flowering plants could thrive and diversify, for the same reasons mammals couldn't diversify until the dinosaurs were gone. The huge herds of plant-eating dinosaurs created the disturbed habitats crucial for the flowering plants to develop. By the time the dinosaurs vanished, the flowering plants had gained a foothold that led to their dominance of the botanical world. Thus, as much as any scientific laws, the haphazard events of natural history led to the existence of today's species, including us.

So when you use the different parts of plants for identification (or for food), consider how they came to be, what they do, and how they fit into the larger picture.

LEAVES

You'll see more leaves on your field trips than any other plant parts, so understanding them will enhance your outdoor experiences. Leaves are solar panels par excellence. Only seven cells thick, they harness sunlight more efficiently than anything we've invented, forming the food chain's first link. Leaves can be oval, elliptical, swordlike, lance-shaped, arrow-shaped, or anything in between. They may be hairy or bald, tasty, bitter, acrid, or toxic. Their different shapes and forms reflect complex adaptive answers to environmental challenges: They must contend with harsh weather,

palmate-compound leaf

twice-compound leaf

feather-compound leaf

toothed

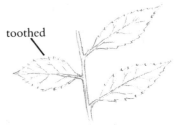

basal rosette

double-toothed leaf

opposite leaves

alternate leaves

hungry vegetarians, and other plants competing for sunlight, while exposing the largest possible surface area to the sun.

So why don't plants with the largest leaves always win out? In the tropics, where the climate is comparatively uniform, they often do. But large upper leaves shade out the lower leaves of the same plant, and farther from the equator, where weather varies more, large leaves are a liability. Storm winds can tear them to shreds. So leaves divide partially into lobes, or separate completely into compound leaves.

How can you tell if you're looking at one compound leaf, or many simple (undivided) leaves? For woody plants it's easy: Leaves come from leaf buds that form the previous summer. A simple leaf already has next year's bud on the twig, next to the base of the leafstalk. A leaflet, the subdivision of a compound leaf, doesn't have a bud near its stem's base, because no new leaflet will form there next

year. Also, twigs are wooden, but the midribs of compound leaves are not.

There are two kinds of compound leaves: Palmate-compound leaves have leaflets emerging from a central location, the way your fingers emanate from your palm. Feather-compound leaves' leaflets emerge along an axis, like the feathers coming from a bird's tail.

Some woody plants have twice-compound leaves. The subdivisions themselves are divided. These leaves are huge, and so are the leaf buds that produce them. Small simple leaves usually develop from the smallest leaf buds.

Leaf configuration is basic to identifying plants. We call paired leaves opposite. Unpaired leaves are alternate. Leaves that encircle a stem are whorled. Don't look at the newest growth to determine leaf configuration, because the stem parts that would separate alternate leaves may have had insufficient time to develop. Look at older growth.

A few plants will try to fool you by having opposite leaves in some places, and alternate leaves in others. So look at a few parts of the plant. Always examine many samples in nature, to get a good understanding of what's going on.

Another important leaf formation is the basal rosette, where all the leaves emerge from the ground at a common point. Many plants take this form in the cold weather, when growing tall would expose them to the elements, there are no taller plants competing for sunlight, and spreading in a circle captures the most sun. One of my first magazine articles was about cold-weather plants and basal rosettes. The editor consistently changed the spelling of "basal" to "basil," leaving the readers wondering where to find the recipe for spaghetti sauce.

Teeth are another feature that helps you identify plants—some leaves have serrated edges. If the serrations are serrated, the leaves are double-toothed. Double-toothed leaves and twice- or thrice-compound leaves are again themes of repetition on different scales, as described above.

The leaves' attachment to the rest of the plant provides further distinctions. Some leafstalks are long, others are shorter. Some, like winged sumac's, have flared edges. We call these stalks winged.

Some leaves have no leafstalks. The leaves that partially surround the plant's stem clasp it. When the leaf base surrounds the stem, it looks as if the stem is perforating the leaf.

The above distinctions are for broad leaves, which include all leaves that aren't needles. Needles, which are narrow, come in different shapes and configurations too, and we'll discuss them as we cover the species that have needles.

FLOWERS

Flowers make plant identification much easier, especially for nonwoody plants. To identify something, you need to group it with its relatives, and separate it from unrelated groups, gradually narrowing your search. Taxonomy, the science of classifying living things, goes hand in hand with identification.

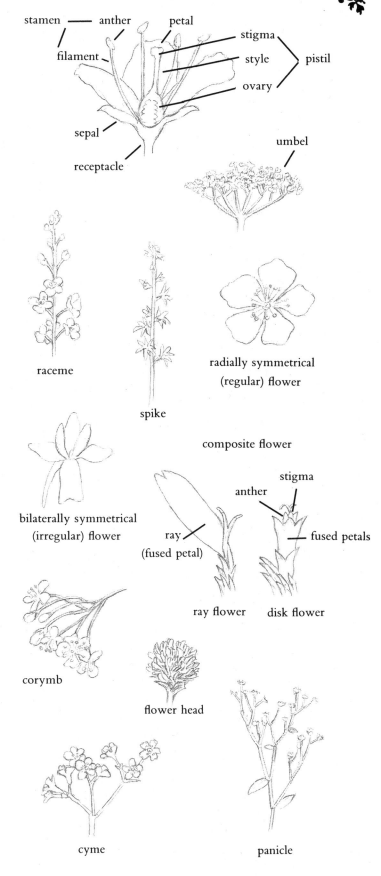

It's hard to determine genealogy with leaves. Because they're around all season, evolution changes them very greatly. Unrelated plants in similar environments, with matching survival strategies, often evolve similar leaves. Flowers don't change as greatly because, as reproductive structures, they're around for a shorter time. Natural selection has less opportunity to work, so flowers retain more of their group's common ancestor's features. Wild roses, for example, evolved different colors and sizes, but they all look like roses. Shakespeare got it right: Smelling them confirms their identity with 100 percent certainty. (It's amazing how many beginners don't think of smelling plants.)

Linnaeus, the eighteenth-century Swedish biologist who invented modern scientific nomenclature, scandalized plants by classifying them according to their sexual parts. Long before Darwin, he didn't know that related plants had evolved from common ancestors, but he saw how anatomical similarities between living things formed natural groups and subgroups: species, genus, family, class, phylum, and kingdom. This was a great stride from medieval botany, where anthropocentric systems competed to classify things according to their usefulness to man. Thus rodents and insect pests were grouped together as vermin.

Common names are still confusing today. The same plants have different names in different regions. Unrelated plants have similar names for no reason. Different edible and poisonous plants may even have the same name!

Universal scientific names clear up this confusion. They're in Latin or ancient Greek—dead, unchanging languages. This makes them hard to remember, so I suggest you use them as references: If you compare information about plants from different sources, check the scientific names to make sure the plants are identical. Eventually, you'll be able to impress people because you'll automatically remember the scientific names of the plants you use the most.

Scientific names are often descriptive, and we've included translations where they're meaningful. Sometimes, they're whimsical. Linnaeus sent scientists around the world to gather specimens. One became severely seasick, sailing across a stormy Mediterranean Sea to collect African specimens. He vowed never to set foot on a ship again. When it was time to return, he spent years walking around the entire Mediterranean Sea. So Linnaeus named a tree after him—a species of arid areas, which never grows near water.

Flowers evolved as modified sets of leaves specializing in reproduction. Resting on a base called the receptacle, they're arranged on a short stem in concentric whorls.

What floral parts do we look at to identify plants? The petals, modified leaves that are collectively called the corolla, are the most obvious feature. Simply noting their color and counting them differentiates your specimen from many other plants.

Surrounding the corolla is another set of modified leaves—the calyx, composed of sepals—usually green. They're often smaller and less conspicuous than the petals—most noticeable when they enclose the flower bud. If the petals are fused into a tube, you can't count them. But you can enumerate a flower's basic divisions by counting the sepals. Sepals also vary. Some flowers have large, colored sepals, often replacing absent petals. Sometimes sepals fuse, forming a calyx tube.

Surrounding the calyx may be another set of modified leaves, the bracts—collectively known as the involucre.

Enclosed by the petals are the reproductive parts of the flower. The male stamens consist of threadlike filaments, each supporting a tiny saclike, pollen-producing anther.

The female part or pistil consists of three parts. The stigma, which catches the pollen, is the expanded tip of the style, a tube for conducting pollen to the ovary. The ovary is the bulging container of the ovules. It may be simple or compound. The ovary and ovules eventually develop into fruit and seeds.

Under selective pressure (the gradual force of natural selection), these basic structures may change radically. Some flowers lack male or female parts. Some reproduce asexually, opening only when the seeds are mature. Some don't set seed at all, relinquishing reproduction to other parts of the plant—asexual vegetative reproduction. Some structures in such nonfunctioning flowers, like the daylily's, are vestigial—they're present only because the genes that create them are still active.

A major distinction among flowers is whether they're small, numerous, and inconspicuous; or large and showy. The former produce much more pollen, depending on wind and pollen volume for pollination. Their parts are often so greatly reduced, we hardly recognize them as flowers. They often cluster as catkins—long stems bearing stalkless male or female flowers. Catkins fall from the plant as a unit. Most wind-pollinated trees, such as oaks and walnuts, bear catkins. Wind-borne pollen often lands in people's noses, causing hay fever.

Showy flowers are the most highly adapted and efficient. They entice insects, birds, or bats to pollinate them. The most advanced are the orchids. Some are so specialized that one wasp species pollinates one orchid species. The pollen-bearing part may resemble the female wasp's genitalia. The male wasp futilely tries to mate with it, gets covered with pollen, and flies off in frustration. Endowed with more libido than brain, he gets fooled over and over, repeatedly pollinating orchids without gratification.

A major distinction among animal-pollinated flowers is whether they're radially symmetrical (regular), like an apple blossom, or bilaterally symmetrical (irregular), like a violet. Flower shape and color are adaptations for selecting efficient pollinators. Insects land on a flower for a meal of nectar or pollen, but not all insects transfer pollen.

Some flowers, like wood sorrel's, are bowl-shaped. Effective and ineffective pollinators alike are admitted. Others, such as jewelweed's, have nectar at the end of a long tube that forms a conspicuous spur. Only hummingbirds and insects with long tongues, such as butterflies, gain access (although you sometimes see a tiny hole at the base of the spur, evidence that a nonpollinating invertebrate cheated by making a hole to get to the nectar). Flower and animal-head shapes evolved in unison, like locks and keys, so pollination and feeding are assured.

Pealike flowers—members of the legume family, such as wisteria and black locust—are bilaterally symmetrical, with a winged keel similar to a boat's, positioned under a prominent flaglike standard, that attracts the pollinators. You must be a robust, sturdy insect, such as a bee, to shoulder your way through the keel to reach the nectar within, picking up pollen in the process.

Flower arrangements provide further distinctions to help you identify plants. When the flower is out of season, the fruit usually retains the configuration. The pattern may even remain on dead, out-of-season plants (plant skeletons).

Stalked flowers often grow on a long central axis, with the lower flowers maturing first. This is a raceme—typical of the mustard family. If the flowers are stalkless, as in the mints, it's a spike. If the central axis branches, like a grape's, it's a panicle.

An umbel resembles an umbrella. Stemmed flowers of equal length originate from one point, as with wild onions. If the lower flowers of the central axis have longer stalks, and the cluster is flat-topped, you have a corymb, like the sweet cherry tree's. When the flowers at the edge of a flat-topped, branched cluster mature first, you have a cyme, like the virbunums and elderberries.

It's also important to note where the flowers originate. The dandelion arises from the middle of its rosette. Some flowers grow in terminal clusters, at the tips of the plants. Others come from the leaf axils, the crotch between the branch and the leafstalk.

A clover looks like one flower, but closer examination reveals many small flowers grouped together into a flower head. Here the recurrent themes of repetition and grouping reap a great advantage. One insect can pollinate all the flower head's flowers at once.

The composite family, which includes burdock, yarrow, and sunflowers, takes this strategy one step further. The radially symmetrical flower head has two kinds of flowers: Sterile, straplike ray flowers, resembling petals, radiate from the circumference. They attract insects to the many tiny, inner, fertile disk flowers. This arrangement is so efficient that composites outnumber all other North American families. As always, there are variations. The dandelion, for example, is a composite lacking disk flowers.

FRUIT

When flowers' ovaries ripen, they form fruit, sometimes along with accessory tissues. The fruit's composition depends on the plant's seed-dispersing strategy. It need not even be fleshy. From a culinary standpoint, nuts aren't fruits, but botanically, they're one-seeded, dry fruits with hard outer walls, like acorns or hazelnuts, that don't open when mature. (Pods open when they mature.)

Sunflower seeds are fruits called achenes. Like nuts, they're one-seeded fruits that don't open when mature, but the seed attaches to the inside of the shell at one point.

Fleshy fruits are also distinctive. True berries, like autumn olives, currants, and ground-cherries, are thin-skinned fruits with seeds loosely embedded in a soft, succulent pulp.

Peaches and plums are drupes: One hard stone or pit encloses the seed. A pome, like an apple or pear, is a many-seeded fruit that doesn't open. Its fleshy portion comes from the receptacle, not the ovary.

Individual fruits may also fuse, the way flowers sometimes do. When the flower cluster's fruits form a dense mass, you get a multiple fruit, like mulberries. If the bunch of fused fruits comes from one flower with multiple ovaries, you get an aggregate fruit, like a raspberry or blackberry.

When a fruit's fleshy part originates from a flower part other than the ovary, it's an accessory fruit. The strawberry's flesh, for example, comes from the receptacle. It's an accessory fruit, while its seeds are achenes.

ROOTS

Things we don't normally consider fruits (nuts, seeds, and pods) are fruits after all, but underground plant parts that we call roots aren't always true roots.

Rhizomes, for example, are horizontal underground stems that give rise to true roots, and to aboveground shoots. They may form dense colonies, like cattails, so colonies indicate that the plants are connected by underground rhizomes. Some rhizomes, like Solomon's seal's, store food underground.

Some rhizomes, like the groundnut's, also produce tubers along their lengths. These short, thick sections of underground stems usually bear minute buds—the familiar eyes of potatoes—that can grow into new plants. Tubers also store food, and many are edible.

A bulb is a single, large, roundish bud, consisting of a short stem surrounded by layers of underground, scalelike leaves. Onions and garlic

fibrous root

rhizome

taproot

corm

bulb

tuber

are typical. Bulbs often store food, and many are edible.

Corms are short, upright, thickened underground stems also specialized for food storage. They're not layered like onions, nor attached to rhizomes or bearing buds like tubers. Sometimes they're edible, but they don't provide any major wild foods in the United States.

Taproots are true roots. They form from the primary root—the first root that emerges from the seed. They're usually large and vertical, with branches. Many, like burdock and common evening primrose, store food and are edible.

Fibrous roots look like they sound. These true roots are tangled masses of wiry fiber. Most are not edible or medicinal. They hold the soil together, keep out other plants, and often hamper you from digging up other roots.

Edible roots are usually in season from fall to early spring, when they store food for spring growth. You can even collect them in the winter; if it's warm enough that the ground doesn't freeze, the basal rosettes that mark the roots' locations still grow, or you can find the "skeletons" of last season's plants. By mid-spring, most roots relinquish their stored food and become tough and woody, although exceptions such as burdock and groundnuts are edible all year.

THORNS AND NATURAL HISTORY

Thorns provide a simple, if sometimes painful, identifying characteristic. Why do some plants have thorns? For protection? Then why do many plants lack thorns? What are their other defenses?

If an animal encounters another animal, it can ignore it, run away, try to devour it, or try to mate with it. Plants interact with the world through structure and chemistry. But their structure necessarily puts them at risk. Plants are highly extended. They must send thin, exposed leaves into the environment, to get sufficient diffuse solar energy. Underground, they must extend their surface area even farther to get scarce water and highly diluted minerals. Animals, by comparison, are compact, for protection. Their most vulnerable extension is often the neck, which many mammalian carnivores target.

To compensate for their greater vulnerability, and to make the most of their poor resources, plants have developed the most sophisticated chemistry on earth. That's why animals like us were able to evolve. We eat plants or the animals that eat plants, and live on their ready-made nutrients. And this is why plants have such great potential for providing new foods, medicines, and biochemical information.

Some plants defend themselves with poisons. This isn't universal, since animals can evolve to tolerate the poisons. Many of our best wild food plants defend themselves by rapid regeneration— a great advantage flowering plants have over their predecessors, especially in disturbed habitats.

This returns us to the question of thorns—an alternative strategy to regeneration. When can't plants regenerate? What do they require for regeneration or growth? Soil, sunlight, and water. Are there habitats that deprive them of these needs? Of course—deserts have inadequate rainfall, and poor, sandy soil. Does this lead to the evolution of thorns there? Yes, deserts are full of thorny plants, such as cacti. The seashore is also desertlike, with sandy soil that won't hold rainwater, plus toxic saltwater. Thorny plants like roses, blackberries, and prickly pears thrive there.

SHOOTS

Another plant part to consider is the shoot. There are two kinds, both involved with new growth. On a perennial woody plant, the shoot is the young, growing tip. The greenbrier vine's shoot, for example, consists of the tender, new stem, and its associated immature leaves and tendrils.

The other kind of shoot is the new growth of an herbaceous plant, as it emerges from the ground, like the shoots of milkweed, pokeweed, and Japanese knotweed. They consist of the tender, young stems with their associated immature leaves.

Herbaceous plants' shoots lack many of the mature plants' identifying characteristics, but the shoots may be the plant's only edible part. If you're not sure what shoot you've found, follow it to maturity, then return the following year to harvest it. Even though we're used to instant gratification, to genuinely attune ourselves to our planet, we must learn to accept natural seasons and cycles, and let the plants teach us patience.

LIFE CYCLES

Wild plants basically have three kinds of life cycles. Annuals usually live one season, flower, and die. But if they're winter annuals, they begin life in the fall, hold out through the winter, and complete their life cycles the following year.

Biennials probably evolved from winter annuals. They normally form basal rosettes the first year, die to the ground in the winter, then form new basal rosettes early the following spring. A flower stalk follows, then the plant blooms, goes to seed, and dies. However, if the rosette is prevented from flowering, it may survive and try again the following year.

Perennials, which include all trees and shrubs, and some herbaceous (nonwoody) plants, are the simplest. They live indefinitely.

PLANT TYPES: HERBACEOUS PLANTS, TREES, SHRUBS, AND VINES

Plants take many easy-to-differentiate basic forms, with some overlap. Some are herbaceous, lacking woody parts. These may be upright or creeping, or they may be climbing vines. Some vines are also woody. Other woody plant forms are trees, which have main trunks with branches, and shrubs (also called bushes), which are usually smaller, with branches coming from the ground. There are intermediate forms—some small trees are sometimes shrubby. For example, people normally prune yews to make shrubs. Left to themselves, yews often grow into trees.

Trees and shrubs have a protective outer bark, consisting of dead tissue, an inner bark or cambium, consisting of living cells, and what we think of as wood inside—more dead cells with fluid-conducting vessels. Plants like brambles and Japanese knotweed are woody, but they lack these tissues.

NATIVE AND EXOTIC SPECIES

The distinction between native and foreign (exotic) plants is not obvious in the field, but it's especially important ecologically. Plants that have been growing somewhere for hundreds of thousands of years fit into the ecosystem, with links to other species. They may be important food sources for animals, and the animals may, in turn, pollinate the plants or disperse the seeds. They may control erosion, and they normally don't escape their bounds and disrupt the balance of species. Foreign species usually lack enemies. They tend to invade and disrupt ecosystems where they don't fit in. Nevertheless, they may still prevent soil erosion, and many are excellent edibles.

Species naturally spread from one continent to another, although the process is much slower without human intervention. When the last Ice Age tied up so much water in glaciers that sea levels lowered, a land bridge developed between North and South America. Much earlier, South America had been connected to Australia, so South American mammals were still marsupials. New northern placental mammals apparently outcompeted the marsupials, and drove most to extinction. One exception was the opossum. This generalist marsupial lives in many habitats, eats all kinds of food, and reproduces rapidly. It migrated to North America, and its range is still expanding northward today.

Because of human activity, the distribution of species is quite rapid. Planted species are constantly escaping cultivation and establishing themselves in the wild. Also, the climate is changing, as a result of a complex combination of natural factors and the greenhouse effect. When I led tours in the Northeast in the early 1980s, watercress always died in late fall and reappeared in spring. Recently, it's gotten so warm that it grows all winter. Also, wild foods come into season much earlier than they used to because of warmer, earlier springs. So wild plants' ranges and seasons are changing. Although we give ranges and seasons for all the plants we cover, they're approximate, and subject to change. Of course, the same plants begin and end their life cycles earlier the farther north you go, and vice versa.

NATURAL HABITATS

People who are new to studying nature and wild foods are surprised that you can forage in and around cities, as well as in the country. They think the country is where plants grow, and the city is all concrete and steel. Nothing could be further from the truth. The country is great, but many cities are full of parks—legacies of past environmentalists. Urban and suburban areas also contain a variety of environments, each with its own communities of plants. As we look at these habitats, be aware that they overlap. Don't put natural habitats into pigeonholes. The only real pigeonholes in nature are usually occupied by pigeons.

LAWNS, MEADOWS, AND HUMANS

Most of us take lawns for granted, but there are lawns and there are lawns. if you put enough destructive chemicals into the environment to annihilate everything but cultivated grass, you'll have the kind of monotonous lawn that is ingrained in our middle-class culture as the ultimate ideal of beauty. As an artist and botanist, I find overcultivated lawns boring.

Lawns were originally practical defenses against dangerous animals and people. Over the centuries, they became a status symbol, associated with beauty. Wild plants, which we couldn't control or subjugate, became ugly "weeds." The peace and beauty of a forest is a modern conception. In the past, the woods were desolate and threatening, as portrayed in fairy tales. (Beauty was found in pastoral settings.)

The problem with lawns is that we force them into ecosystems where they don't belong. The most natural lawns occur in Great Britain. As any tourist will tell you, it rains there all the time, so they don't need watering. Until recent times, herds of sheep served as low-tech lawn mowers.

Golf courses in arid regions, on the other hand, squander scarce water resources to maintain lawns where they'd never grow naturally. In other regions, the chemicals that suppress natural forests pollute the ecosystem, and vast energy resources are wasted mowing down edible plants.

If a lawn or meadow is subjected to only occasional mowing, an interesting habitat with a variety of beautiful plants—many renewable wild foods—results. Such a habitat also provides food and shelter for many small creatures.

What kind of wild foods grow on lawns? Those

that grow, mature, and reproduce most rapidly. These traits favor survival in a temporary habitat, where tall plants may shade out the pioneers within one or two seasons. Of course, if humans continually mow down the taller plants, the dandelions, peppergrasses, plantains, purslane, common mallow, and other species will have a permanent home where they can proliferate.

Many plants of open, sunny habitats, such as shepherd's purse and chickweed, are annuals. They use all their solar energy to make seeds. Other survival strategies also work here. Dandelions are stubborn perennials that grow faster if you cut them down, and reproduce prodigiously.

OVERGROWN FIELDS

If you leave a meadow or lawn to its own devices too long, you eventually get an overgrown field, with taller plants, some shrubs, and even a few pioneer trees. Many of the herbaceous biennials, such as wild carrot and curly dock, grow here. This is a good habitat for annuals like lamb's-quarters and amaranth, and perennials like daylily, spearmint, stinging nettle, bayberry, and many others.

Succession is still uneven. A local overgrown field that was probably a farm before I was born still has open, sunny sections filled with strawberries. Other parts are thick with black-cherry trees mature enough to produce fantastic fruit. Field garlic and other intermediate-size plants grow there too.

THICKETS

Small woody plants, such as shrubs and small trees, dominate thickets. They're also great habitats for berry bushes, such as raspberries and blackberries, and vines, such as greenbrier. A thicket in a local golf course has blackberry shrubs ten feet tall behind a row of stinging nettles. In its most dense section, there's a stand of Hercules'-club fifteen feet tall, mixed with staghorn sumac. Again, nature doesn't respect human distinctions, and parts of this area are more like an overgrown field,

with stands of mugwort and milkweed. I keep my students busy studying this wonderful setting.

WOODLANDS

For unknown reasons, beginning foragers assume you have to go to a forest to find wild foods, and that forests don't grow in cities. They're misinformed on both counts. As we've already seen, other natural habitats are also great places to forage. Also, many large urban and suburban parks maintain forested areas, often with sections of varying maturity, including wet and dry regions, and plenty of biodiversity.

The first time I used a field guide to identify edible wild plants, a friend and I went into a local forest in early spring. We were in the wrong place at the wrong time. The woods were still mostly bare, and this secondary-growth forest had little understory. (Old-growth forests have a richer ground cover. Early spring wildflowers such as ramps use the sunlight to grow, flower, and set seed before the trees leaf out and shade the forest floor.)

Today I know I could have found field garlic, garlic mustard, false Solomon's seal, Japanese knotweed, and sassafras in these woods in early spring, but I didn't know these plants, or any others. Miraculously, we found a plant we thought was wild lettuce. It hadn't flowered, so we couldn't be sure, and neither of us could persuade the other to be the guinea pig. Fortunately, we didn't give up, and little by little, we began to learn the local plants and habitats.

Forests are also great in mid-spring. You may find redbuds, wood sorrel, violets, and ramps. The habitat is usually not uniform—there are thickets with berries, and openings where you can harvest greenbrier, Asiatic or Virginia dayflower, daylilies, jewelweed, and wood sorrel in the summer. Then, in the fall, you'll get the best nut crop along sunny edges. As you become more familiar with wild plants and your local natural areas, you'll even begin harvesting in the forest during warm spells in the winter, when you can gather sassafras, black birch, garlic mustard, field garlic, winter cress, and wintergreen.

DISTURBED AREAS

When humans or other factors destroy the plants growing in a location, new species move in very quickly. When I was eight, I had my appendix removed in a local hospital. About twenty-five years later, the hospital was demolished, leaving a field of rubble. The site was too polluted for foraging, but it was great watching new plants beginning from scratch.

The first edible pioneers were shepherd's purse, dandelions, and mugwort. Poor man's pepper, red and white clover, white and yellow sweet clover, and foxtail grass soon followed. Various nonedible plants (see page 2) also entered. The sandy, brick-covered soil was so poor, many plants associated with overgrown fields and empty lots, like pokeweed, burdock, curly dock, amaranth, and lamb's-quarters never established themselves, although broadleaf dock thrived.

Unhappily, there was a final step in the succession process. Everything was displaced by a new black, glass-and-steel office building, one of the ugliest in the neighborhood, and I lost interest in the site completely.

DESERTS

Deserts are dry and usually sandy. Plants and animals have a hard time, resulting in fewer species and fewer individuals per unit area than other habitats. Within these limits, there is still plenty of life, and the habitat can be quite beautiful, especially after rare rainfalls, when the desert blooms. Edible plants include cacti such as prickly pear, and yuccas. There are other edible desert plants, but because these specialized plants have limited ranges, and most Americans don't often visit this habitat, we'll leave them to books specializing in desert flora.

MOUNTAINS

Mountains are the results of volcanoes, or of continental plates colliding and the earth buckling. The process takes millions of years, and the subse-quent erosion takes more millions of years to tear them down. Meanwhile, many of the same plants that grow elsewhere grow on mountains, at least on the lower slopes. I find wild garlic, spicebush, blueberries, huckleberries, black birch, sheep sorrel, Hercules'-club, and countless other species on mountains. Mountains have forests, thickets, overgrown fields, streams, lakes, and meadows.

Mountains, like the seashore, have habitats arranged in zones. Different communities of plants and animals live at different altitudes. Going up and down a mountain can be like going hundreds of miles north or south. As you get higher, it gets colder and windier, and the trees get shorter. New growth is killed where conditions are beyond the species' tolerance, so the wind and cold prune the trees and shrubs, making them shorter and shorter. Above the tree line, only herbaceous plants grow. These regions have very poor soil, harsh weather, high winds, late springs, and early winters. Only the hardiest plants survive here. The habitat is very similar to the Arctic tundra.

THE SEASHORE

The seashore is another zoned habitat, where plant communities change as you get closer to the sea. Because salt water is toxic to land plants, only the most salt-tolerant species, such as sea rocket, grow close to the sea. Trees and such plants as beach plums, bayberries, autumn olives, black locust, and black cherry grow much farther back. In between are salt marshes, dunes, and sandy fields—all rich with life. You can find asparagus, lamb's-quarters, wild carrots, parsnips, and much more. Even when the terrain looks monotonous, it's not. Subtle variations of soil chemistry, climate, and the area's history (such as which species arrived first) create different plant and animal communities.

Zones continue even as you enter the sea. Some plants and animals thrive only in the intertidal zone—submerged at high tide, dry at low tide. This is where you find rockweed and glasswort. Some seaweeds, like dulse, grow beneath the low-tide mark, while others, such as kelp, are more deeply submerged.

OTHER WETLANDS

In addition to the seashore and salt marshes, there are other interesting wet areas. Swamps are wetland areas with trees, while marshes or wet meadows are wetlands with grasses. Bogs are acidic wetlands, where dead plant material doesn't decay. It forms peat, which may be transformed into coal over millions of years. These habitats have plants like cattails, wild onions, mints, blueberries, elderberries, cranberries, and wintergreen.

Lakes and rivers are also great places for foraging. The water and extra sunlight both favor plant growth. I find curly dock, wild ginger, nettles, watercress, violets, toothwort, and many other species in such habitats.

SEASONAL CYCLES

Ultimately, it's your responsibility to use this book to discover what grows in your region, and where and when to look for it. Abundance and quality also vary greatly from year to year, depending on rainfall, sunshine, temperature, insect population, and other less obvious factors. As you learn to recognize more plants, you'll find that there are plenty of bountiful harvests and crop failures every year. No two years are identical. We're now going to take you through the seasons to look at the best, most practical wild food plants growing in a variety of habitats.

CHAPTER 2

EDIBLE WILD PLANTS OF MID-SPRING

Mid-spring is a great time for foragers to get started. The cold weather is finally over, and the wild plants are mature enough to make identification easy, especially since many are flowering. Newly emerging edible shoots, greens, and flowers make this season exciting and colorful. You'll want to gather shoots and greens while they're sweet and tender. Later on, some species defend themselves by getting tough, bitter, or even toxic, although others remain edible much longer. Flowers usually have very short seasons, and this is the best time for them. Identifying spring flowers also prepares you for summer and fall, because many of these blossoms will give way to edible fruit, berries, nuts, and seeds.

Why do most plants flower in the spring? That's been a hot science topic for over one hundred years. Plants' physiology is regulated by hormones, just like ours. However, plants don't have endocrine glands, so it's more difficult to determine where the hormones are produced, what they are, and how they work.

What stimulates plants to make these substances, called auxins? Water and temperature are important, but so is light. If you put plants in a windowless room and manipulate the length of day and night, it turns out that short nights, not long days, stimulate many plants to flower. In nature, that means more plants flower toward the summer solstice than at any other time. So foraging throughout the spring up to the summer solstice normally brings you the greatest variety of wildflowers (except for the desert, where flowers bloom after rare rainfalls).

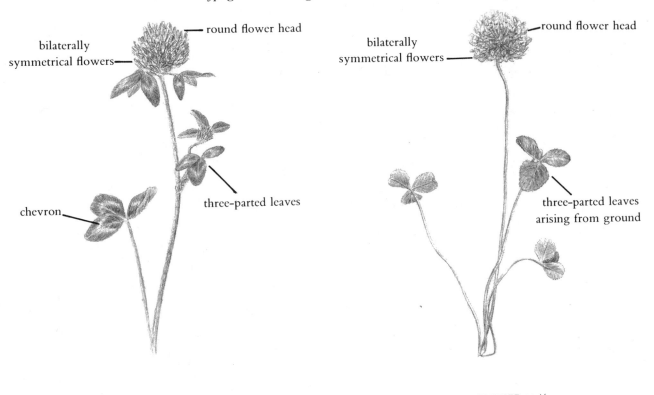

bilaterally symmetrical flowers — round flower head — three-parted leaves — chevron

RED CLOVER × ⅓

bilaterally symmetrical flowers — round flower head — three-parted leaves arising from ground

WHITE CLOVER × ⅓

PLANTS OF LAWNS AND MEADOWS IN MID- TO LATE SPRING

CLOVERS

(*Trifolium* and *Melilotus* species)

Not all clovers have three leaves. When we visit any overgrown field on my educational tours, and a certain woman attends, she always finds so many four-leaf clovers (a mutation that breeds true), you'd think they outnumbered the three-leaf variety.

Trifolium means "three leaves," and all clovers have leaves in sets of three. Clovers usually have compact flower heads consisting of many tiny, pea-like, bilaterally symmetrical flowers. Alfalfa, which has edible sprouts, also shares these characteristics, but there are no poisonous look-alikes.

The most common and conspicuous species are red clover (*Trifolium pratense*), the state flower of Vermont, imported from Europe as a hay crop;

and white clover (*Trifolium repens*), a native. Their round flowerheads are purple to pink and white respectively. Both grow to under 1½ feet tall, although the red clover is the taller. Red clover's leaflets arise from the flower stalks, and are marked with prominent V marks, called chevrons. White clover forms large, dense stands of underground spreading stolons—specialized branches along which new plants can spring to life. *Repens*, in fact, means "creeping." The leafstalks arise from the stolons, not from the flower stalks.

You can find clovers on lawns, fields, and disturbed areas. Both species bloom from spring to fall, but the best time to collect the flowers is in late spring, when the most flowers bloom.

The flowers are the sweetest and most enriching part of these plants. Select only the best flower heads. The ones that are turning brown taste awful. When you get home, sort the flowers on a tray and

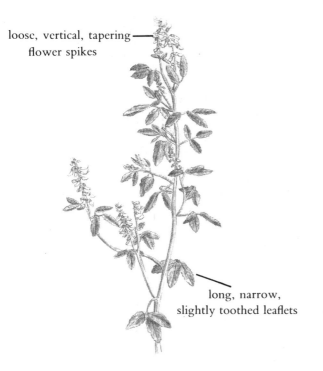

loose, vertical, tapering flower spikes

long, narrow, slightly toothed leaflets

WHITE OR YELLOW SWEET CLOVER × ⅖

discard any brown ones you missed before. Fresh or dried clover flowers make one of the best-tasting herb teas. You can also use finely chopped flowers in salads.

The American Indians, who often ate the entire plant, sometimes included the dried roots in stews. I also dry the flower heads. They grind into an excellent flour—good mixed with whole-grain flour in breads, muffins, and pancakes. This adds a chewy texture and natural sweetness.

A long-standing student of mine took a job as a wild-foods teacher at a wilderness survival school. She found herself on a barren mountaintop with dozens of hungry military men, and the smallest fraction of the common wild edibles that grow in more accessible places. Fortunately there were plenty of clover flowers. Her clover-flower granola was a big hit. She pan-roasted the flowers over a fire until they were sweet, dry, and crunchy. Hop and white clover worked better than the heavier red clover flowers.

Clover leaves are most edible when they're very young, in the early spring. Different sources give

varying accounts of their digestibility. In *Edible Wild Plants of Northeastern North America*, Lee Peterson recommends soaking them in salt water for several hours, or boiling them. This destroys much of the nutrition.

Wild-foods teacher Linda Runyon eats up to half a cup of raw clover leaves at a time in salads all season, and does well with them. But American Indians who were forced to live in the mountains and subsist on clover leaves supposedly died of a disease called "the bloat," because they couldn't digest the clover. On the other hand, "the bloat" is supposed to be a disease of cattle only. Linda counters with the claim that clover's high-protein leaves are a staple in China, where overpopulation makes protein scarce, and that people haggle over the price and quality. Personally, I don't care for clover leaves' flavor or texture, raw or cooked. I like using them for making vegetable stock, and they're acceptable for making tea.

Since clovers are legumes, they provide a protein that complements that of whole grains. Combining whole grains and legumes makes for a better-quality protein than either alone, with a more complete array of essential amino acids, the constituents of protein, which the body uses to assemble its own proteins. Clover-flower infusion is full of nutrients, providing beta carotene, vitamins C, B_1, B_2, B_3, B_5, B_6, B_9, B_{12}, biotin, choline, inositol, and bioflavonoids. It's a good source of the minerals magnesium, manganese, zinc, copper, and selenium.

A strong infusion of flowers and leaves is good for detoxification and rebuilding. It stimulates and cleanses the liver and gallbladder, and creates a gradual sense of all-over strengthening and nourishment that helps with a variety of maladies. People use it for gout, arthritis, skin disorders, and AIDS.

The tea is anti-inflammatory, calming, expectorant, and antispasmodic. People use it to reduce the severity of bronchial coughs, whooping cough, and even tuberculosis. Before vasodilators, people smoked anti-asthma cigarettes containing red clover. Forcing the bronchial tubes open with drugs ignores the causes of asthma, which include allergies, food sensitivities, psychological factors, and undiscovered factors. Asthma deaths mysteriously increased during the 1970s and 1980s. Now it's been discovered that the pharmaceutical drugs, which act in the short term, increase the severity and frequency of asthma in the long run, creating greater dependency on the drugs.

Red-clover-flower tea and compresses have been used for cancer for centuries. They were part of an early natural alternative cancer therapy called the Hoxsey formula, attacked by the cancer establishment. However, research by University of Alabama scientist Stephen Barnes shows that the genistein in red clover may inhibit breast-cancer growth by blocking estrogen receptors. Genistein is much safer than the very similar chemotherapeutic agent tamoxifen, which has recently come under fire for causing serious liver damage in some patients.

Other species of clover are also interesting and useful. The hop clover (*Trifolium agrarium*) stands almost 15 inches tall, with yellow flower heads ½ to ¾ inch across. Use the flowers the same as red and white clover. Look for it in overgrown fields, where the dead flower heads overwinter and resemble dried hops.

White and yellow sweet clover (*Melilotus alba* and *Melilotus officinalis*) are nearly identical, except that one has white, the other yellow, flowers. These straggly biennials, 2 to 6 feet tall, have long, narrow, slightly toothed leaflets in threes—typical of clovers. The small white or yellow flowers grow in loose, vertical, tapering spikes. They contain coumarin, which smells a little like vanilla.

You can find these species in overgrown fields, poor and disturbed soil, and near the seashore. The young leaves first come up in early spring.

Use the flowers the same way you'd use the other species, or scent your clothing by putting the fragrant flowers in your dresser drawers. You can eat the sweet, tender, young leaves in small quantities in early spring. **Caution:** Use only completely fresh leaves. Avoid any that are spoiling. Some time ago, someone noticed that cattle got sick and died if they ate decomposing sweet clover leaves. Upon investigation, it turned out that fermentation transformed harmless coumarin into dicoumarin, which stops blood from clotting. The cattle were dying of internal hemorrhages. This substance became the first medical blood-thinning drug, dicumerol, as well as warfarin, a rat poison.

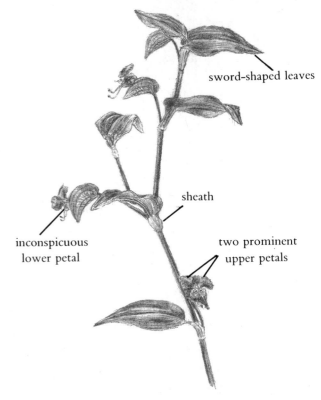

ASIATIC DAYFLOWER × ⅖

I learned of this when a friend who liked the flavor of sweet clover planned on making a wine by *fermenting* the leaves! Of all the hundreds of edible plants to use, this must be one of the worst— that would produce a poisonous wine. Fortunately, my friend did some research beforehand. I would have loved to see his face when he learned that his first glass of sweet-clover wine might also be his last.

ASIATIC DAYFLOWER

(Commelina communis)

This beautiful Asian annual has two upper sky-blue petals about ½ inch long, and one lower, smaller, inconspicuous, translucent petal. The grasslike leaves are sword-shaped, 3 to 5 inches long. The leaves' bases wrap around the stem to form a sheath. The hairless plant usually grows only 1 foot high, although under very favorable circumstances, it can reach a height of 3 feet.

There are close relatives—all edible. Virginia dayflower (*Commelina virginica*), for example, is a very similar, larger plant, with all three petals blue. You usually find dense stands of dayflowers in open places in woods, on rocky soil, and in cultivated and disturbed areas. Asiatic dayflower grows in the Northeast, while Virginia dayflower grows in the Southeast, with partially overlapping ranges. Other relatives grow in the Southeast, ranging westward to Texas or New Mexico. They bloom in spring, summer, and fall.

The flowers are only about an inch across, distinguishing the plants from poisonous irises or blue flags (*Iris* species), cultivated garden plants and wildflowers with much larger, showy flowers. Some coarse, hairy grasses resemble immature dayflowers, but you'd never be able to swallow them.

Spiderwort (*Tradescantia virginiana*) has three round, violet petals, golden stamens, and very long, narrow leaves that make the plant look spiderlike. This 3-foot-tall plant bears only a slight resemblance to the dayflowers. Its very young leaves and stems, as well as its flowers, are edible. The same is true of the two western species of spiderwort (*Tradescantia pinetorum* and *Tradescantia occidentalis*).

People often confuse the dayflowers with the daylily (page 232), but only because these very dissimilar plants have the word *day* in their names.

I eat the tender young leaves and the flowers of dayflowers raw in salads, or simmered ten minutes in soups, from mid-spring to mid-fall. These very common wild vegetables taste somewhat like string beans.

If you open the flower in the summer or fall, you'll find a seed that looks and tastes like a tiny green pea. It's not worth the time to collect in quantity, but children, who prefer fun to efficiency, love them.

Linnaeus, the father of taxonomy, named the genus after the three eighteenth-century Commelin brothers. Two became well-known botanists, represented by the two prominent petals. The third was a ne'er-do-well who died young, represented by the insignificant, translucent petal.

dense, terminal raceme of pink flowers

lance-shaped leaves

fringed sheath

occasional faint chevrons

branch

LADY'S THUMB × ⅓

LADY'S THUMB, REDLEG

(*Polygonum persicaria*)

This upright or spreading herbaceous annual grows over 2 feet tall, usually branching. The plant is so common, people know it from the corner of their minds without ever having noticed it.

The narrow, pointed, lance-shaped leaves are about 1½ to over 4½ inches long. There often is a slightly darkened triangular spot (chevron) toward the center, supposedly resembling a lady's thumbprint. On tours, I ask a lady to press her finger against a leaf, showing everyone the thumbprint she made. Then I announce that all the leaves now bear her thumbprint.

Lady's thumb has a conspicuous terminal cluster of tiny, pink flowers growing in dense racemes a few inches long.

Lady's thumb's stem has swollen joints at regular intervals—typical of the buckwheat family—accounting for the generic name, *Polygonum*, which means "many joints." At each node, a sheath connected to the short leafstalk wraps around the stem.

Tiny hairs grow from the top of the sheath, forming a fringe.

No poisonous plants resemble lady's thumb. However, lady's thumb belongs to a subgroup of the buckwheat family called the smartweeds. Other smartweeds have similar flowers and jointed stems, although some of them are larger. None has lady's thumb's fringe of hairs on the stem.

Smartweeds, also called smartass and water pepper, make you smarter: Many species are so acrid, and they make your mouth smart so badly, that you become smart enough never to eat them again. If you're dumb enough to touch your eyes with smartweed juice on your fingers, you'll smarten up even more.

Nevertheless, you can try using minced smartweed leaves like hot pepper, or use smartweed tincture as a liniment for bruises and sprains (pungent herbs often increase circulation). And some non-acrid smartweeds are mild-tasting and good to eat.

Lady's thumb grows on lawns, meadows, open sunny areas, and disturbed soil. It especially favors low, wet ground.

Lady's thumb is in season from mid-spring to late fall, flowering most of the time. The leaves and flowers are edible, with a mild flavor, somewhat like lettuce. You can add them raw to salads, or enjoy them steamed, sautéed, or in soups. Cook them 5 to 10 minutes. Because they're so mild, they're best mixed with stronger-tasting vegetables, especially those too strong to use alone.

It's so common, you can collect it nearly anywhere you go. Once you've learned it, it becomes one of the most easily recognized and safest wild edible plants. In addition, it improves the quality of soil where it grows.

PURSLANE, PURSLEY, PUSSLEY

(*Portulaca oleracea*)

Oleracea means "eaten as a cultivated herb." This annual was esteemed as food in its native India, and in Persia. Archaeologists find purslane seeds and pollen in excavations dating back thousands of years. For two thousand years it was a well-known cultivated garden vegetable in Europe. Suddenly, in the twentieth century, it was forgotten, and just as quickly it's making a comeback. Farmers remove it from their tomato or corn patches, and set it behind inflated price tags at farmers' markets, confident that unusual, expensive fare is irresistible to yuppies.

Purslane is a smooth, reclining annual European herbaceous plant that can cover your yard with a doilylike mat. It produces many prostrate and erect thick, smooth, succulent, creeping, reddish-green stems, with branches 4 to 10 inches long. The fleshy, toothless, opposite or alternate, stalkless, paddle-shaped leaves grow from ½ to 2 inches long, at the end of the branches.

Purslane's tiny, pale yellow flowers, with five petals and two sepals, bloom from mid-summer to fall. They're hard to spot, only ⅕ inch wide, hiding between the branches, and usually opening only in the morning sun. The fruit capsules are ¼ inch or less long. Beginning in late summer, their tops fall off, and tiny, dark red to black mustard-sized seeds fall out.

There are other very similar *Portulaca* species,

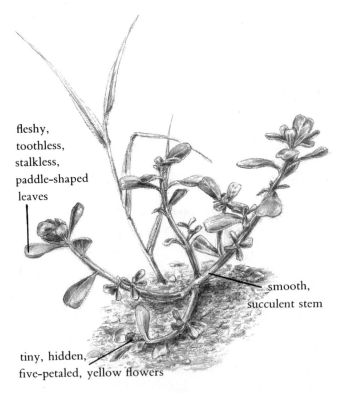

fleshy, toothless, stalkless, paddle-shaped leaves

smooth, succulent stem

tiny, hidden, five-petaled, yellow flowers

PURSLANE × ⅖

all edible, and no poisonous look-alikes. However, spurge (*Chamaesyce* species), which is poisonous, sometimes grows with purslane. This reclining plant has thin, stringy stems, and white, milky sap. Purslane has no colored sap. Although the two plants don't closely resemble one another, watch out that you don't carelessly include any spurge in your salad.

Purslane grows in sunny, sandy soil in fields, vacant lots, disturbed soil, and lawns and cultivated areas across the United States. I gather most of my crop in local front yards. The neighbors, who know me from past and present media attention, always welcome me, thinking I'm going to devour all their weeds. Little do they know that I always leave the roots alone, so the wild edibles will grow back even faster. Let's hope they won't read this book, or I'll no longer be welcome in my own neighborhood.

Notwithstanding my predation, most gardeners attack purslane relentlessly. Nonetheless, it has effective survival strategies: Most of its energy goes toward making seeds (which are good to eat), manufactured when the plant is only a few weeks old. The succulent stems retain enough water to ripen the seeds, even after they've been removed from the earth. Gardeners who uproot purslane but leave it in the garden help it reproduce.

Purslane appears in late spring and dies in the fall. If the seeds don't germinate, or the competition is too stiff, it doesn't come back. I've scheduled many field walks in parks where purslane was superabundant the previous year, only to find none whatsoever. I'm sure some readers are laughing at me, because they live in areas where the plant is so plentiful you couldn't eat it all if you tried.

The stems and leaves of purslane have a wonderful sweet-sour flavor—great raw in salads. Purslane is mucilaginous, slightly thickening soups, the same as okra. You can also bread purslane stems and put them in casseroles, and purslane stems make great pickles. One thing I haven't been able to do is dry purslane in my food dryer—the stems are too waterproof. For long-range storage, precook it and freeze it, or pickle and can it.

If you find purslane in very large quantities, put some in paper bags for a few weeks. Then remove the seeds, strain out the coarser debris, and winnow the remainder on a windy day by tossing it into the air from a sheet (or use a fan). The seeds, which are heavy, fall back, while the chaff flies off. I use the tasty seeds as a cereal, in granola, as a substitute for poppy seeds, or ground into flour and used with whole-grain flour.

Purslane is very nutritious, providing iron, beta carotene, vitamin C, calcium, phosphorus, and riboflavin. It's a great source of omega-3 fatty acids, which prevent heart disease and nourish the immune system. People spend a fortune on capsules of fish oil from the health stores to get small amounts of this nutrient, then try to destroy all the purslane in their gardens. But purslane is delicious and free; fish oil only tastes good if you're a cat.

WOOD SORREL, SHAMROCK

(*Oxalis* species)

Wood sorrel has a three-parted, palmate-compound leaf, consisting of three heart-shaped leaflets suspended on a slender stem, usually no more than 8 inches tall. The smooth leaflets fold along their seams under adverse conditions. This herbaceous plant forms colonies, which arise from tough, underground rhizomes.

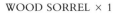

small, five-petaled, radially symmetrical flowers

slender stem

three-parted leaves

heart-shaped leaflets

WOOD SORREL × 1

Wood sorrel's small, five-petaled, radially symmetrical flowers' colors vary from species to species. It blooms from spring to fall. Common wood sorrel's (*Oxalis montana*) flower is a soft yellow, while violet wood sorrel's (*Oxalis violacea*) is violet, of course.

Although wood sorrel and sheep sorrel (page 198) are unrelated, their similar sour flavors (*Oxalis* means "sour") and names forever link them. Although they taste alike, they look different, and many people prefer one or the other. Some like sheep sorrel's soft texture and hardiness; others are more attracted to wood sorrel's delicate heart-shaped loveliness.

There are no poisonous look-alikes. People commonly confuse wood sorrel with clover (page 24), which is not poisonous, until you show them that clover's three leaflets are oval, not heart-shaped, and that the flowers differ.

For a common weed, wood sorrel is dainty and slender, even though there's no such thing as a delicate weed. It adapts to the city and country alike, tolerating difficult situations and thriving where other plants fall by the wayside. Wood sorrel grows in moist, partially shaded areas such as woods, disturbed areas, lawns, and lawn edges. You can find various species throughout most of the United States.

Also known as the shamrock, it's important in Ireland—associated with the missionary Saint Patrick. A tribal chief asked him to explain the doctrine of the Trinity. Picking up a wood-sorrel leaf, Saint Patrick said, "Here in this leaf, three in one, is a symbol of my faith, three Gods in one." The chief was so impressed, he converted to Christianity.

The plant is so distinct, it's one of the first wild edibles I teach young children. They love the lemon flavor, and they're fascinated by eating "hearts." Use it the same way you would sheep sorrel, in salads, soups, et cetera. The Indians seasoned beaver and muskrat with it. It's a wonderful trail nibble, and thirst-quenching on a hot day. Like sheep sorrel, wood sorrel has oxalic acid (see Sheep Sorrel), but not enough to cause problems with normal use.

Wood sorrel has lots of vitamin C, potassium oxalate, and mucilage. Use the infusion the same as sheep-sorrel tea: to cool, as a diuretic, and an astringent. Wood sorrel is supposed to have stronger blood cleansing abilities, and herbalists give it to cancer patients. Avoid foods containing oxalic acid if you have kidney stones, rheumatism, or gout.

OTHER PLANTS OF LAWNS AND MEADOWS IN MID- TO LATE SPRING

Edible or Medicinal

Chickweed, common mallow, dandelion flowers and roots, daylily tubers, wild onion leaves and bulbs, ground ivy, peppergrasses, plantain leaves, sheep sorrel, storksbill, strawberry leaves, thistle stalk, violets, yarrow.

For Observation Only

Plantain flowers, shepherd's purse seeds, sow thistle flowers, strawberry flowers.

PLANTS OF CULTIVATED AREAS IN MID- TO LATE SPRING

Edible or Medicinal

Amaranth leaves, burdock root and immature flower stalk, chickweed, clover flowers, dandelion flowers and roots, dayflowers, daylily tubers, Hercules'-club shoots, lady's thumb, lamb's-quarters, linden flowers, common mallow, orpine leaves, plantain leaves, purslane, redbud flowers, sassafras leaves, flowers, and roots, violet leaves and flowers, rose flowers, wisteria flowers, wood sorrel.

For Observation Only

Autumn olive flowers, Russian olive flowers, black walnut catkins, blueberry flowers, burdock flowers, sweet cherry flowers, Kentucky coffee tree flowers, gingko leaves and flowers, hackberry flowers, juniper, Kousa dogwood flowers, mulberry flowers, plantain flowers, sow thistle flowers, viburnum flowers.

PLANTS OF DISTURBED AREAS IN MID- TO LATE SPRING

BRACKEN, BRAKE, PASTURE FERN

(Pteridium aquilinum)

Arising from a velvety base, mature bracken is 1 to 3 feet tall. This fern is silvery at first, grass-green when mature, with brownish stems. Individual fronds have three branches, each with rows of feathery, blunt-tipped leaflets extending 1 to 2 feet, creating a triangular or fan shape. The upper leaflets aren't completely cut away from the midrib. There are usually purplish or brownish black spots called nectaries, where the leaflets join the stalk. The mature fronds' backs are dotted with the sori—reproductive structures. According to legend, if you gather their spores on St. John's Eve and hold them in your hand, you'll become invisible at the moment of St. John's birth.

The shoot or fiddlehead looks like an eagle's claw: The three curled-up prongs resemble an arthritic bird's talons. The whole fiddlehead is covered with woolly, silver-gray hair. You often see mature ferns growing alongside the fiddleheads, as well as dried-out fronds from last year. Bracken fiddleheads are easy to distinguish from other unopened ferns because of the woolly, grayish-white color, and the lack of scaly coverings or long hairs. When you find one fiddlehead, you usually find a whole colony, because the bracken propagates underground by long, branched, woody, hairy, spreading rhizomes.

People who study hundreds of ferns still appreciate bracken for its strange hardiness amid a delicate, shy family. Most ferns can't tolerate pollution, poor soil, or disturbed habitats, but bracken encroaches on grassy areas next to highways and parking lots. Any plant that can prevail over grass's domineering roots must be a robust survivor. In addition, it grows in woods, dry, open places, and old fields.

Bracken grows throughout the world, except for very cold and hot regions. From Siberia south

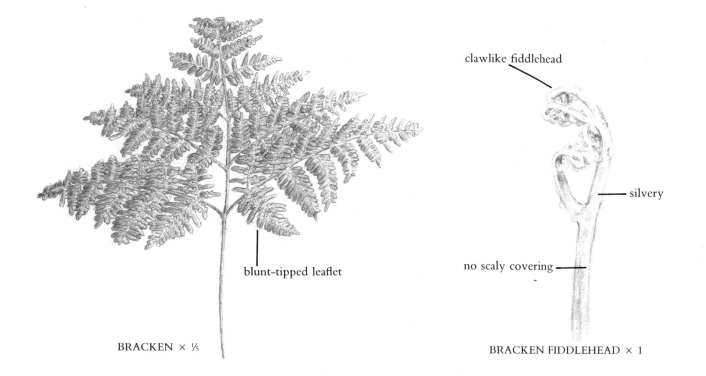

blunt-tipped leaflet

clawlike fiddlehead

silvery

no scaly covering

BRACKEN × ⅕

BRACKEN FIDDLEHEAD × 1

to Australia, east to Japan, and on the British Isles, bracken's fiddleheads are on foragers' minds. This famed gourmet delicacy is loved by the Japanese, tolerated in times of hunger by the Western palate, and much appreciated by New Zealand and Australian aborigines, who use the rhizomes as well as the shoots, although the underground parts are somewhat toxic. (Don't try to use them.)

All the ferns living today are long-term survivors from the Carboniferous Era, the Age of Ferns, some 345 million years ago. Back then, ferns came in all sizes and shapes. There were tree ferns with woody trunks as tall as 50 feet. Many of these extinct species are still with us today, transformed by slow geochemical processes into fossil fuels.

Late spring is the best time to search for bracken. Compared to other ferns, it's a slowpoke. You may sometimes even see its little heads slowly unfurling from the soil in the summer.

For foragers, bracken is a beautiful, delicious spring woodland treat. Break off the young fiddlehead tip close to the ground, so you have a sprout about 6 to 8 inches long. Older individuals become too fibrous to snap. Take only a very small portion of the fiddleheads in each spot. Unlike flowering plants, the ferns won't grow back.

Rub off the white-green wool, and cook like asparagus, 20 to 30 minutes. Tannin turns the cooking water brown and bitter. Discard the water, or use it for medicine (see Oaks, page 173, for tannin's medicinal uses) or dye. A friend of mine gathers baskets of fiddleheads, which she simmers in a large enamel pot. It provides a clear yellow-green coloring for cotton fabric.

Caution: The fiddleheads are safe only before their fronds uncurl. Don't eat them afterward. Not only are they very bitter but they've poisoned livestock, causing hemorrhages. Also, it's best to cook the fiddleheads, to destroy the enzyme thiaminase. If you eat too many raw fiddleheads, they'll deplete your body of vitamin B_1. Some people have adverse gastrointestinal reactions to bracken. Try a small sample the first time. Furthermore, this fern contains three carcinogens: Japanese people who eat commercially available bracken in quantity all year have elevated rates of stomach cancer. Nevertheless, I wouldn't be afraid of eating reasonable quantities of wild fiddleheads during their short season.

BURDOCK

(*Arctium species*)

Burdock's large leaves remind me of elephants' ears. They're toothless, wedge-shaped, very rough and coarse, with very wavy edges—up to 2 feet long and 1 foot broad. The undersides are so densely woolly that under magnification, it looks as if someone had dipped a spider in LSD and turned it loose to spin its web.

Each leaf arises from a purplish-green leafstalk up to 1½ feet long and ¾ inch thick. There are two equally edible species: Great burdock (*Arctium lappa*) has a solid, round leafstalk, except for the upper surface, which is trimmed by a single, long groove. The leafstalks of common burdock (*Arctium minus*) are hollow, not furrowed.

Baby burdock's taproot is pencil-sized, but the massive, mature root may be 4 feet long and 3 inches across. The root is whitish after you wash it. Its surface texture resembles tree bark. Cut the root horizontally, and you'll see two concentric,

stout, single, central, immature flower stalk of second-year plant

GREAT BURDOCK IN PRE-FLOWER STAGE
(CARDONE) × ⅓

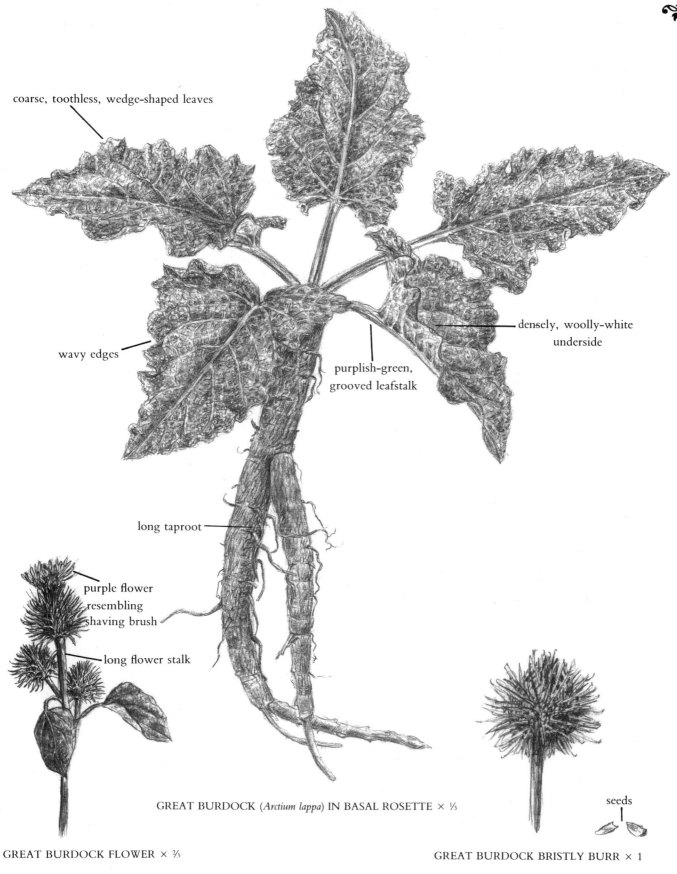

coarse, toothless, wedge-shaped leaves

wavy edges

densely, woolly-white underside

purplish-green, grooved leafstalk

long taproot

purple flower resembling shaving brush

long flower stalk

GREAT BURDOCK (*Arctium lappa*) IN BASAL ROSETTE × ⅓

GREAT BURDOCK FLOWER × ⅔

seeds

GREAT BURDOCK BRISTLY BURR × 1

beige-white rings. Note the mild, starchy-sweet fragrance, a combination of potato and artichoke.

There are no poisonous look-alikes in the wild. Rhubarb, a garden vegetable I've never seen growing wild, has similar-looking poisonous leaves, but the flowers are different. Nonpoisonous bitter or broad-leaf dock leaves also resemble burdock's, but they're hairless underneath. Curly dock's (page 236) edible leaves are hairless and narrow. The taproots of the latter two plants are yellow inside.

Burdock is a Eurasian biennial. Collect the root in spring, summer, or fall of the first year, when the leaves form a basal rosette. You can still dig up the root in the beginning of the plant's second year of life, when the leaves again form a basal rosette. Use it only before the flower stalk appears. One of my students wasn't paying attention and didn't want to dig up the root herself. So she paid a farmer from a local green market to dig up the burdock growing as a weed in his fields. The longer she cooked it, the tougher it got. She finally phoned me, and we determined that the farmer had dug up all inedible second-year burdock.

Greater burdock's stout, single central flower stalk, which appears in mid- to late spring, grows from 2 to 9 feet tall, while common burdock's flower stalk grows from 2 to 5 feet tall. The thicker the flower stalk the better, because under the tough skin is a tender, celerylike core that tastes like artichoke heart. I'm often surprised at its size, because it lies hidden by the rough, heart-shaped to oval leaves that surround it. These leaves are much smaller and less crinkly and wavy than the basal leaves.

Burdock blooms in late spring and early summer. Greater burdock's long-stemmed flowers are 1 to 1½ inches across. Common burdock's similar short-stalked flowers are ¾ inch across. These composite flowers resemble shaving brushes, and correspond to the "choke" of burdock's close relative, the artichoke. The flowers' centers are pink to purple. The prickly bases are green. When they appear, the stalks become too tough to eat.

The flower heads give way to globular, bristly, clinging brown burrs, which give the plant its name. They stick to your clothing if you brush against them—the Skagit Indians called burdock "sticks to everything." Dog owners and parents curse this plant as they painstakingly remove the burrs, which fragment when you pull on them. America is graced with this plant because the burrs attached themselves to immigrants' clothing, or to livestock, releasing their tiny, dark brown, crescent seeds upon our shores. Under magnification, you'll notice hundreds of tiny hooks, formed from the bracts. Someone became a millionaire because the burrs inspired him to invent Velcro.

The burrs make good toys. Children wear them as "nature pins," or make animals or other objects by sticking them together. Sometimes they throw these "itchy-balls" at each other. One group of particularly unruly, poorly supervised day campers delighted in throwing them into my hair whenever my back was turned—easy, since I was leading.

Look for burdock on disturbed soil, in backyards, empty lots, untended gardens, overgrown fields, and urban parks. It grows in full sun or partial shade, in rich or poor soil, but not in the sand or near the seashore. It flourishes throughout the northern half of North America, from coast to coast, but not in the Deep South. You can also find it in ethnic produce markets, selling for four dollars per pound, although the wild root is much better.

First- and second-year plants usually grow together. Although burdock dies after the second year, some dead flower stalks persist over the winter, so you can locate the stands any time of year.

Burdock's root is the part most often mentioned in cookbooks and herbals. Its food and medicinal uses make it invaluable. Not only does it build the system when you eat it, it gives you a week's worth of exercise when you dig it up. After a long, strenuous shoveling, when you finally start to pull up the deep, massive taproot, you can swear that there's a hungry Asian person on the other end pulling in the opposite direction.

Dig up the first-year taproot with a spade or fork, soon after a rainstorm, where the soil is the softest and least rocky, and the digging is easiest. I usually shovel for about forty minutes, or until my arms feel like they're going to fall off, returning home with about four pounds.

I used to say that the best implement for collecting it is the bulldozer, until it finally happened on my tour. A construction crew had dug a trench by a park building surrounded by burdock, to put in new pipes, and dozens of three-foot-long first-year burdock roots were exposed—a once-in-a-lifetime occurrence. I dove into the abyss, pulled root after root into the trench, sideways, and tossed them to the surface. When I finally emerged to claim my treasure, there was none to be found. My students had taken every last root!

The easiest way to clean the root is under running water, with a soapless, coarse-wire scouring pad. Slice the root diagonally, bark and all, as thinly as possible, to break up the fibers. Or save time and effort using the finest slicing blade of a food processor.

Burdock root's mild, nutty, somewhat sweet flavor makes it quite versatile. Its firm consistency makes it perfect for water-based cooking. I've steamed it, or mixed it into sauces, soups, chilis and curries, where it adds little bits of soothing, starchy texture. In Oriental dishes, it serves as a filling, satisfying main ingredient, sautéed with sesame oil, tamari soy sauce, and ginger. It's also great steamed or baked in a covered casserole dish or Dutch oven, like a potato, 20 minutes to 1 hour. When a fork penetrates easily, it's done. For a more chewy texture, finely grate it and sauté in oil. Some people like it very tender, so they slice and pre-steam before sautéing. Sweet peppers, carrots, turnips, parsnips, mushrooms, and mustard greens all complement burdock.

An elderly Italian man I met in the woods taught me another way to eat burdock. He'd never heard of eating the root, but said the Italians use the immature flower stalk, which they call cardone. Cut it off relatively late in the spring, before the flowers appear, while the stalks are still relatively tender. Don't worry, they'll soon grow back.

Peel them, parboil 1 minute to dispel the bitterness, add them to soups, stews, or casseroles, or enjoy them plain. You can do the same with the long leafstalks, but they're less substantial and harder to peel. Try cardones in recipes that call for commercial stalks like celery or Chinese cabbage.

The Italians dip the peeled, parboiled flower stalks in eggs and bread crumbs, and fry them. They taste like artichoke hearts, only better. You can layer a casserole dish with tomato sauce, breaded, fried cardones, and mozzarella and Parmesan cheese, then bake this Cardone Parmesan until bubbly.

Burdock leaves are also good to eat, but only if you're a goat or an iguana. For people, they're horribly bitter. When burdock leaves first appear in early spring, some people boil them in several changes of water to remove the bitterness, but there are so many better-tasting spring greens, it's hardly worth the effort. I once led a frustrating field walk, just after it had rained, where everyone inexplicably hated the flavor of all the mulberries they tried. It turned out we had handled wet burdock leaves earlier. Moisture from the berries was transferring bitter burdock-leaf residue from people's hands and ruining everything else.

Burdock root is very nutritious, providing vitamin C, biotin, vitamins B_1, B_6, B_{12}, vitamin E, potassium, sulfur, silica, and manganese. It provides inulin, a helpful sugar for diabetics and hypoglycemics (sufferers from low blood sugar) because it doesn't elicit rapid insulin production.

Burdock is one of the best and safest herbal medicines, with a wide range of uses, dating back to the days of Hippocrates. People pay ridiculous prices in health-food stores to make burdock decoction, then throw away the roots. You can simmer fresh or dried wild burdock, eat the root, and use the cooking water medicinally. You can also tincture the root.

Many people swear by this herb, relating miracles that would put many doctors to shame. Burdock is used for liver dysfunction, urinary tract disorders, and weight loss. It's a thorough diuretic that cleanses the entire body as it tones and soothes. It's most effective as an alterative in chronic conditions that require subtle, slow, and safe long-term nourishment and support. As a general detoxifier and immune-system stimulant, it helps people without clear-cut pathologies who don't feel well, and for biochemical imbalances where someone is easily upset and needs grounding.

People with sugar metabolism problems or chronic weakness often have stomach, kidney, liver and/or lymphatic problems. Burdock root, with its high mineral levels, will build, stabilize, and detoxify these organs. The tea is mucilaginous (gooey), so it soothes and tonifies the stomach muscles and mucous membranes, promoting digestion and eliminating toxins from the digestive tract.

It's helpful for colds, with antimicrobial action attributed to its polyacetylenes. Its demulcent action soothes the upper respiratory tract's mucous membranes, while its diaphoretic action makes you perspire, stimulating the immune system and promoting detoxification. Burdock root's diuretic effect also stimulates the kidneys to eliminate excess fluids, along with unwanted concentrates. The combination of diuretic and antibiotic effects make this an excellent herb for cystitis.

Burdock helps alleviate many acute and chronic skin problems. I've seen the resulting clear, vibrant skin. Severe eczema and psoriasis may disappear over a period of six months to a year, especially if burdock decoction is combined with a good diet and exercise. A compress of concentrated burdock leaf infusion is also good for eczema. You can make burdock oil by soaking the root in olive oil for a month. People also use this externally, while drinking the decoction, for a variety of skin problems, from rashes to acne.

An infusion of the seeds also soothes external sores, cleanses the skin, soothes the mucous membranes and the kidneys, and promotes urine elimination. Collect them in autumn, when the burrs are dry and brown. Crush them in a bag with a hammer to free the seeds, and put them in water. The seeds sink, the trash floats.

A student got a big kick in the ribs a few days after collecting burdock with me. There were no broken bones, just a gigantic, painful bruise. I advised him to chop up the burdock leaves, mix them with clay, and bandage this poultice to his injury. In a few days, the bruise was gone. This poultice is also good for gouty swellings and inflamed surfaces.

EPAZOTE, MEXICAN TEA, WORMSEED, JERUSALEM OAK

(*Teloxys ambrosioides*)

This naturalized Mesoamerican annual usually grows from 3 to 5 feet tall. Its leaves are long-oval, 2 to 4 inches long, tapering at the base and tip, with coarse, wavy teeth. It branches, with clusters of tiny, greenish flowers, arising in spikes from the leaf axils. The seeds are glandular-dotted. The whole plant is always very fragrant—like pine or turpentine.

flower spikes arising from leaf axils

long, oval, wavy, toothed leaves

EPAZOTE × ½

Epazote grows on disturbed soil throughout the country.

Epazote looks a little like its edible relative, lamb's-quarters (page 45), which has diamond-shaped leaves and is always odorless.

This book discusses plants that are edible, not edible, poisonous, and medicinal. This plant is all four simultaneously: Poisoning symptoms include vomiting, diarrhea, abdominal cramps, dizziness, and headache. But every few years, the *New York Times* food section has an article on Mexican cuisine, touting epazote as an indispensable ingredient for authentic Mexican cooking. New Yorkers race past Central Park, ignoring volumes of epazote choking the walkways, and buy the same plant, imported from Mexico, at Macy's gourmet basement for $4.99 an ounce.

How can this plant be both edible and poisonous? Toxicity is often a matter of dosage. People who consume large quantities of epazote, mistaking it for lamb's-quarters, have gotten ill, while Hispanic people who use it sparingly, like parsley, thrive, especially if they're living in the tropics, where the plant's antiparasitic effects are beneficial. The active substance is called oil of chenopodium. (Epazote was in the genus *Chenopodium* until very recently.)

Epazote is covered with resin glands, which produce a mildly toxic resin. Fortunately, it's easy to avoid them, since resins never dissolve in water. If you don't believe me, try mixing water with turpentine. It can't be done outside of a zero-gravity environment because the liquids have different densities.

Avoid the resins altogether by putting fresh or dried epazote—leaves, stems, or flowers—in a tea bag or a tea ball. (You're supposed to use bay leaves the same way—see discussion under Bayberries, page 79.) Cook your recipe, remove the herb, and discard it. The flavor goes into the food, but the resins remain on the plant.

This herb is a superb wild seasoning. I sometimes use it with bayberry leaves. Epazote adds flavor to Mesoamerican cuisine, and it's great in beans, soups, and tomato sauces. I never make chili without it. It also helps digestion, and it reduces gas from beans.

Until recently, the dangerously toxic lab-extracted essential oil was used to treat roundworms, hookworms, dwarf tapeworms, and intestinal amoebas. Now it's been replaced by safer synthetics.

Moderate doses of epazote infusion are safe, but I never knew how effective they were until an elderly Cuban woman attended a tour. As a child in her native country, she had often observed a senior herbal healer using epazote for parasites. It was effective, without side effects.

The tea is also reputedly an antispasmodic, a cardiac stimulant, a diuretic, and a diaphoretic. It's also supposed to induce menstruation (although it's never had that effect on me), and reduce menstrual cramps. A poultice is supposed to be good for arthritis.

Some people enjoy the tea as a beverage; others don't like it. It seems this is an herb you either like or dislike, with no middle ground.

PINEAPPLE WEED

(*Matricaria matricarioides*)

This fragrant European annual provides one of my favorite herb teas.

It grows 4 to 18 inches tall and has fernlike leaves that are very finely divided (dissected) into linear segments. It has globular, yellow-green composite flower heads under ½ inch across. Under magnification, you'll see the many tiny individual buttonlike flowers. Unlike its well-known relative, chamomile (*Matricaria chamomilla*), it has no petallike ray flowers. Also, the whole plant smells like pineapple when you crush it. Nevertheless, people attending my tours are often certain that pineapple weed is chamomile.

Pineapple weed grows on disturbed sites and on poor soil, where there's plenty of sun. I find it in heavily trodden fields and cracks in the sidewalk. My groups are always skirting the edges of baseball diamonds, a preferred habitat, trying to gather this plant without getting conked by a fly ball. It comes up in mid-spring, flowers from late spring to fall, and grows throughout most of North America.

Use it fresh or dried. An infusion of one tablespoon of fresh leaves, stems, and flowers per cup makes a wonderful chamomilelike beverage. A

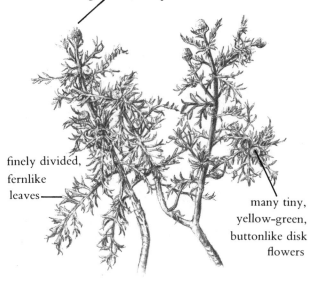

globular, composite flower heads

finely divided, fernlike leaves

many tiny, yellow-green, buttonlike disk flowers

PINEAPPLE WEED × ⅓

handful per cup is medicinal strength. This tea is one of the best for relaxation after a stressful day. It's a nervine, with a gentle calming effect on the nervous system, also traditional for menstrual cramps. It's also a carminative, improving digestion and relieving gas. It's also supposed to be an antiseptic, used for colds and urinary tract infections, such as cystitis. **Caution:** Some people, especially hay-fever sufferers, may be allergic to this plant.

POISON IVY AND ITS RELATIVES

(*Toxicodendron* species)

These relatives of the cashew, the mango, and the sumacs cause people more problems than any other plants, yet they're simple to avoid. After I've pointed out their different forms and asked people to locate them, even the kids are poison ivy experts by the end of one tour. The straightforward generic name comes from the Greek, where *toxico* means "poisonous," and *dendron* means "plant" or "tree."

Poison ivy (*Toxicodendron radicans*) has long-stalked, alternate, three-parted palmate-compound leaves. One leaflet points to the left, one to the right, and one has a stem and points straight ahead.

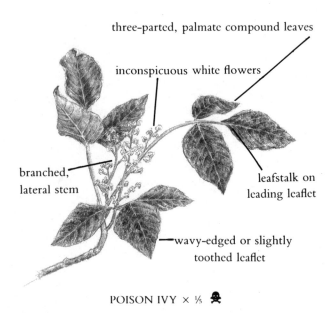

three-parted, palmate compound leaves

inconspicuous white flowers

branched, lateral stem

leafstalk on leading leaflet

wavy-edged or slightly toothed leaflet

POISON IVY × ⅕

The leaflets have some indentations on the edges that you could almost call teeth. The leaflets range from 4 to 14 inches long, with pointed tips and more rounded bases. Their leafstalks are reddish near the leaf's base.

These variable leaves are dark glossy-green most of the season, although they have red overtones when they first appear in the spring, and they turn scarlet in the fall. The plant is so beautiful in autumn, someone brought it to his garden in England, and now the British Isles are blessed with this plant.

Poison oak (*Toxicodendron quercifolium* on the East Coast, or *Toxicodendron diversilobum* on the West Coast) is nearly identical, except that its leaflets are partially subdivided into lobes. Botanists still argue over whether all these variants are really different species.

The plants' general forms are also highly variable. They may grow as herbaceous plants, upright shrubs, or woody vines. The vine has aerial roots that anchor it to the tree or fence but don't absorb nutrients. Dark, dense, hairy-looking aerial roots are a certain identifying characteristic in the winter.

Inconspicuous white flowers growing on branched, lateral stems bloom in late spring. They're followed by small, round, cream-white berries in the summer and fall. The berries are edible, but only if you're a bird. (The downy woodpecker eats the berries and spreads them through its excrement.) They're as poisonous to us as the rest of the plant: "Leaves of three, let it be; berries white, take flight." Of course, other plants, like blackberries, raspberries, wild beans, and hog peanuts also bear three-parted palmate-compound leaves.

One of the good things about poison ivy and its relatives is that, unlike many other plants, it's always easy to find. It grows throughout North America, favoring disturbed habitats such as edges of trails, on fences, in fields, in marshes, in thickets and woods. Because it needs plenty of sunlight, it doesn't grow in virgin forests or very old, undisturbed woods. It's especially common at the seashore, where its roots help prevent beach erosion and provide cover for small animals.

The plant is poisonous all year. I heard of a man who randomly picked up a poison-ivy twig to chew on in the middle of the winter. His throat got so swollen that the hospital had to insert a breathing tube to save his life.

The poison is the yellow oil urushiol, a lacquerlike phenolic compound. It doesn't affect animals, but four out of five people are allergic to it. Exposure leads to severe skin blistering—contact dermatitis—usually within one to twelve hours. Washing with soap containing oils spreads the urushiol, although washing with an oilless soap (see your pharmacist) helps prevent the rash. The oil also spreads through the blood, so you may even break out where you haven't touched the plant. Even contact with a dog that has run through poison ivy can cause a rash.

Avoid poison ivy even if you know you're not allergic to it, since repeated exposure may initiate allergy. Every time I pointed out poison ivy, one older woman who regularly attended my tours would shock everyone by holding a piece of poison ivy in her bare hands, innocently asking, in a thick German accent: "Mr. Brill, is this the plant you mean?" She was immune, until one day she showed up with bandages on her arms. She had flirted with danger once too often.

The best way to prevent the rash is to rub juice from the broken stem of jewelweed (page 73) on the affected area. Juice from the crushed leaves of plantain species (page 227) also help.

A colleague is so sensitive that oil carried by the wind sets off a reaction. She tried a radical solution: eating poison ivy, starting with one very tiny leaflet in early spring, when the urushiol content is minimal. The next day she ate two, then three. I don't advise this risky approach—it may backfire, and I wouldn't try it myself. However, she was desperate, and it worked. She desensitized herself, and continues to consume tiny quantities for maintenance. I've heard that Pacific Northwest lumberjacks routinely protect themselves from poison ivy in this way, but you can also get a bad rash where the poison ivy leaves enter the body— as well as where they exit!

The worst thing to do with poison ivy is to burn it. (If it becomes necessary to eradicate this plant, uproot it in late fall, wearing protective clothing, when it has a minimum of poison.) The Boy Scouts handbook forbids the use of any vine in campfires, to avoid such accidents. Smoke carries the oil, producing a rash over 100 percent of the body. If you inhale the smoke, you can get the rash in your throat, bronchial tubes, and lungs. This can be fatal, especially if you're camping out, where there's no hospital. Poison ivy inspired early-twentieth-century scientists to create weapons with substances as irritating to the human body as this plant. That's how the mustard gas of World War I, and subsequent biological warfare, originated.

Nevertheless, poison ivy has medicinal uses. An ointment of equal parts poison ivy vine, prickly ash bark, and alfalfa seeds is supposed to be good, applied externally, for arthritis. I don't know if it works, and I suggest you don't try to find out yourself. There are safer herbal and nutritional treatments. Poison ivy is also used in homeopathic medicine. Here, herbs are repeatedly diluted so many times that there are literally no molecules left in the medicine. This makes homeopathy difficult to understand from a scientific basis, but it's supposed to stimulate the body's defenses against the symptoms the offending substance causes. The patient's constitution is more important than the symptoms in choosing the best homeopathic treatment, but for people with skin problems, poison ivy is often included in the regimen. Of course, there are conflicting claims about whether homeopathy works. Some people benefit from it, others don't, but it's safer than drugs.

POKEWEED, POKE SALLET, INKBERRY, SCOKE

(Phytolacca americana)

The stout, branching stem of this nonwoody, native herbaceous plant grows 4 to 12 feet tall, and dies to the ground in the winter. It's tinged with purple-red: *Lac* means "red," and *phyto* means "plant," so *Phytolacca americana* means "American red plant." Pokeweed has oval, untoothed, stalked, alternate, emerald-green leaves 8 to 12 inches long and 2½ to 5 inches wide. The leaves, subject to a plant mosaic virus, are sometimes faintly mottled.

The radially symmetrical flowers, about ½ inch

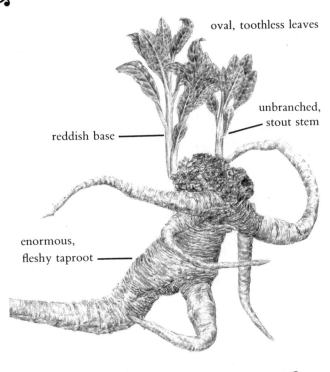

oval, toothless leaves

unbranched, stout stem

reddish base

enormous, fleshy taproot

POKEWEED SHOOTS AND TAPROOT × ⅛

across, have green centers and five white sepals resembling petals—no true petals. They bloom in the summer and fall, growing on 6- to 8-inch-long, bright magenta terminal racemes.

After the sepals fall off, in late summer or fall, the ovary at the flower's center enlarges and becomes a dark purple berry ⅓ inch across, filled with tiny seeds. Each berry is indented, as though poked, accounting for the plant's common name. The fruit stalk droops with the weight of the berries.

In the mid- to early spring, the young shoot's stout, unbranched stem is about 1 inch thick, with a reddish base. Most of the oval, toothless, alternating leaves still wrap around the stem, pointing upward.

The fleshy, perennial taproot can become enormous over the years.

This aggressive plant grows in disturbed and partially sunny areas, in fields and thickets, and at the edges of trails. It grows in eastern North America, from forest to seashore, and in the South all the way to the West Coast. It has escaped cultivation and become naturalized in Europe, and it's probably doing the same elsewhere.

Caution: Pokeweed is one of the most widely used edible wild plants in North America, even though it has poisonous parts. Because the toxins

can kill, this is not a plant for unsupervised beginners to eat. Study it over several seasons, build experiences with other wild foods, and work with an experienced forager before attempting to survive a pokeweed dinner.

Never include any pieces of poisonous root in your collection, and leave the poisonous mature plant, flowers, berries, and seeds alone. The berries are important energy sources for birds, as they were for the extinct passenger pigeons. They contain less poison than the stem, leaves, and roots, but the seeds inside are very poisonous.

Only the young shoots that emerge in mid-spring are safe. Collect shoots 6 to 8 inches tall. Larger shoots are dangerous, especially those with red stems. Shoots growing in the summer are poisonous, so collect only in springtime. Mature plants are very poisonous.

When I first found this plant, I could identify only mature pokeweed. Working backward every year, I identified younger and younger plants growing by the older plants, until I could recognize the youngest shoots.

To make the young shoots safe, always boil them in two changes of water, as described in the cooking section. Boiling vegetables usually destroys vitamins, minerals, and flavor, but pokeweed contains fat-soluble betacarotene, which is not destroyed by boiling, and there's still plenty of flavor. Although the boiling may destroy pokeweed's vitamin C and dissipate the iron, it's absolutely necessary to remove all traces of poison.

Properly cooked pokeweed tastes a little like asparagus. The flavor is strong, without being bit-

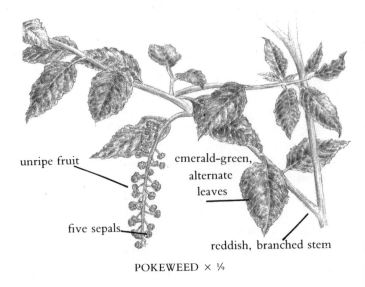

unripe fruit

emerald-green, alternate leaves

five sepals

reddish, branched stem

POKEWEED × ⅑

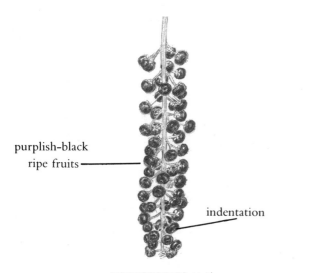

purplish-black
ripe fruits

indentation

POKEBERRIES × ⅓

ter or pungent. Season with oil and vinegar, if desired, and serve as a side dish. Add properly precooked pokeweed to soups and stews. I dry the cooked shoots, and reconstitute them in soups in the winter.

After learning about it from the Indians, the vitamin-starved pioneers celebrated poke's first appearance every year. We still have springtime pokeweed festivals in the South today, and some Southerners say no dish of greens is quite complete that doesn't contain poke. It's cultivated there, and you can buy it, canned, in the supermarket. It's also been imported for cultivation to Europe and Africa, where it sometimes escapes and becomes a "weed."

One of pokeweed's poisons, the alkaloid phytolaccine, causes increasingly severe vomiting, diarrhea, and intestinal cramps, as well as a burning mouth. There is usually visual impairment, and weakened respiration and pulse. Convulsions and death may follow. The folk remedy includes drinking lots of vinegar and eating a pound of lard. I doubt this works, although with proper medical treatment, victims usually recover within twenty-four hours.

Pokeweed also contains triterpenoid saponins, resins, phytolaccic acid, and tannin. A mitogen affects the DNA (genes) in the cells, affecting their ability to divide. The cells with the fastest turnover, those in the bone marrow, are most affected, so pokeweed poisoning may eventually lead to reduced red blood cell counts (anemia), or abnormalities of the white blood cells.

On the other hand, there are research reports that mitogens may be developed for making T- and B-lymphocyte (white blood cells) mature and stop dividing—the goal in treating leukemia, where immature white blood cells proliferate uncontrollably. (Mature white blood cells can't divide.) Pokeweed is also being studied for an antiviral substance that may be used to treat herpes and influenza. Antibacterial substances in the leaves are also under investigation.

Pokeroot has been used as medicine since the days of the Indians, but it's much too dangerous for any but the most skilled and experienced practitioners. People used to treat breast cancer with pokeweed poultices, to "burn out" the tumor and allow its removal. This causes severe burns, and probably doesn't work, but it's interesting that the folk culture detected pokeweed's anticancer potential, as discussed above.

People have used pokeberry tea for arthritis. In low dosages, it's supposed to act as an alterative to stimulate metabolism and decongest the digestive tract, the lymphatic system, and various other organs. The elderly mother of one of my students, well versed in Southern folk traditions, drank one cup of tea made from one pokeberry every day for decades, claiming it cured her arthritis and prevented its recurrence.

A pokeroot ointment is used externally for many skin diseases, including scabies, ringworm, and intermittent fungus infections. A lotion and tincture have also been used for vaginal yeast infections. Finally, presidential candidate James K. Polk and his supporters wore pokeweed leaves as a campaign symbol in 1844.

🌿 OTHER PLANTS OF DISTURBED AREAS IN MID- TO LATE SPRING

Edible or Medicinal Plants

Amaranth leaves, asparagus shoots, brassica leaves, catnip, chickweed, chicory roots, clover flowers, coltsfoot leaves, curly dock flower stalks, dandelion flowers and roots, dayflowers, daylily tubers, wild onion leaves and bulbs, garlic mustard leaves, flowerbuds, and flowers, ground ivy, hedge mustard leaves and flowers, horseradish leaves and roots, lady's thumb, lamb's-quarters leaves, orache leaves, peppergrasses, strawberry blite leaves, povertyweed leaves, milkweed shoots, mullein leaves, nettles, orpine leaves, plantain leaves, purslane, raspberry leaves, sheep sorrel leaves, spearmint, storksbill, thistle stalk, thyme, wood sorrel.

For Observation Only

Blackberry flowers, passion flower flowers, pin cherry flowers, plantain flowers, raspberry flowers, rose flowers, goatsbeard flowers and seeds, shepherd's purse seeds, sow thistle flowers, sumac flowers, thistle flowers, wild potato vine flowers, winter cress seeds.

PLANTS OF FIELDS IN MID- TO LATE SPRING

BLACK LOCUST

(*Robinia pseudoacacia*)

This thorny native tree grows up to 80 feet tall, with dark brown-gray, deeply furrowed bark. It's one of the worst trees to stand under during a thunderstorm, when torrents of electrically conductive rainwater flow down the grooves. The tree has feather-compound leaves 6 to 12 inches long with seven to twenty-one toothless, elliptical to oval leaflets 2 inches long. Pairs of small spines protect the leaf axils.

The white, fragrant, pealike flowers are bilaterally symmetrical. They bloom in mid- to late spring, arranged in drooping, dense, showy clusters. In late summer, the tree bears flat, hairless, black seedpods, 2 to 6 inches long and about ¾ inch across, containing four to seven flat, rounded seeds of less than ¼ inch. Good imagination and poor vision just might make these pods resemble locusts.

The black locust grows in dry woods, old fields, and near the seashore in eastern North America, and it's been planted throughout the rest of the country. A nearly identical relative with the

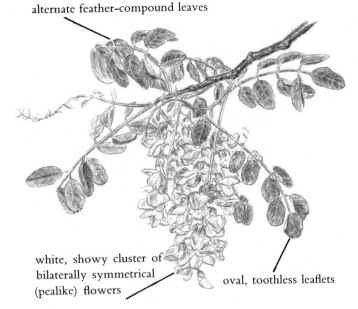

alternate feather-compound leaves

white, showy cluster of bilaterally symmetrical (pealike) flowers

oval, toothless leaflets

BLACK LOCUST TREE × ¼

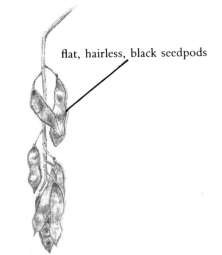

flat, hairless, black seedpods

BLACK LOCUST PODS × ¼

same food uses, the New Mexican locust (*Robinia neomexicana*), grows in the Southwest.

The generic name comes from John and Vespasian Robin, who first cultivated locust trees in Europe back in the 1500s. This legume superficially resembles the related acacia trees, so the specific name, *pseudoacacia*, means "false acacia."

There are no similar trees in North America with poisonous flowers. Some related herbaceous plants and vines have poisonous *pods*, but no trees with showy, pealike flowers and short, flat, dark-brown pods have poisonous *flowers*. A cultivated Asian species, the pagoda tree, has feather-compound leaves and similar flowers (I know of one person who mistakenly ate some, with no ill effects) but they bloom in the summer, not spring, and the pods are knobby and translucent green.

The legume or bean family includes both trees and nonwoody plants. This large, diverse, successful family has representatives around the world, having originated before the continents separated.

Legumes do something most other plants can't. They have swellings called nodules along their roots, where nitrogen-fixing bacteria live. The tree gives the bacteria a home, and the microorganisms take otherwise inert nitrogen from the atmosphere and make it available to the tree.

Carbohydrates and fats consist of oxygen, hydrogen, and carbon. Protein contains these plus nitrogen. All life requires protein, and most of the nitrogen in the food chain on land got there because of the legumes. That's why the black locust is a pioneer tree, appearing in overgrown fields before most other trees, and thriving in soil too nitrogen-poor for most of its competitors.

Black-locust wood is strong and durable. It's used for fence posts and for making wooden nails, but the best parts of this tree, from my point of view, are the edible blossoms. They're as sweet-tasting as they are beautiful and fragrant, with a flavor like perfumed peanuts. I use them raw in salads, simmer them five minutes in soups, and include them in pancake batter to make fritters. A student discovered that they're terrific stirred into oatmeal when you turn off the heat, at the stage where you let it sit covered, before serving.

The best way to preserve these gourmet flowers is freezing them raw in airtight containers or plastic bags. You can freeze locust-fritter batter or soup.

I once dried locust flowers in a food dehydrator, destroying all their texture and flavor, but leaving them otherwise intact.

Caution: Black locust leaves and inner bark contain a poisonous phytotoxin (plant toxin) that causes digestive system distress. Many sources also condemn the tiny seeds as poisonous without substantiating evidence, while the late foraging guru Euell Gibbons claimed them harmless, and fed them to his family. Because there are so many deadly legumes, and these seeds are so small, I've never tried them, and I advise you to follow my example.

At the other extreme, my colleague Jim Duke, writing in *Peterson's Field Guide to Herbs*, thinks the flowers are harmful even if you sniff them. Having sniffed, eaten, and fed the flowers to hundreds of adults and children sine the early 1980s without ill effects, I must disagree.

COW PARSNIP

(Heracleum maximum)

This is a huge, conspicuous herbaceous (nonwoody) perennial of the umbelliferous family, related to carrots and parsnips. Coarse-looking and woolly, it grows from 3 to 10 feet tall.

The woolly, toothed compound leaves are huge, often 3 feet long—partially divided into three sections, like maple leaves. The leafstalks' swollen bases clasp the stem, making the plant distinctive.

The umbrellalike flower head, which can be up to 8 inches across, consists of numerous tiny white flowers, with notched petals often purple-tinged. The outer flowers are larger than the more central ones. They give way to green seeds in summer. The large taproot smells spicy.

Caution: Don't confuse the cow parsnip with water hemlock, a deadly plant with purple-streaked stems and without swollen leafstalk bases. Also, cow parsnip contains furanocoumarins: Touching the plant in conjunction with sunlight and wet, sweaty skin may give sensitive people rashes.

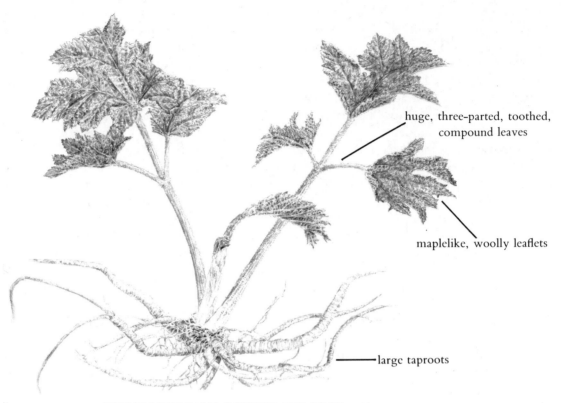

huge, three-parted, toothed, compound leaves

maplelike, woolly leaflets

large taproots

COW PARSNIP BASAL ROSETTE AND ROOT × 1/10

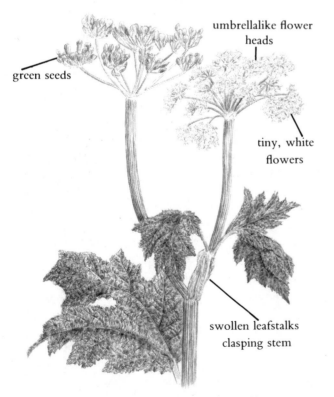

umbrellalike flower heads

green seeds

tiny, white flowers

swollen leafstalks clasping stem

COW PARSNIP MATURE PLANT × 1/10

The cow parsnip grows from Alaska through the northern half of the eastern and western United States, although its range persists in mountainous areas south to Georgia. I find it in wet areas—old meadows, open, partially wooded regions, and along the edges between meadows and woods—as long as the ground is moist.

If they're tender, you can eat the large roots of the young plant in spring, before the flower stalk develops, and in the fall. They taste like a combination of commercial parsnips and rutabagas, but stronger, so you may prefer them peeled and boiled.

The very young leafstalks and very young flower stalks, which taste like celery, are the best parts of this plant. You can peel them and eat them raw, simmer them in soups, or boil them in a couple of changes of water, depending on how strong they taste to you. The longer you cook them, the milder they get. You can also use the green seeds as seasoning in the summer, although I don't think they taste very good. Add them to soups, stews, and breads the way you'd use celery or caraway seeds.

Various Indian tribes used a decoction of the roots for colds, coughs, sore throats, flu, headaches, and cramps. They applied them as a poultice for sores, bruises, swellings, boils, and arthritis. An infusion of the roots is supposed to be good for indigestion and asthma, and people used to use it for epilepsy.

LAMB'S-QUARTERS, GOOSEFOOT, PIGWEED

(Chenopodium album)

Once immigrants brought lamb's-quarters from Europe, the seeds were spread throughout the continent. Lamb's-quarters is a branching, herbaceous annual that usually grows from 3 to 5 feet tall, although I've seen it 10 feet tall under especially favorable circumstances. Its slender, grooved stems are often tinged with red—especially next to the leaf joints.

The long-stalked, alternate, simple leaves are mealy white underneath. People ask if the white powder is mold or pesticide spray, but it's a protec-

tiny, ball-like, inconspicuous, green flowers

flowers in dense spikes in leaf axils

mealy, white underside

long-stalked alternate leaves

branching stems

slender grooves

diamond-shaped leaves

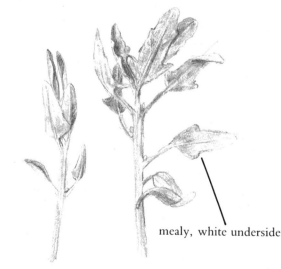

mealy, white underside

YOUNG LAMB'S-QUARTERS SHOOT × 1

LAMB'S-QUARTERS × ½

tive, waxy bloom. The young leaves are linear, while the older leaves, which can grow up to 4 inches long, are diamond-shaped.

Tiny, green, inconspicuous, edible, balllike flowers grow in short, dense spikes in the upper leaf axils and at the plant's tip, in summer and fall, eventually turning reddish-brown. By late autumn, each plant may have as many as 75,000 tiny, shiny, black seeds inside.

Lamb's-quarters has virtually no odor. There are similar *Chenopodium* species that you'll recognize once you're thoroughly familiar with lamb's-quarters. You can eat the odorless species. Use those that are resinous-smelling the way you would epazote (*Teloxys ambrosioides*, page 36), but don't eat them—large quantities are poisonous.

Orache (*Atriplex patula*) is a salty-tasting edible relative that grows near the seashore and in salty, alkaline soil throughout the country. Its leaves are arrow-shaped instead of diamond-shaped, and while other wild-food authorities praise the virtue of this salty-tasting species, my students and I usually reject it as unpleasantly bitter. It's better in some locations than others.

Strawberry blite (*Chenopodium capitatum*) is another edible, odorless relative. Its leaves are more triangular than lamb's-quarters—even arrow-shaped, like orache. In the summer, the flowers growing in the leaf axils are more rounded than lamb's-quarters. The insipid-tasting fruits that subsequently develop look like tiny bright-red strawberries.

Povertyweed (*Monolepis nuttalliana*) is a fleshy, branching, sprawling annual relative that grows up to 1 foot tall, with lance-shaped alternate, mealy-textured leaves about ½ to 2½ inches long. They're usually irregularly toothed, sometimes smooth, with a pair of lobes at the bases. Many clusters of tiny, petalless, greenish flowers grow in the upper leaf axils, followed by brown, flattened fruits. Look for it in alkaline or dry soil from the eastern edge of the Great Plains to California, and from southern Canada to Mexico. Use it like lamb's-quarters.

Where does lamb's-quarters get its common name? I've heard two stories: 1. The mature leaf is supposed to look like a cut of lamb meat, the quarter. 2. There's an ancient English festival called "Lammas Quarter," associated with the lamb's-quarters relative orache, which was also confusingly called lamb's-quarters in England. Plants in this group may have gotten their common name from the festival.

The plant's other names are obvious. It's called pigweed because pigs eat it. The leaf is shaped like a goose's foot, and *cheno* means "goose," while *podium* means "foot." *Album* means "white," referring to the leaf's underside. So the translation of *Chenopodium album* is "white goose's foot."

Use the tender, young shoots under 10 inches tall when they first appear in mid-spring, and continue collecting the leaves, especially the smaller, more tender ones at the tips, until the plant is killed by frosts in late fall.

Lamb's-quarters is very common throughout the country. You'll find it in backyards and vacant lots, in overgrown fields, in urban parks, and along roadsides. It does especially well in poor or disturbed soil, but it isn't particular, and will just as soon take over the best parts of your garden as thrive in the sandy soil near the seashore.

If you begin learning wild foods with only a few plants, as I recommend, this widely distributed, easy-to-identify, tasty, nutritious, long-in-season plant should be one of the first on your list.

Caution: Lamb's-quarters will absorb nitrates from contaminated soil, and make you sick. (See the warning under Amaranth, page 145.) It will also pick up the carcinogenic pesticide 2-4 D.

Lamb's-quarters sometimes falls victim to a leaf miner, a very tiny insect larva that lives inside the leaf. It leaves a visible, red path where it's eaten its way through the leaf. If one stand of the vegetable is infested, seek your supply elsewhere.

Lamb's-quarters is delicious any way you prepare it. It tastes like its relative spinach (it's also related to beets), only better, and it's never bitter, as spinach sometimes is. Include it in salads, steam it (see the instructions on waterless cooking, page 281), throw it in soups, stews, or casseroles, sauté it, or dry it to store. It tastes much better than spinach in quiches. However, it shrinks by about two thirds when you cook it, so be sure to collect enough.

Lamb's-quarters dries well. You can easily reconstitute the leaves for any recipe, or powder them and use them to enrich and flavor main-course dishes. It's especially good in tofu and cheese dishes.

Even the tiny black seeds are edible, and very nutritious. They provide protein, calcium, phosphorus, potassium, and niacin. Napoleon used them to make bread for his army when other food was scarce. Collect them in late autumn, as described under Amaranth (page 145). Note: Amaranth is also called pigweed, although it's a different plant.

Along with dandelions and watercress, lamb's-quarters is one of the most nutritious of foods. The leaves are a super source of beta carotene, calcium, potassium, and iron—superior to spinach. It also provides trace minerals, B-complex vitamins, vitamin C, and fiber.

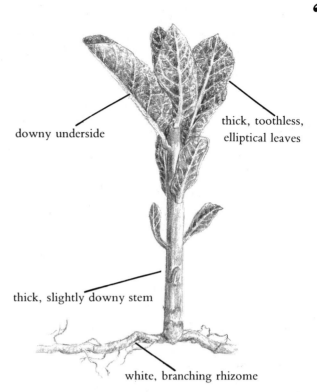

downy underside

thick, toothless, elliptical leaves

thick, slightly downy stem

white, branching rhizome

MILKWEED SHOOT AND RHIZOME × 1/3

COMMON MILKWEED

(Asclepias syriaca)

If you identify and prepare this herbaceous annual properly, it's a most delicious wild vegetable. However, there are serious dangers involved in foraging for this beautiful plant that set it off limits to unsupervised beginners.

Common milkweed is a thick-stemmed, unbranched, upright herb that grows 3 to 5 feet tall. It contains a bitter, white, sticky latex or milk—protection against insects. It has thick, velvety, toothless, elliptical, opposite leaves 4 to 9 inches long, 2 to 4½ inches wide, downy underneath, tapering to a rounded point at each end. Milkweed's stem is hairy under magnification—an important difference from poisonous look-alikes. Underground, there's a white rhizome or rhizomes, giving rise to true roots and small or large stands of plants.

The flower buds look like loose heads of broccoli, 1½ to 3 inches across. The long-stemmed, purple-pink, attractive flowers are set in umbrellalike globular clusters, often drooping. Each individual flower is ½ inch tall and less than ¼ inch across, divided into five radially symmetrical parts.

warty, cucumberlike seedpods

MILKWEED SEEDPODS × 1/4

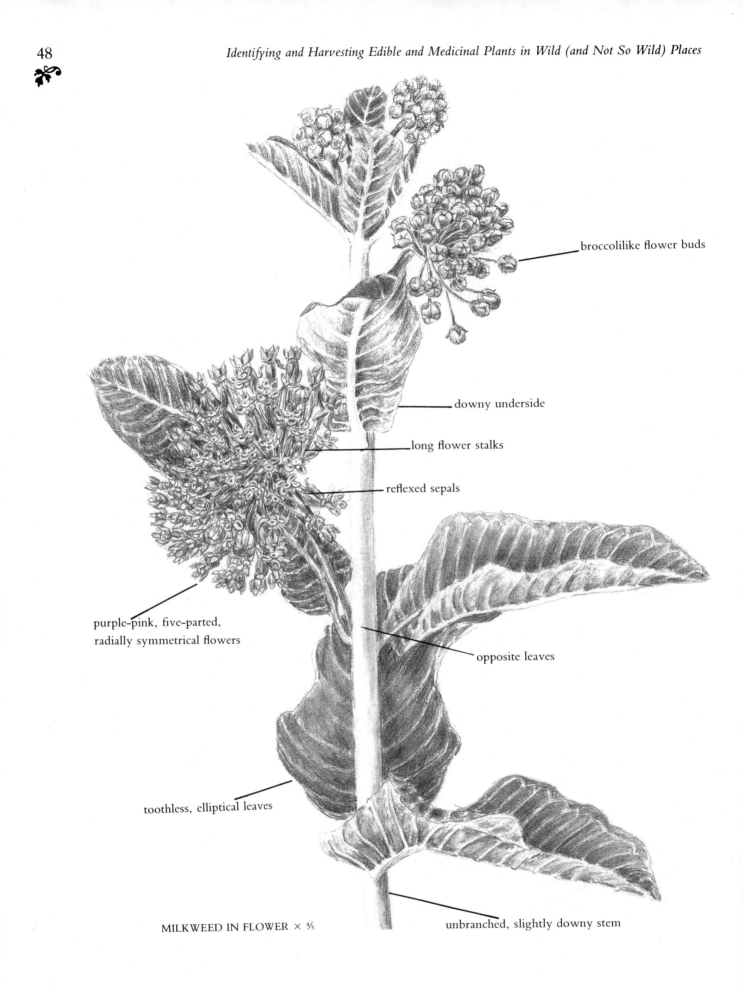

broccolilike flower buds

downy underside

long flower stalks

reflexed sepals

purple-pink, five-parted,
radially symmetrical flowers

opposite leaves

toothless, elliptical leaves

unbranched, slightly downy stem

MILKWEED IN FLOWER × ⅕

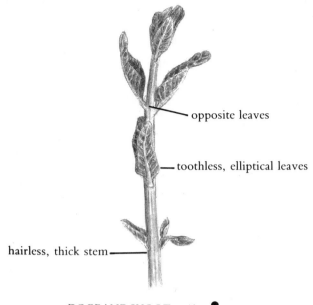

DOGBANE SHOOT × ¼ 🕱

- opposite leaves
- toothless, elliptical leaves

hairless, thick stem

The sepals are about as long as the flower, and they're reflexed (bent backwards). The fragrance attracts human and insect foragers alike.

The mature seedpods look like warty cucumbers, about 4 inches long, filled with silky seeds. I once distributed them to some high-spirited grade-school children in a special education class. They had a great time taking them apart. Then I made the mistake of showing them how the seeds propagate: I gently puffed on one, and its built-in parachute carried it aloft. Within a few seconds, the children had blown all the seeds into the air, and the classroom looked like the scene of a blizzard.

Look for common milkweed in old fields and on roadsides. It grows in disturbed habitats, and in the poor, sandy soil near the seashore. It ranges through most of the eastern United States.

BUTTERFLY WEED × ¼ 🕱

Caution: Common milkweed's shoots closely resemble dogbane's (*Apocynum* species). This toxic relative has the same leaves, and white, milky sap. Tiny hairs on milkweed's hairy stem, visible only under magnification, provide the sole distinction. Dogbane's stem is bald, although its *leaves* are hairy under magnification, as are common milkweed's. When they mature, dogbane's branching habit (milkweed doesn't branch), its tiny grouping of small, white flowers, and its long, thin, curved and paired seedpods set them apart.

There are more dangers. A milkweed species called butterfly weed or pleurisy root (*Asclepias tuberosa*) is also poisonous, although people used it to treat pleurisy, an infection of the membranes surrounding the lungs, before antibiotics were discovered. Unlike common milkweed, butterfly weed's sap is clear.

Other milkweed species on the West Coast and Southwest with white, milky saps are poisonous. Only use common milkweed (*A. syriaca*), which grows in the eastern United States. Avoid all other species there, and don't use milkweeds elsewhere.

Milkweed shoots are in season in mid-spring. The flower buds and flowers are ready in early summer, and the immature pods are edible by mid-summer.

You must prepare milkweed properly. The white, milky sap is mildly toxic. I know two people who get skin rashes on contact, although most people are unaffected. To get rid of the bitter sap, boil the plant in two changes of water. (See the cooking section, page 272.)

Follow this procedure for shoots 8 inches tall or less, when they first appear in mid-spring, with the tender tops of older plants, and the young, immature leaves near the tip of the plant later on. In these stages, the vegetable tastes a little like sweet string beans. Use the same procedure for the unopened flower buds, which look and taste like broccoli.

The sweet, open flowers have much less sap. Parboil them one minute, then add them to pancake batter to make fritters, or include them in soups, stews, casseroles, or other vegetable dishes. They have a wonderfully refreshing, perfumed flavor. After the flowers fall off, but before the seedpods mature, while they're still firm and under 1½ inches long, they're also good boiled in two changes of water. But I find collecting them too labor-intensive.

Be sure the plant is very abundant. There are areas where it's so common it's considered a nuisance, but it may be rare elsewhere. Milkweed is also a favorite of other species. Don't get stung by bees when you're gathering milkweed flowers. Bees love milkweed, and they're a good sign, indicating your dinner isn't contaminated—they're the first insects killed by insecticides.

Monarch butterfly larvae also use milkweed. Whereas the bees collect nectar from the flowers, monarch caterpillars eat the leaves. (Note: Transfer any accidentally collected caterpillars onto other milkweed plants.) They don't eat enough to harm the plant. Since milkweed sap is bitter, so are the monarchs. A bird may eat this bright orange and black insect once, determine that it tastes like the food I had to eat in summer camp, and never attack another monarch again.

There's even a monarch mimic, the viceroy butterfly, that resembles the monarch so closely that predators avoid it. For generations, textbooks pronounced it without bitter flavor, although the writers never tasted one. When scientists recently fed birds dewinged viceroys, they spat them out. They're just as unpalatable as the monarchs. In fact, poisonous and bad-tasting species worldwide, including a newly discovered New Guinea genus of poisonous birds, appropriately called the *Pitohui*, are all colored orange and black. This may be evolutionary truth-in-advertisment.

Milkweed's edible parts provide water-soluble vitamin C, probably dissipated by boiling, and fat-soluble beta carotene, probably retained. The root contains many biologically active substances.

Named after Asklepios, the Greek god of healing, milkweed's long list of medicinal applications goes back centuries. Most important, a root decoction strengthens the heart in a different way from digitalis, and without the foxglove derivative's toxicity. It's up to medical researchers to determine how to use it, and in what doses. Even though the herb is safe, don't use it for dangerous conditions without competent supervision.

The root decoction also soothes the nerves. It's listed as an emetic, anthelmintic (it kills worms), and stomach tonic. It helps relieve edema (water retention), probably by strengthening the heart.

It's also a diaphoretic and expectorant. People use it for coughs, colds, arthritis aggravated by the cold, threatened inflammation of the lungs, asthma, bronchitis, female disorders, diarrhea, and gastric mucus.

During World War II, scientists unsuccessfully tried to make rubber from the sap, and schoolchildren had to collect bags of the fluffy seeds to fill life preservers.

MINTS (many genera)

This is one of the easiest groups of wild foods to recognize, and various species grow throughout the United States. All mints have square stems and opposite leaves, although not all plants with square stems and opposite leaves are mints. Some mints are odorless, but the fragrant species are useful to foragers. Any plant with a square stem, opposite leaves, and a minty fragrance is a mint, and all mints are relatively safe. **Caution:** Commercially available essential oils of mints are toxic if you take them internally, but these substances are too diluted in the plants to make them dangerous.

If a mint species smells good, it makes a tasty tea and a good seasoning. Chop up the fresh leaves finely, or crumble the dried leaves, and include them in fruit dishes, sauces, desserts, or salad dressings.

The infusion facilitates digestion, and it's a mild astringent—good for diarrhea. Large doses are diaphoretic—good for colds or flu. Modern medicine uses menthol, the main active ingredient, as an antiprurient (to stop itching). Applied externally, it mildly anesthetizes pain receptors, and mildly stimulates cold receptors, creating a pleasant cooling sensation.

Mints have been used since ancient times, at least since the time Hades, the Greek god of the underworld, tried to rape the nymph Minthe. Hades' wife, Persephone, caught him in the act, and foiled her husband by turning Minthe into a sweet herb. After that, the Greeks placed mint around dead bodies, reminding Hades of his defeat, and driving him away. Mint also covers up the smell of the dead.

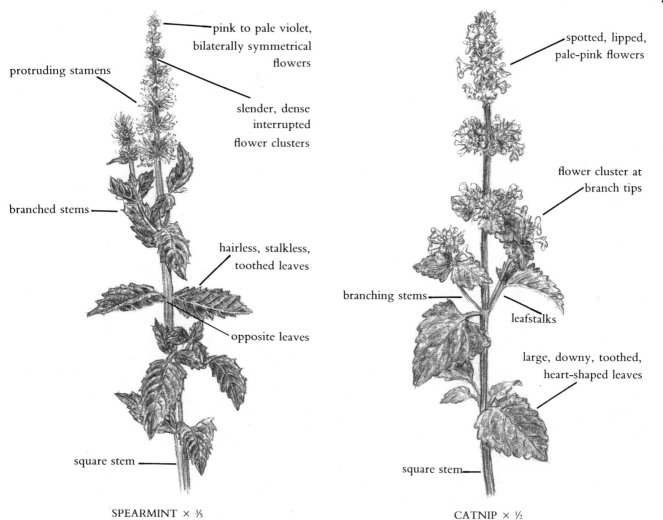

pink to pale violet, bilaterally symmetrical flowers

protruding stamens

slender, dense interrupted flower clusters

branched stems

hairless, stalkless, toothed leaves

opposite leaves

square stem

SPEARMINT × ⅗

spotted, lipped, pale-pink flowers

flower cluster at branch tips

branching stems

leafstalks

large, downy, toothed, heart-shaped leaves

square stem

CATNIP × ½

Fragrant mints are easy to identify as to genus, but harder to identify as to species. They're very prolific, so new, unlisted foreign cultivated species often make their way into the wild.

It's best to identify mints in flower, anytime from spring to fall, although the majority flower in summer. The small flowers grow in the leaf axils, or in terminal clusters. They're bilaterally symmetrical, with the lower petal expanded to form a "lip." Flower colors may be white, pink, yellow, violet, and blue. Here's a sampling of aromatic mints:

Spearmint (*Mentha spicata*) is a European species with branched stems, often tinged with purple, that grows up to 10 to 20 inches tall. The toothed, stalkless leaves are hairless. The tiny flowers, which bloom in summer and fall, are arranged in slender, dense, interrupted clusters. They're pink to pale violet, with protruding stamens. The odor and flavor are familiar from many commercial products that exploit the plant's essential oil, although the mint essence of the commercial products is always synthetic. Look for spearmint in wet areas along roads, in ditches, and in wet meadows throughout the United States. Collect from early spring to late fall.

The leaves are high in beta carotene and vitamin C. Spearmint is one of the best remedies for flatulence. It's a mild astringent, good for children's diarrhea and stomachaches. Unlike peppermint, it's also a mild diuretic.

Catnip (*Nepeta cataria*) is famous for its remarkable intoxicating effects on all members of the cat family. One sprig will transport any feline into a state of sheer ecstasy.

This branching European perennial can reach

over 3 feet in height. The large, toothed, downy, heart-shaped, stalked leaves are up to 3 inches long. The whole plant has a grayish appearance because of the heavy whitish down that covers it. The lipped flowers clustered at the tips of the branches are pale pink, with tiny purple spots. Inside each little flower are red anthers. The calyx is soft-hairy. Look for it in wet areas along roads, in ditches, and in wet meadows throughout the United States. Collect from spring to fall.

Catnip contains tannins and a volatile oil including nepetalactone, a mild sedative. The herb also acts as a diaphoretic, so it helps you sweat out and sleep—a blessing for people with infections such as colds, flu, measles, or other childhood diseases. It's one of the best herbs for children with infections. It also helps calm unsettled stomachs, and alleviates colic, gas, and diarrhea. I know one mother who got good results giving this herb to her colicky baby. It's also good for convalescents in weakened states.

I've found water mint (*Mentha aquatica*) growing in only one stream, where it was probably planted. It favors very wet places, flowering in late summer and fall, and it tastes wonderful. It grows 18 to 30 inches tall. The leaves are ovate-round. It's covered with distinctive curved hairs. The flower heads—crowded in globular terminal spikes—are wider than those of peppermint, and the stems and leaves are often purplish. Collect from early spring to fall.

Peppermint (*Mentha piperita*), which is similar, has a familiar odor and hot taste. Its dark green, toothed leaves are hairless and short-stalked, and its branching stems are purplish. This perennial grows up to 2½ feet tall, or reclines. The flowers are pink to violet. They grow on short or interrupted spikes, blooming from summer to fall. This European herb is a common cultivated mint, and it grows wild throughout the United States, preferring wet places, such as ditches, along streams, and in wet meadows. Collect from spring to fall.

Peppermint is about 0.4 percent volatile oil, consisting of menthol, azulene, vitamin C, bioflavonoids, vitamin E, and many other substances. It's the most powerful mint for digestive disorders. It increases stomach acidity, essential for digestion. The flavonoids stimulate the gallbladder to con-

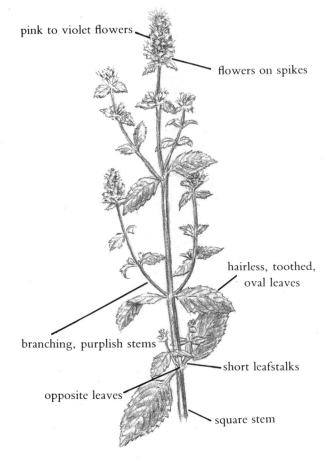

pink to violet flowers

flowers on spikes

hairless, toothed, oval leaves

branching, purplish stems

short leafstalks

opposite leaves

square stem

PEPPERMINT × ½

tract and secrete bile. It also normalizes gastrointestinal activity, reduces gas, and lessens cramps. There's also evidence that its menthol inhibits or kills many kinds of pathogenic microbes, including herpes simplex and other viruses. Its azulene has been shown to reduce inflammation and heal ulcers.

Peppermint is the mint of choice for dyspeptic adults, good for nausea and diarrhea, while spearmint, which is gentler, is more beneficial for children. Peppermint is used for fevers and chills, heart disorders, arthritis, convulsions and spasms, and headaches. Oil of peppermint, applied externally, is also supposed to relieve the pain of shingles, and calm nerves. It's often included in analgesic massage oils and liniments.

American pennyroyal (*Hedeoma pulegioides*) is the most pungent mint I've ever smelled, even stronger than peppermint. It has small, lance-

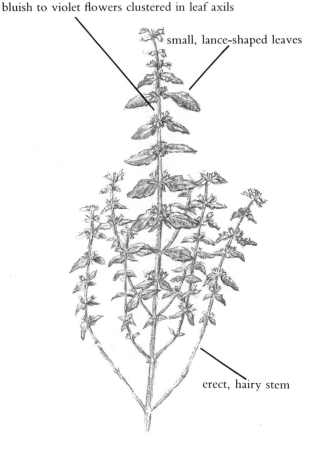

bluish to violet flowers clustered in leaf axils

small, lance-shaped leaves

erect, hairy stem

PENNYROYAL × ⅕

tight clusters of tiny flowers in leaf axils

downy, unbranched stem

pale violet to white, bell-shaped flowers

WILD MINT × ½

shaped, toothed or toothless leaves, ½ to 1½ inches long. The erect, hairy stem grows from 6 to 18 inches tall. The pale bluish-violet (sometimes pinkish) flowers grow in clusters in the leaf axils. This European herb is naturalized in the eastern half of the United States, and escapes cultivation elsewhere. It grows in dry woods, in fields, and on mountains—especially common along paths in the woods. Pennyroyal is best when it's just flowering, although you can collect from spring to fall.

Dry at a very low temperature so the oils don't evaporate. As with many other mints, the tea is a diaphoretic, carminative, aromatic, and stimulant. It's also an emmenagogue: Some Indians called it "squaw mint" because it helps suppress menstruation. Pennyroyal is great for the flu—especially stimulating to the sinuses and bronchial tubes, because its oils seem more concentrated than other species'. They say that pennyroyal is a great insect

repellent, but it doesn't work for me. Sachets of dried pennyroyal create a refreshing scent in closets, and supposedly repel moths.

Wild mint (*Mentha arvensis*) has downy, unbranched stems and tight clusters of tiny, pale violet to white, bell-shaped flowers in the leaf axils. In a group of great-tasting herbs, it has its own wonderful flavor. The menthol used in commercial flavoring originally came from a cultivated variety of this species. This herb grows on damp soil and shores in eastern North America and California. Collect from spring to fall.

Oswego-tea or bee-balm (*Monarda didyma*) is a very fragrant, tasty mint that can grow up to 3 feet tall, with large, toothed leaves and very hairy stems. Scarlet tubular flowers grow in a dense head. Large red or purplish bracts surround the flower head. Other similar *Monarda* species' flowers have different colors. Look for them along streams, in wet meadows, and in other wet places through-out eastern North America from spring to fall. It's a great tea and seasoning. Like wild thyme (below), it contains thymol, which has antibacterial, anti-fungal, and antihelminthic properties. Indians boiled the roots and leaves together for worms and parasites. Use it medicinally like wild thyme.

Horse balm, stoneroot, or citronella (*Collinsonia canadensis*) is another favorite mint with strongly lemon-scented flowers and fruits. Called stoneroot because the perennial root is large, knobby, and very hard, this branching native grows from 2 to 5 feet tall. The oval, pointed, toothed leaves are huge, up to 10 inches long, probably to get more sun in the plant's shady habitat. The yellow flowers have especially long lower lips and fringed tips. They grow in large, loose, branching clusters (panicles), unusual for the mint family. I find it in rich moist woodland areas, espe-

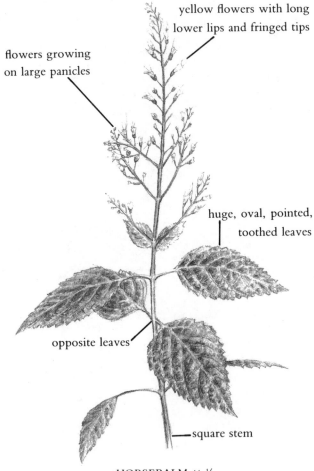

flowers growing on large panicles

yellow flowers with long lower lips and fringed tips

huge, oval, pointed, toothed leaves

opposite leaves

square stem

HORSEBALM × ⅓

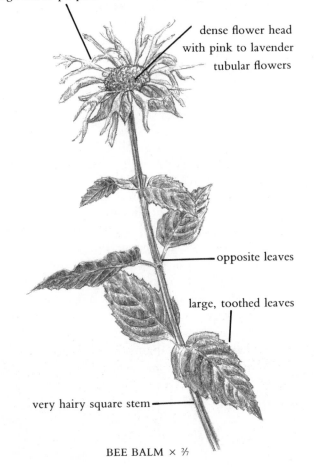

large red or purplish bracts around flower head

dense flower head with pink to lavender tubular flowers

opposite leaves

large, toothed leaves

very hairy square stem

BEE BALM × ⅔

cially on woodland roads. Collect the flowers and seeds in late summer and fall. The root is good all year.

An infusion of the root is antispasmodic, tonic, astringent, diaphoretic, and diuretic. It's used for cystitis, bladder infections, constipation, indigestion, cramps, sore throat, and congestion. Although eating the fresh leaves may cause vomiting, a tea of the flowers and fruits is delicious.

The first time I found wild thyme (*Thymus serpyllum*), I had been walking on it and smelling it for hours, without realizing what I was doing. I finally looked down, and there was the same herb I'd occasionally bought fresh, for cooking. Wild thyme is closely related to the garden variety. This European plant is especially abundant in the Alps, and it was brought here from Eurasia.

It's a weedy, sprawling plant, growing from 4

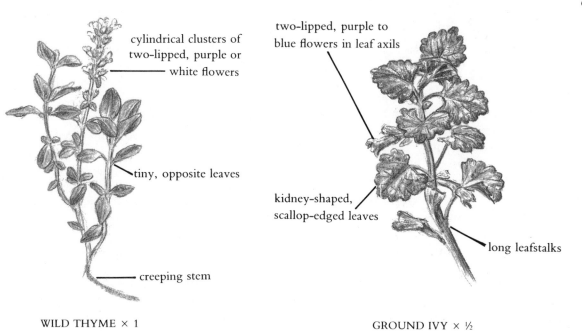

cylindrical clusters of
two-lipped, purple or
white flowers

tiny, opposite leaves

creeping stem

WILD THYME × 1

two-lipped, purple to
blue flowers in leaf axils

kidney-shaped,
scallop-edged leaves

long leafstalks

GROUND IVY × ½

inches to 1 foot long, and forming dense mats. The leaves are a little larger than garden thyme's but still tiny (about ⅛ inch wide and ½ inch long). Cylindrical clusters of flowers grow on top of the plant, or in the leaf axils. They're two-lipped and bilaterally symmetrical—purple or, rarely, white. It grows in fields, along roadsides, and in disturbed habitats. It's in season from spring to fall, flowering in the summer.

Thyme is a standard culinary seasoning for salads, stuffing, eggs, fish, and vegetables. To make thyme vinegar, simply cover a jar of thyme with vinegar, leave it for a few days or weeks, and strain it out. It's great for cooking, and inhaling the fumes is good for headaches caused by nervousness.

I was long familiar with this plant as a culinary herb when I discovered that it also has important medicinal properties. Thyme contains B-complex vitamins, vitamins C and D, trace minerals, borneol, fluorine, triterpenic acids, and gum. It also contains thyme oil, which consists of thymol and carvacrol. Thymol is a proven antibiotic, antiseptic, antifungal, and anthelmintic (it expels or kills worms).

An infusion of the leaves, stems, and flowers is good for colds, flu, coughs, asthma, chronic respiratory problems, sinusitis, mucus congestion, headaches, stomach problems, and gas. It helps

relax the nervous system—useful for insomnia, nervous disorders, and depression. The infusion makes a good underarm deodorant, mouthwash, and after-shave lotion. It even lowers cholesterol levels.

Honey of bees feeding on thyme is great for relieving coughs and sore throats (honey is antiseptic). Thyme honey, a traditional remedy for headaches and upset stomachs, has a real thyme flavor. Use it sparingly.

Caution: Although using medicinal dosages of thyme is safe, the isolated oil is toxic, and only competent professionals should use it. Also, avoid using thyme medicinally if you're pregnant.

There's also a common trailing mint called gill-over-the-ground or ground ivy (*Glecoma hederacea*). This fragrant perennial's long-stalked, round to kidney-shaped, scallop-edged leaves are ½ to 2½ inches across. They're often tinged with purple. Two-lipped purple-blue flowers appear in whorls in the leaf axils, just before mid-spring. Each joint or node in the creeping stem produces roots. You can find it in meadows, on lawns, along roadsides, and in sunny places throughout most of the United States. It grows from very early spring to very late fall.

Beginners often confuse this European plant with the common blue violet (page 229), which is odorless, or garlic mustard (page 254), which smells like garlic. However, ground ivy smells like basil. It makes an excellent tea, providing vitamin C. **Caution:** Lifestock has reportedly been poisoned by eating the leaves, so use it only as tea, especially if you're a cow or horse.

Drying drives off its essential oils, so gather it fresh—you can't buy it; it's therefore not in most herb books. It's been used as a cough remedy, and a treatment for jaundice, poisoning, sciatica, gout, and the plague. A poultice of bruised fresh leaves was applied to wounds and sores. The plant is called gill-over-the-ground because it was used to clarify and add a bitter flavor to beer—gill is a technical term from beer making.

It's also an effective diuretic. I once had a date with a woman who lived about five miles from me. We spent the afternoon foraging, shared some delicious ground ivy tea, and off I went, bicycling home. From one cup of tea, I needed to make five pit stops. Unfortunately, there were no pits.

❧ OTHER PLANTS OF FIELDS IN MID- TO LATE SPRING

Edible or Medicinal Plants

Amaranth leaves, asparagus stalks, bayberry leaves, bracken fiddleheads, burdock roots and flower stalks, chicory root, clover flowers, curly dock stalks, wild onion leaves and bulbs, groundnut tubers, lamb's-quarters leaves, strawberry blite, povertyweed, milkweed shoots, mugwort, mullein, brassicas, peppergrasses, field pennycress, orpine leaves, pineapple weed, pokeweed shoots, purslane, sassafras leaves, flowers, and roots, sheep sorrel leaves and flowers, thistle flower stalk, strawberry leaves, yarrow.

For Observation Only

Autumn olive flowers, Russian olive flowers, blackberry flowers, burdock flowers, caraway flowers, chokecherry flowers, pin cherry flowers, sweet cherry flowers, goatsbeard flowers and seeds, groundnut flowers, hawthorn flowers, mulberry flowers, passionflower flowers, rose flowers, sow thistle flowers, strawberry flowers, sumac flowers.

PLANTS OF THICKETS IN MID- TO LATE SPRING

❧

GREENBRIER, CATBRIER, BLASPHEMY VINE, STRETCHBERRY

(*Smilax* species)

This is a group of long, woody, climbing, perennial vines with round to heart-shaped leaves, slender, paired tendrils, and nasty thorns. The shiny-green, leathery leaves are about 4 inches long, with pointy tips and parallel veins. In the South, they're nearly evergreen.

Several small, inconspicuous, yellow or green six-petaled flowers grow together in small, flat-topped clusters in the spring. Male and female flowers grow on separate vines. The fruits are small globular bluish-black berries, ¼ inch or less across,

ripening in the fall and often persisting into the winter.

It's called greenbrier because the vine is green, or catbrier because the tendrils resemble cats' whiskers and the thorns scratch like a cat. It's also called blasphemy vine because you commit blasphemy after it scratches you. All species are good to eat, and there are no poisonous look-alikes. Carrion flower (*Smilax renifolia*) is a related nonpoisonous vine that resembles greenbrier, but with rank-smelling flowers and no tendrils.

This atypical member of the lily family prefers partially shaded areas, such as the edges of trails and openings in the woods. Different species grow throughout most of the United States.

Greenbrier is best in mid-spring, when you can

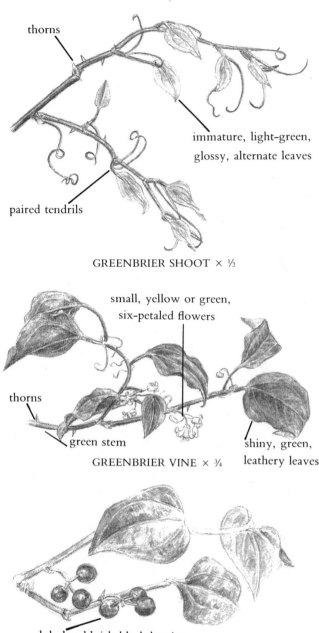

thorns

immature, light-green, glossy, alternate leaves

paired tendrils

GREENBRIER SHOOT × ½

small, yellow or green, six-petaled flowers

thorns

green stem

GREENBRIER VINE × ¾

shiny, green, leathery leaves

globular, bluish-black berries

GREENBRIER BERRIES × ¾

especially if you're a vegetarian. Also, beware of poison ivy (page 38), which grows in the same habitats.

Greenbrier has a piercing, sweet-sour flavor. It's one of my favorite salad greens—the leaves are more delicate than any other wild vegetable. It's great in soups, or steamed or sautéed. Cook no more than five minutes.

It's quick and easy to collect in quantity, especially when all the leaves are young and tender, but it's more perishable than other greens, spoiling after a few days, even refrigerated in sealed bags.

There are reports that you can dig up the gelatinous roots of the larger greenbrier species, dry and powder them, and add them to soups or stews, but other sources say this labor-intensive use doesn't work.

Also, greenbrier's dry, seedy, tasteless berries aren't worth collecting. The vine is sometimes called stretchberry because the berries are like rubber. Some birds or mammals dine on them in the winter, and they're welcome to them. The vine itself, including the sharp thorns, makes an excellent winter survival food, but only if you're a deer.

HERCULES'-CLUB, DEVIL'S-WALKING-STICK, ANGELICA TREE

(Aralia spinosa)

You can identify this small, branchless shrub or small tree by its trunk alone: All 6 to 30 feet are covered with such nasty-looking, sharp, sometimes stout spines, only the Devil could use it as a walking stick. If Hercules were to swat you on the butt with it, you'd take your meals standing for a very long time. Even the twice-compound alternate leaves, which can be up to 6 feet long, are thorny. The many medium-sized oval leaflets are toothed and pointed.

In late summer, the tree has halos of tiny, white flowers in umbels, accounting for the name: angelica tree. This member of the ginseng family is not related to angelica (page 66), and the tiny, black berries that follow the flowers are certainly not

eat all the tender, young, shiny, emerald-green leaves, as well as the shoots. (The shoot includes the tender tip of the stem, unfolding leaves, and associated young tendrils.) Even the thorns on the shoots are edible—they're tender, not sharp. The mature leaves of late spring and early summer are too tough to eat, but there are still plenty of tender shoots until midsummer. Note: Ants love greenbrier, so shake off your shoots before you eat them,

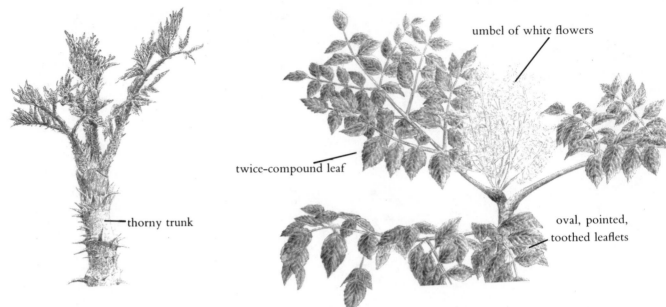

thorny trunk

umbel of white flowers

twice-compound leaf

oval, pointed,
toothed leaflets

HERCULES'-CLUB SHOOT × ¼

HERCULES'-CLUB × ¹⁄₂₀

angelic. They're poisonous if eaten in quantity and guinea pigs have died after eating the seeds.

As if the poisonous fruit and spiny thorns were not defense enough, some people develop contact dermatitis, a poison ivylike rash, with inflammation and blisters, from handling the bark and roots.

Spreading underground by rhizomes, Hercules'-club establishes large stands along streambanks. I've found it growing wild in open areas of moist woods. It's also planted as an ornamental in urban parks. You can find this native plant, which is also native to Asia, throughout most of eastern North America, and on the West Coast wherever people have introduced it. Despite its formidable armaments, Hercules'-club is an excellent edible. The developing shoot and very young, tender leaves are delicious cooked.

Cut off the shoots before the leaves are fully unfurled. They're in season in mid-spring. Steam or sauté them, or simmer them 10 to 15 minutes in soups. They're a little like asparagus spears, but more meaty.

The Japanese marinate this vegetable in vinegar and serve it, cooked, with rice. **Caution:** It may be unsafe to eat raw. Also, don't take more than a few shoots from any one tree. I have observed that they regenerate, but we shouldn't push its recuperative powers too far.

I once came across a man using this species as a walking stick. I was quite wary, lest he tempt me to sign a contract in blood that was too good to be true. It turns out that you can make a walking stick out of this forbidding-looking tree. The trunk is straight and strong, and just the right size. Just be sure to trim off all the thorns.

ring of tiny,
black berries

HERCULES'-CLUB BERRIES × ⅓

WISTERIA

(*Wisteria* species)

This is a smooth-barked, twining vine with beautiful, edible blossoms and poisonous seeds. It lives hundreds of years, so the woody base may be as thick as a tree trunk. The alternate leaves are

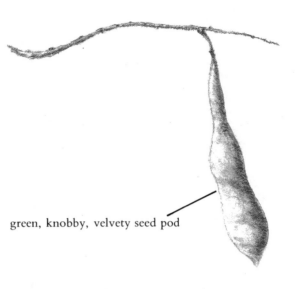

green, knobby, velvety seed pod

WISTERIA POD × ½

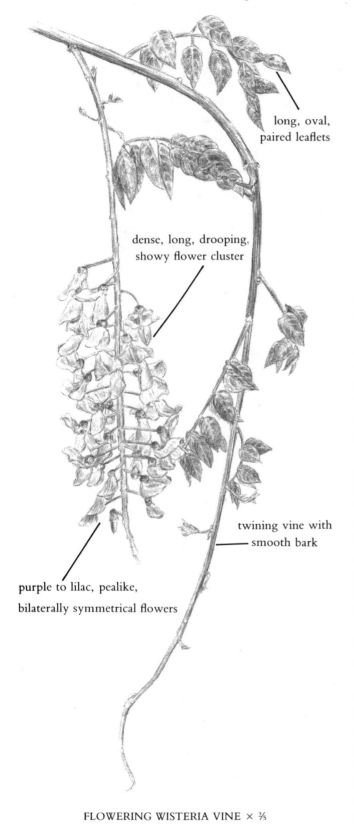

alternate, feather-compound leaves

long, oval, paired leaflets

dense, long, drooping, showy flower cluster

twining vine with smooth bark

purple to lilac, pealike, bilaterally symmetrical flowers

FLOWERING WISTERIA VINE × ⅖

feather-compound, 2 feet long, with long-oval, toothless pairs of leaflets.

The dense, long, drooping clusters of showy, fragrant flowers are purple to lilac, sometimes even white. They're pealike, bilaterally symmetrical, about 1 inch long—adapted for bee pollination. The bees have to crawl into the flower to get the nectar. The old pollen brushes off, and new pollen for the next flower collects.

There are native and Asian species. The plant gets its name from a man named Wister, who first brought the Asian species to America. The cultivated Asian species, which readily escapes into the wild, blooms in mid-spring. Other species may bloom earlier. Look for wisteria in cultivated gardens and adjacent thickets, and along streams and floodplains, in partially shaded areas in woods. Unfortunately, it often shades out and kills the trees that support it.

Wisteria flowers have a sweet, perfumed flavor. Use them the same way as black locust flowers (page 42).

Caution: In the summer, wisteria produces green, knobby seedpods 4 to 6 inches long. Asian species' pods are velvety; native species' pods are smooth. The seeds within reportedly contain toxic glycosides, lectins, or both. They're so poisonous that two seeds can cause severe illness. Symptoms include mild to severe nausea, repeated vomiting, abdominal pains, and diarrhea. In extreme cases, dehydration, collapse, and death may occur. If you recover, you usually do so within twenty-four hours.

In the 1960s, a friend was tempted to eat wisteria seeds by her girlfriend's boyfriend, who was trying to poke fun at her because she was a vegetarian and ate sprouts: "Here, have some of these; they're good. I eat them all the time," he said, and unknowingly fed her poison. She got so sick, she could hardly crawl to the bathroom to throw up. She eventually recovered completely, but her doctors were surprised that she survived.

No animal is stupid enough to eat and scatter the seeds, so the plant has a different way of spreading. The pods have two layers of tissue, which dry unevenly. Since organic materials tend to contract when they dry, this creates great tension. The pod finally splits and the seeds are ejected explosively. People who take the pods home may be rudely awakened late at night by botanical "gunfire."

OTHER PLANTS OF THICKETS IN MID- TO LATE SPRING

Other Edible or Medicinal Plants

Bayberry leaves, elderberry flowers, grape leaves, groundnut tubers, mugwort, nettle leaves, pokeweed shoots, raspberry leaves, redbud flowers, sassafras leaves, roots, and flowers, spicebush leaves.

For Observation Only

Autumn olive flowers, Russian olive flowers, blackberry flowers, blueberry flowers, currant and gooseberry flowers, grape flowers, groundnut flowers, hawthorn flowers, passionflower flowers, poison ivy flowers, raspberry flowers, rose flowers, sumac flowers, viburnum flowers, wild potato vine flowers.

PLANTS OF WOODLANDS IN MID- TO LATE SPRING

HONEWORT, WILD CHERVIL

(Cryptotaenia canadensis)

Honewort is an erect, freely branching perennial, growing from 1 to 3 feet high, with a smooth, hairless, faintly ridged stem. The three-parted palmate compound leaves are hairless, dark green and shiny above, and whitish-green below. The oval, pointed leaflets are sometimes lobed, with sharp, uneven teeth.

Like other members of the parsley family, it has tiny white flowers growing in loose clusters, like a leaky umbrella. The flowers' stems are different lengths. The cup-shaped, sepalless flowers appear in late spring. By midsummer, they give way to oblong, flattened, ribbed green seeds that are often curved. Goutweed (page 259) looks somewhat similar, but the leaves and flowers are more dense.

Honewort grows in rich, moist soil, in light shade. You can find it near streams and in woods. It's a native of eastern North America and Japan,

showing that these regions were much closer together in the geological past.

Honewort has a fresh herby flavor, like goutweed, celery, or chervil. Collect it in spring and use the leaves as a parsleylike seasoning, raw or cooked for 5 to 15 minutes. Look for new growth in summer and fall, when the smallest leaves are the best. In Japan, it is used in sushi and miso soup.

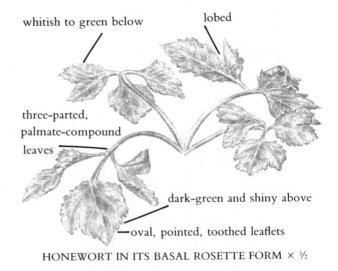

whitish to green below

lobed

three-parted, palmate-compound leaves

dark-green and shiny above

oval, pointed, toothed leaflets

HONEWORT IN ITS BASAL ROSETTE FORM × ½

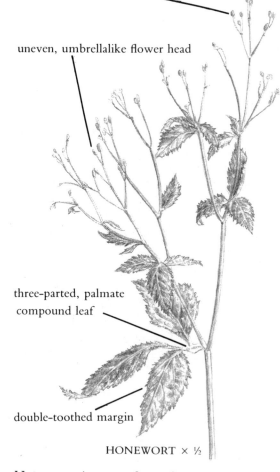

tiny, white, radially symmetrical flowers

uneven, umbrellalike flower head

three-parted, palmate compound leaf

double-toothed margin

HONEWORT × ½

Honewort is one of my favorite plants to dry. I crumble it in almost any recipe; it's especially nice in bean soups. Since it grows abundantly, it's quick collecting.

Use the seeds like caraway seeds. The roots of the young plant are supposed to be like parsnips or carrots, but I always find them small and tough. Like its relatives, angelica and cow parsnip, you can use the stem like celery. It's mild and crunchy.

Traditional herbalists in New England use an infusion as a diuretic and urinary tract tonic, to strengthen and cleanse the kidneys, and to relieve frequent urination.

I'd been hunting for honewort for years, and kept missing it, even though it's common. One of my students finally taught it to me. She transplanted some of this as-yet-unidentified plant into her weed garden, where any cultivated plant that dares appear is summarily destroyed. When it flowered, she identified it with a field guide, and showed it to me.

KENTUCKY COFFEE TREE

(Gymnocladus dioica)

This native tree grows from 40 to 100 feet tall, with dark, scaly, grooved bark. It bears gigantic, twice-compound leaves—each up to 3 feet long, with seven to thirteen oval, toothless, paired leaflets. In the spring, white flowers grow in clusters from the leaf axils. The best identifying characteristic is the flattened, curved, brown seedpod, 5 to 10 inches long, 1 to 1½ inches wide, and 1 to 2 inches thick. Inside are six to ten flattened seeds under 1 inch in diameter, ¼ inch thick, set in a sticky, green pulp.

Caution: The pods somewhat resemble those of poisonous legumes. Don't use them without expert supervision. The longer, flat, twisted green pods of the honey locust have edible green seeds, but those of the green pods of the knobby wisteria vine are deadly.

You can find the Kentucky coffee tree growing in rich soil and woods in the Southeast, especially west of the Appalachians. Farther north, and throughout the rest of the country, landscapers often plant this large tree in parks, where it does well but doesn't normally reproduce.

The pods ripen in autumn, but unless there's

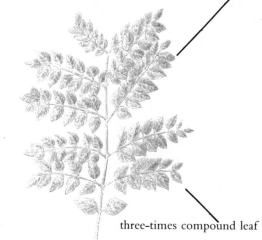

7 to 13 oval, toothless, paired leaflets

three-times compound leaf

KENTUCKY COFFEE TREE LEAF × ⅟₁₅

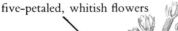

five-petaled, whitish flowers

KENTUCKY COFFEE TREE FLOWERS × ⅔

an exceptional windstorm, they don't fall to the ground until the following spring. The relatively imperishable seeds eventually fall out of the disintegrating pods, and you can often collect them from the ground all year.

This tree's seeds provide an excellent caffeine-free coffee substitute, even though it's a legume, unrelated to coffee. **Caution:** The green pulp and *raw* seeds are poisonous. They contain hydrocyanic acid, the alkaloid cytosine, and toxic saponins. People reportedly have been poisoned by eating the raw pulp, and livestock have been poisoned after eating the sprouts, leaves, and pods. Symptoms include severe gastrointestinal irritation and nervous disorders.

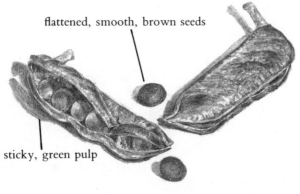

flattened, smooth, brown seeds

sticky, green pulp

KENTUCKY COFFEE TREE PODS × ⅓

To make a safe wild coffee, wash all the pulp off the seeds, and bake them at 300°F for three hours, in a covered roasting pan. Cooking destroys the poison, and the cover stops the seeds, which pop like popcorn, from making a mess in your oven.

Grind the seeds to the consistency of coffee in the blender, one-half cup at a time, and brew or percolate in your coffeemaker, as you would ground commercial coffee. Experiment with how strong to make your wild coffee—it's a matter of taste. Speaking of taste, my first experience making wild coffee from this tree was a disaster. It tasted just like coffee, which I can't stand. People who like coffee, on the other hand, enjoy this wild caffeine-free substitute.

I use the seeds as a seasoning, the way you sometimes use coffee to flavor chocolate or mocha cakes. Use one or two ground-up roasted seeds in carob or chocolate cake or carob-chip cookie recipes, to provide a coffee overtone. (Carob or St. John's bread is a legume with seeds that taste like chocolate. Unfortunately, it grows only in Asia.)

There's also a folk remedy for radiation poisoning using Kentucky coffee tree seeds, cornsilk, linden flowers, and the seaweeds Irish moss, kelp, and dulse. But I can't elicit any volunteers to take a tour of Chernobyl to see if it works.

LINDEN, LIME TREE, AMERICAN BASSWOOD, MONKEY-NUT TREE

(*Tilia* species)

The lindens are attractive, medium-sized trees 60 to 100 feet tall. The alternate, simple, heart-shaped, toothed leaves have pointed tips and uneven bases.

When the fragrant, yellow-white flowers are in season, you can find the tree by simply following your nose. The flowers attract bees, other insects, and people. They're about ½ inch wide, with five petals and sepals. Racemes of stalked flowers hang from a striplike bract, about 5 inches long and 1 inch wide. The strip with attached fruits subsequently catches the wind and spins off like a heli-

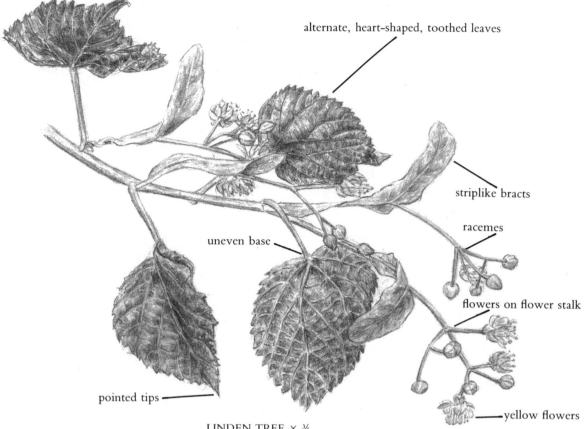

alternate, heart-shaped, toothed leaves

striplike bracts

racemes

uneven base

flowers on flower stalk

pointed tips

yellow flowers

LINDEN TREE × ⅓

copter. The fruits are hard, green, spherical, and stalked. Because they're spherical, small—only ¼ inch across—and paired, they're commonly called "monkey nuts."

There are a few similar species in the United States, all good to use: The imported European linden (*Tilia europaea*), has leaves 2 to 4 inches wide. It's been planted in parks and on residential streets throughout the United States. The American basswood (*Tilia americana*) is larger, usually growing 60 to 80 feet or taller. It has leaves 4 to 10 inches wide. It's native to the eastern United States, growing in the moist soil of valleys and woods. It's sometimes planted in other regions, especially in parks and botanical gardens. Lindens have no poisonous look-alikes. Avoid the bees that are also attracted to the flowers.

Linden flowers bloom in late spring and early summer. They produce a volatile oil, including farnesol, which creates the attractive fragrance.

Steep a handful of linden flowers in a cup of water for one of the best-tasting herb teas in the world. This beverage is very popular in Europe, but virtually unknown in the United States. A strong infusion is relaxing. People use it for nervousness, stress, high blood pressure, headache, and insomnia. It's even been recommended for spasmodic conditions such as asthma and epilepsy.

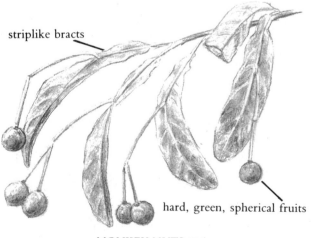

striplike bracts

hard, green, spherical fruits

MONKEY NUTS × 1

It's a diaphoretic, great for the flu, colds, and accompanying mucus congestion. It's also supposed to be good for hoarseness.

A tea of the inner bark is a diuretic. It's used for kidney stones and gout. It's also antispasmodic, and it dilates the coronary arteries. Under medical supervision, it may be good for coronary heart disease.

The white wood is valued for carving, and the Indians made a strong rope from the tough, fibrous inner bark of the native species. I love this tree, even though it causes problems on my walks. It's hard to get people to leave when it's time to look for the next plant.

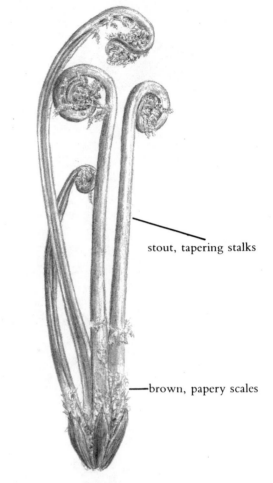

stout, tapering stalks

brown, papery scales

OSTRICH FERN FIDDLEHEAD × ⅕

OSTRICH FERN

(*Matteuccia struthiopteris*)

All ferns have two kinds of fronds (leaves)—larger sterile (vegetative) and smaller fertile (reproductive) ones. The ostrich fern is one of the best edible species. It's easiest to recognize when mature: The large, dark-green sterile fronds grow 2 to 6 feet tall. They're featherlike, with one main stem running through the frond. Toothed leaflets arise in a curved arch from the main stem, so they do actually resemble ostrich plumes. They come out of the ground as though arising from a vase. Unlike bracken's (page 31) frond, the ostrich fern is unbranched.

The fertile fronds are much smaller, at most 2 feet tall—leathery and stiff, with rounded podlike sori (spore-containers). They grow surrounded by the much larger sterile fronds.

Caution: You eat only fern fiddleheads—they become poisonous once they've unfurled. The ostrich fern's fiddlehead has a curled head, like a violin. The stalk is firm and stout near the base, then rapidly tapers. It's covered with large, brown, papery scales. Thick rhizomes creep along and sprout the new fronds from a black, rough tangle of branches. Last year's tough, small fertile fronds persist through the winter.

The cinnamon fern's fiddlehead looks similar, but its fertile frond is reddish and knobby, not feathery like the ostrich fern's. It's often difficult to identify one fiddlehead from another in the spring, and many species aren't edible. Some are bitter, others are bound in a thick, papery membrane that's nearly impossible to remove, but no fiddlehead is poisonous.

If you're uncertain of your identification, come back to confirm your ID when the fern is mature, and harvest the following spring. I had to wait more than ten years to locate ostrich ferns in my region, and another two years to pinpoint the right time to collect.

Ostrich ferns grow in moist woods, and near riverbanks, where there's some sunlight. They occur in Canada and the Northeast, ranging westward through South Dakota and northern Washington State. When you locate them, you'll probably find them in abundance.

Ferns are primitive plants that reproduce via spores instead of flowers and seeds. While most ferns today grow in undisturbed woodland habi-

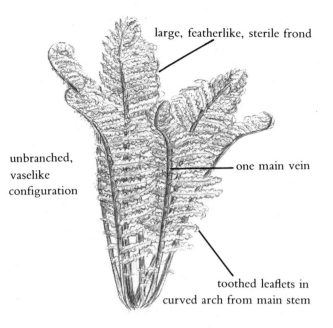

large, featherlike, sterile frond

unbranched, vaselike configuration

one main vein

toothed leaflets in curved arch from main stem

OSTRICH FERN × ¹⁄₁₂

name because it comes covered with showy clusters of small, long-stalked, fragrant, purple-red flowers in the spring, before the leaves appear. It's a legume, so the blossoms are pealike—bilaterally symmetrical. The fruit is a flattened, pinkish-green pod 2 to 3 inches long.

The redbud is supposed to grow in moist woods and thickets throughout eastern North America, and in the South through Texas, but it's one of the few native species favored by landscapers, who plant it in cultivated parks throughout the country. Another redbud species (*Cercis occidentalis*) grows in the foothills, dry slopes, and canyons of California, Nevada, Utah, and Arizona. It's nearly identical to the eastern species, and you use it the same way.

The flowers have a sharp, sweet-sour flavor everyone loves, and they provide vitamin C. They're gorgeous when added raw in salads. You can toss them into oatmeal when it's finished cooking, put them in pancake batter to make fritters, add them to bread dough, or use them in cooked vegetable dishes. Freeze them in ice cubes and serve

tats, there are always exceptions: When a brushfire occurs in the woods, ferns are often the first to recolonize. So a worldwide layer of dense fossil fern spores in sediments from just after the age of the dinosaurs is evidence that their extinction was due to an impact with a comet or asteroid. After worldwide forest fires caused by the catastrophe, ferns recovered first.

Use them like bracken. You need not peel off the brown scales, and you can eat them in quantity without the cautions associated with bracken. They're less bitter than bracken. You can slice them up raw and put them into salads, or cook them 10 to 15 minutes. They taste a little like asparagus.

Remember, don't eat unfurled fronds, which are poisonous. Collect them only where they're very abundant, and take less than half of any bunch, so you don't deplete the rhizomes' energy reserves.

REDBUD, JUDAS TREE

(*Cercis canadensis*)

The redbud is a native tree, often shrubby, growing up to 50 feet tall, with a rounded crown. The toothless, alternate, pointy-tipped, heart-shaped leaves are 2 to 6 inches broad. It gets its

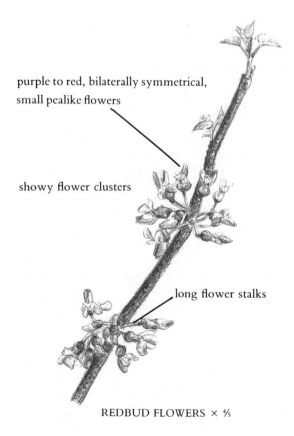

purple to red, bilaterally symmetrical, small pealike flowers

showy flower clusters

long flower stalks

REDBUD FLOWERS × ⅖

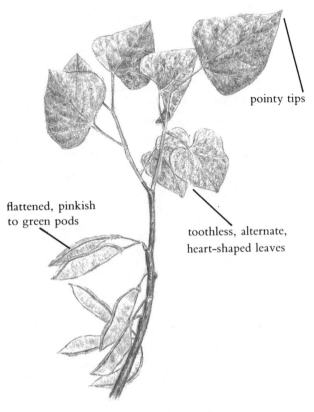

pointy tips

flattened, pinkish
to green pods

toothless, alternate,
heart-shaped leaves

REDBUD PODS × ⅓

them in drinks, or use them in winemaking. You can also pickle the unopened flowerbuds.

Although this is one of the best-tasting flowers you'll ever experience, you should always leave some flowers to mature. In the very beginning of the summer, you can supposedly use the tender, young pods, lightly cooked, like snow peas. Because they're inedibly tough when I find them, I'm either finding them too late, or the information is incorrect.

OTHER PLANTS OF WOODLANDS IN MID- TO LATE SPRING

Other Edible or Medicinal Plants

Aniseroot leaves, flowers, and seeds, black birch, bracken fiddleheads, cow parsnip flower stalks, dayflowers, daylily tubers, elderberry flowers, garlic mustard leaves, flower buds, and flowers, goutweed leaves and flowers, grape leaves, greenbrier shoots and leaves, groundnut tubers, Hercules'-club shoots, horse balm root, jewelweed, nettles, pennyroyal, pine needles, shoots, and pollen, pokeweed shoots, raspberry leaves, sassafras leaves, flowers, and roots, spicebush leaves, strawberry leaves, storksbill leaves and flowers, violet leaves and flowers, wild gingerroots, wild leek leaves and bulbs, wintergreen leaves and berries, wood sorrel.

For Observation Only

Black birch catkins, black cherry flowers, black walnut catkins, butternut catkins, blackberry flowers, blueberry flowers, currant and gooseberry flowers, false Solomon seal's flowers, ostrich fern fronds, grape flowers, hackberry flowers, hawthorn flowers, Kentucky coffee tree flowers, mayapple flowers, mulberry flowers, poison ivy leaves and flowers, pawpaw flowers, raspberry flowers, rose flowers, Solomon's seal flowers, sweet cherry flowers, viburnum flowers, waterleaf flowers, wild ginger flowers, wild potato vine flowers.

PLANTS OF FRESHWATER WETLANDS IN MID- TO LATE SPRING

ANGELICA, ALEXANDERS

(Angelica atropurpurea)

This fragrant nonwoody biennial grows 4 to 9 feet tall. Its alternate, three-parted compound leaves again divide into three or five jagged-toothed leaflets. A swollen sheath forms at the base of the leaf and wraps around the stem. The stout, smooth dark-purple stem lends it its specific name, *atropurpurea*. It is a member of the carrot (umbelliferous) family, and its large, globular flower clusters consist of many tiny greenish-white flowers that bloom from June to October. A taproot grows underground. This native plant grows along streambeds and in swamps.

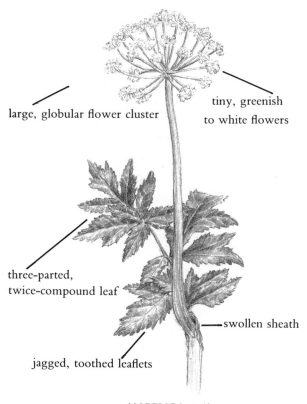

large, globular flower cluster

tiny, greenish to white flowers

three-parted, twice-compound leaf

swollen sheath

jagged, toothed leaflets

ANGELICA × ⅙

Angelica is like a blend of juniper, celery, and asparagus. From mid-spring to early summer, you can enjoy the leaves as a seasoning, or use the peeled stem like asparagus. Peel the very young leafstalks and flower stalks and eat them raw, simmer them in soups, or boil them in a couple of changes of water if they're too strong-tasting. Use the green seeds in late summer and fall as a seasoning in soups, stews, and breads, like celery seeds or caraway seeds. With a taste like juniper, they're used commercially to flavor gin, soft drinks, and confections.

Caution: This is not a plant for beginning or even intermediate foragers to use without expert supervision. It closely resembles its deadly poisonous relative, water hemlock (page 128). Although water hemlock's fingerlike roots have hollow chambers, people have died after confusing the two.

Don't eat fresh angelica roots. They may be poisonous, although preparations of the root, stem, and seeds are used medicinally. The plant's common name comes from the archangel Michael, who instructed a monk, in a dream, to have people

chew European angelica root (*Angelica archangelica*) to protect themselves from the plague. The constituent pinene is both antimicrobial and an expectorant.

You can make a decoction or tincture from the wild root, or get it in health food or herb stores. Indians and colonists used the native species as a general tonic, and for bronchitis, colds, pleurisy, flatulence, colic, liver disorders, rheumatism, anemia, urinary problems, indigestion, and gout. It's supposed to create an aversion to alcohol and help treat anorexia. Laboratory research has demonstrated the validity of many of these claims.

This herb is contraindicated for pregnant women, although a Chinese species, dang gui (*Angelica sinensis*) is touted as "woman's ginseng." Angelica is not good for diabetics, although it's supposed to help hypoglycemia (low blood sugar).

CATTAIL

(*Typha* species)

The cattail is one of the most important and common wild foods, with a variety of uses at different times of the year. A stand of cattails is as close as you'll get to finding a wild supermarket.

You can easily recognize a cattail stand: White, dense, furry, cigar-shaped overwintered seedheads stand atop very long, stout stalks, even as the young shoots first emerge in early spring. The immature swordlike, pointed leaves, with parallel veins, resemble other wetland plants, but last year's stalks provide positive identification.

By late spring, the light-green leaves reach nearly 9 feet tall, forming a sheath where they tightly embrace the stalk's base. The leaves hide the new flower head until it nears maturity. Peel them back to reveal it. The plant is so primitive—dating back to the time of the dinosaurs—that male and female flowers are separate on the stiff, two-parted flower head. The pollen-producing male is always on top, while the seed-bearing female is forever relegated to the bottom. Clearly, this species evolved long before the Sexual Revolution.

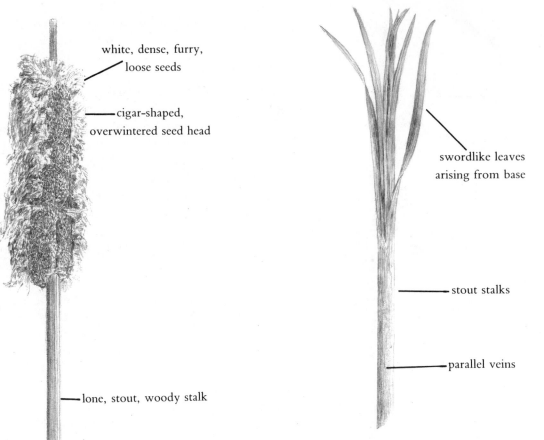

white, dense, furry, loose seeds

cigar-shaped, overwintered seed head

lone, stout, woody stalk

OVERWINTERED FLOWER HEAD OF CATTAIL × ⅖

swordlike leaves arising from base

stout stalks

parallel veins

CATTAIL SHOOT × ⅙

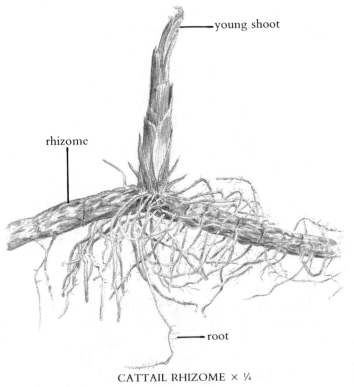

young shoot

rhizome

root

CATTAIL RHIZOME × ¼

(Biologically speaking, this arrangement is effective because the male part withers away when its job is done, whereas the female part must remain connected to the rest of the plant until the seeds have matured and dispersed.)

Once fertilized, the female flowers transform into the familiar brown "cigars"—also called candlewicks, punks, ducktails, and marsh beetles—consisting of thousands of tiny developing seeds. They whiten over the winter after the leaves die, and the cycle repeats.

Cattails grow in dense stands. Like most colonial plants, they arise from rhizomes—thick stems growing in the mud, usually connecting all the stalks. A cattail stand is like a branching shrub lying on its side under the mud, with only the leaves and

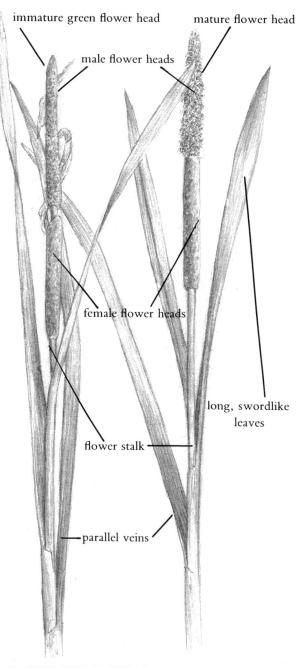

immature green flower head

mature flower head

male flower heads

female flower heads

long, swordlike leaves

flower stalk

parallel veins

CATTAIL STALKS WITH FLOWER HEADS × ¼

reeds have flaglike flowers, and leaves originating along the stalks. When the two species compete, reeds tolerate more salt and win out on land. But they can't grow in shallow water, like cattails.

The reed's young shoot is barely edible in early spring. After lots of peeling, the small yield tastes so bad, you'll be glad there's no more. When this was one of the few plants I could identify, I also wasted time and effort trying to extract very scant starch from reed rhizomes, and searching for the "edible" seeds as rare as hen's teeth. Following the advice of authors who hadn't tried their own suggestions, I learned the hard way to leave the reeds to the wildlife and saxophone players.

Caution: Young cattail shoots resemble non-poisonous calamus (*Acorus calamus*), and poisonous daffodil (*Amaryllidaceae*) and iris (*Iris* species) shoots, which have similar leaves. If a stand is still topped by last year's cottony seedheads, you know you have the right plant. In spring, the cattail shoot has an odorless, tender, white inner core that tastes sweet, mild, and pleasant—a far cry from the bitter poisonous plants, or the spicy, fragrant calamus. None of the look-alikes grows more than a few feet tall, so by mid-spring, the much larger cattail becomes unmistakable, even for beginners.

Cattails grow in marshes, swamps, ditches, and stagnant water—fresh or slightly brackish—worldwide. Finding them is a sure sign of water. Military survival specialist and author Tom Squier once found them completely out of habitat, in a dry, sandy pine forest. A short search revealed an open manhole from an abandoned storm-sewer system, full of water.

The cattail's every part has uses. It's easy to harvest, very tasty, and highly nutritious. It was a major staple for the American Indians, who found it in such great supply, they didn't need to cultivate it. The settlers missed out when they ignored this great food and destroyed its habitats, instead of cultivating it.

Before the flower forms, the shoots—prized as "Cossack's asparagus" in Russia—are fantastic. You can peel and eat them well into the summer. They're like a combination of tender zucchini and cucumbers, adding a refreshing texture and flavor to salads. I love mixing them with pungent mustard greens to balance their mildness. Added to soup toward the end of cooking, they retain a re-

blossoms visible. The two most widespread species in the United States are the common cattail (*Typha latifolia*), which is larger and bears more food, and the narrow-leaved cattail (*Typha angustifolia*), also quite good.

People sometimes confuse cattails with the very common grasslike nonpoisonous reeds (*Phragmites* species), which form dense stands 12 feet tall. But freshing crunchiness. They're superb in stir-fry dishes, more than suitable for sandwiches, and excellent in virtually any context. I love sliced cattail hearts, sautéed in sesame oil with wild carrots and ginger.

Harvest cattail shoots after some dry weather, when the ground is solid, in the least muddy locations. Select the largest shoots that haven't begun to flower, and use both hands to separate the outer leaves from the core, all the way to the base of the plant. Now grab the inner core with both hands, as close to the base as possible, and pull it out. Completely cut off the tough upper parts with a pocket knife or garden shears in the field, so you'll have less to carry. When you get home, peel and discard the outermost layers of leaves from the top down, until you reach the edible part, which is soft enough to pinch through with your thumbnail (the rule-of-thumb). There are more layers to discard toward the top, so you must do more peeling there.

Note: Collecting shoots will cover your hands with a sticky, mucilaginous jelly. Scrape it off the plant into a plastic bag, and use it to impart a slight okralike thickening effect to soups. The shoots provide beta carotene, niacin, riboflavin, thiamin, potassium, phosphorus, and vitamin C.

The proportion of food to waste varies with the size of the shoot. You'll get the best yield just before the flowers begin to develop. A few huge, late-spring stalks provide enough delicious food for a meal. Some stalks grow tall and become inedibly fibrous with developing flowers by late spring, although just before the summer solstice, you can often gather tender shoots, immature flower heads, and pollen at the same time.

You can clip off and eat the male portions of the immature, green flower head. Steam or simmer it for ten minutes. It tastes vaguely like its distant relative, corn, and there's even a central coblike core. Because it's dry, serve it with a topping of sauce, seasoned oil, or butter. Sometimes I also gnaw on the cooked female portions, but there's very little to them. It's easier to remove the flesh from the woody core, if desired, after steaming. This adds a rich, filling element to any dish, and it's one of the best wild vegetarian sources of protein, unsaturated fat, and calories. It also provides beta carotene and minerals.

When the male flowers ripen, just before the summer solstice, they produce considerable quantities of golden pollen. People pay outrageous prices in health stores for tiny capsules of the bee pollen—a source of minerals, enzymes, protein, and energy. Cattail pollen beats the commercial variety in flavor, energy content, freshness, nutrition, and price. To collect the pollen in its short season, wait for a few calm days, so your harvest isn't scattered by wind. Bend the flower heads into a large paper bag and shake it gently. Keep the bag's opening as narrow as possible, so the pollen won't blow away. Sift out the trash, and use the pollen as golden flour in baking breads, muffins, pancakes, or waffles. It doesn't rise, and it's time-consuming to collect in quantity, so I generally mix it with at least three times as much whole-grain flour. You can also eat the pollen raw, sprinkled on yogurt, fruit shakes, oatmeal, and salads.

During fall, winter, and early spring, the cattail rhizomes store food. Digging up the thick, matted rhizomes from the muck, especially in cold weather, is not easy. After years of procrastination, I determined that, as a foraging teacher, the time had come to experiment with cattail rhizomes. Late one autumn, a friend and I went to gather cattail rhizomes from Central Park. It was so messy, I emerged from the park splattered with muck, looking more like a "Wildman" than I had ever intended. We hauled two dripping shopping bags across Manhattan, into her apartment, and onto her balcony. It took half an hour to hose down our harvest, and the mud clogged the drain.

We peeled off the rhizomes' outer layers, still imbued with mud, then worked the starch from the fibers with our fingers, in a large bowl of water. The water became cloudy with the starch. We waited an hour to let the starch settle, and poured off the water, getting enough sweet, tasty starch to thicken a small pot of soup—hardly worth the effort.

An alternate method is to tear apart the washed rhizomes and let them dry, pound the fibers to free the starch, and sift. This yields about as much starch as the previous method. However, I've received reliable reports that people in other parts of the country had better results. Perhaps rhizome quality varies.

I've also tried chewing on the fibers inside the cleaned rhizomes and swallowing the starch, which is very tasty. However, the digging and cleaning is so much work, I'd have to be starving in the winter to bother. Furthermore, there are reports that eating the starch of some species raw may cause vomiting.

The buds of the following year's shoots, attached to the rhizomes, are also edible. Although they make a tasty cooked vegetable, I find them too small to be worth digging up and cleaning, although their size may also vary.

Collecting the flower heads and pollen doesn't harm the plant, because cattails spread locally by their rhizomes—the seeds are for establishing new colonies, and each flower head makes thousands of these. Collecting a small fraction of the shoots also does no damage, since the colony continually regenerates new shoots. Since nobody wants to sink into the mud, people normally collect at the periphery of the stand. Of course, if the stand is small, it's already struggling to survive adverse conditions. Finding a larger stand elsewhere will increase your harvest and give the embattled plants a chance.

The Indians also used cattails medicinally by applying the jelly from between the young leaves to wounds, sores, carbuncles, external inflammations, and boils, to soothe pain.

Besides its medicinal uses, the dried leaves were also twisted into dolls and toy animals for children, much like corn-husk dolls found today. Cattail leaves can be used to thatch roofs, weave beautiful baskets, as seating for the backs of chairs, and to make mats. Archaeologists have excavated cattail mats over 10,000 years old from a Nevada cave.

No longer edible once the pollen is gone, the brown flower heads make good "punks," supporting a slowly burning flame with a smoke that drives insects away. The fluffy white seeds were once used for stuffing blankets, pillows, and toys. The Indians put them inside moccasins and around cradles, for additional warmth. After hours of collecting, I once made a cattail-fluff pillow, but something went wrong. My girlfriend sewed shut the pillowcase with the white seeds inside, and I went to sleep happy, on the softest pillow I'd ever felt. This mood quickly vanished when I awoke at 2:00 A.M. My head was on top of the pillow, but my right arm had "disappeared." I discovered it beneath the pillow. It had "fallen asleep" so badly, it seemed disembodied. After I shook it awake, I wished I hadn't. There was a row of hives from one end of the arm to the other, wherever it had pressed into the pillow. They itched so badly, it seemed to require a forest of jewelweed (page 73) to quell the torment. I had never heard of this reaction in any of the books that lavishly praise cattail fluff as stuffing! Confident that my "allergy" was unique, I handed the pillow to my girlfriend the next evening. She agreed that it was the best pillow ever—until the next morning, when the hives marred her once-beautiful face. She was so angry, she wouldn't talk to me until the hives healed. I learned later that my mistake had been not using thick batting material to enclose the stuffing.

Cattails and their associated microorganisms improve water and soil quality. They render organic pollution harmless, and fix atmospheric nitrogen, bringing it back into the food chain. They've even been planted along the Nile River to reduce soil salinity.

COLTSFOOT, SON-BEFORE-FATHER, COUGHWORT

(Tussilago farfara)

The dandelion has composite flowers with straplike ray flowers, but no tubelike disk flowers. Ever wonder how they would look were they to regain the absent disk flowers? The dandelion's relative, coltsfoot, fills the bill: This perennial herbaceous composite looks like a dandelion, except that its fertile ray flowers encircle many sterile disk flowers. (Daisies and sunflowers are composites that also share this form, although they don't closely resemble dandelions.) Coltsfoot's yellow flowers are 1 inch across, growing singly on a flower stalk, like the dandelion. Unlike the dandelion's smooth flower stalk, coltsfoot's has reddish scales.

The long-stalked, rounded, toothed leaves are broadly heart-shaped, reminiscent of a colt's hoof—10 to 12 inches across, with cottony down

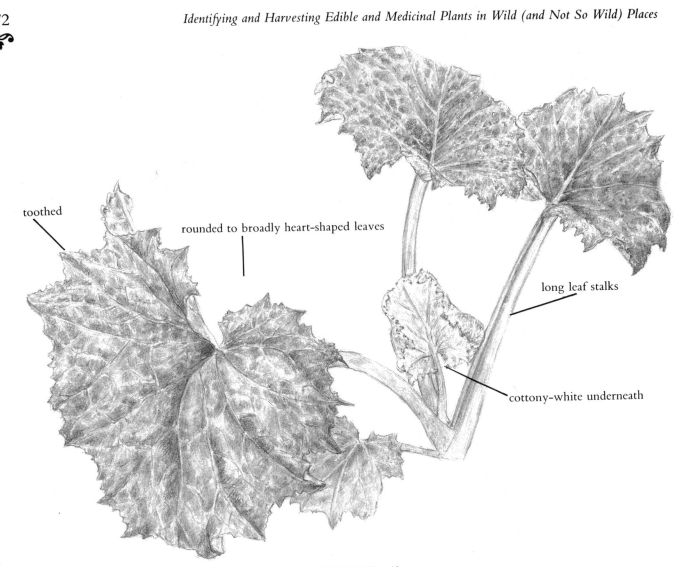

toothed

rounded to broadly heart-shaped leaves

long leaf stalks

cottony-white underneath

COLTSFOOT × ⅓

on the undersurfaces. The flowers often die before the leaves even appear. Medieval botanists thought the European species' flowers and leaves came from two different plants. Mired in dogmatism, nobody ever bothered to examine the roots or experiment. More observant Native Americans named the American species "son-before-father."

Coltsfoot grows along streambanks and riverbanks, in wet places, and on disturbed soil throughout most of eastern North America. The flowers come up in early spring, and you can use the leaves from mid-spring to fall. The flowers, flower stalks, and young leaves are good cooked vegetables—steamed, sautéed, in soups, or coated with batter and deep-fried.

The mature leaf is too tough to eat, but important medicinally: Coltsfoot-leaf emblems were mounted above the doors of Parisian pharmacies, symbolizing efficacious medicine. A strong infusion makes a good cough and cold medicine. It's an expectorant, and the mucilage makes it soothing. The Indians burned the dried leaves in pipes, and inhaled the smoke to dry out coughs, and to treat chronic bronchitis and mucus congestion. I used a combination of coltsfoot and mullein (page 247) tea and smoke to help a friend who couldn't sleep due to a severe cough caused by bronchitis. It worked wonders. The same combination is also supposed to be good for asthma. Scientists think the smoke inhibits impulses along the parasympathetic nerves, and that it works as an antihistamine. Coltsfoot is also used for inflammations, fever, diarrhea, external ulcers, and burns. It contains the nutrients zinc, potassium, and calcium, as well as

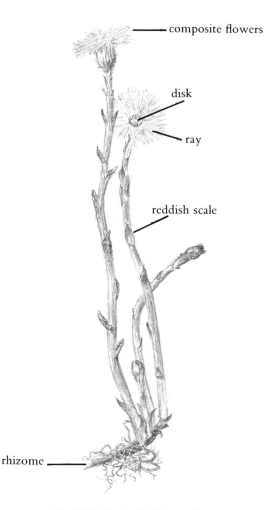

composite flowers

disk

ray

reddish scale

rhizome

COLTSFOOT FLOWERS × ⅖

have confused the two plants without ever testing the foreign species' ashes.

Caution: Use coltsfoot in moderation. Although no one has ever gotten sick from using coltsfoot, it contains traces of pyrrolizidine alkaloids, potentially toxic to the liver in large doses.

JEWELWEED, TOUCH-ME-NOT

(*Impatiens* species)

Although this isn't one of my favorite wild foods, it's one of our most important herbs. I call jewelweed the forager's American Express, because I never leave home without it. It's common, widespread, easy to recognize, and invaluable to anyone venturing out-of-doors, because it's a virtual panacea for skin irritation.

This herbaceous native plant has a distinctive succulent, translucent, hollow stem, powdered with a pale blue-green, waxy bloom and partitioned by nodes, making the plant easy to identify. Jewelweed grows up to 5 feet tall, branching

caoutchouc, volatile oils, pectin, resin, mucilage, and tannins.

The Indians were also supposed to have burned dried coltsfoot leaves and added the ashes to stews, as a source of salt, in areas where salt was unavailable. They even supposedly fought wars over stands of coltsfoot. I once tried burning coltsfoot leaves and adding the ashes to a stew. It didn't taste the slightest bit salty, but like I had added ashes to the stew!

I discovered the same claims for western coltsfoot or sweet coltsfoot (*Petasites speciosa*). This is a similar western perennial of damp, shaded ground with thick, creeping leafstalks, and long-stalked, woolly, deeply-lobed basal leaves—probably the species the Indians really used. The species I used, *T. farfara*, is a European import. Someone may

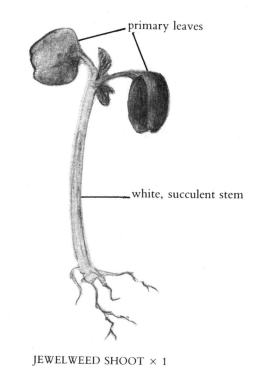

primary leaves

white, succulent stem

JEWELWEED SHOOT × 1

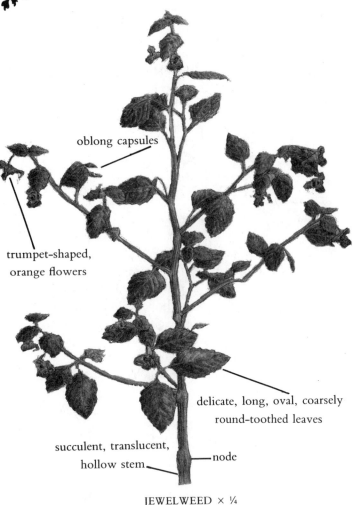

oblong capsules

trumpet-shaped,
orange flowers

delicate, long, oval, coarsely
round-toothed leaves

succulent, translucent,
hollow stem

node

JEWELWEED × ¼

toward the top and toughening with age. There's a clear, watery liquid inside, especially in the nodes.

The delicate, long-oval, long-stalked, leaves are 1 to 4½ inches long, with a few rounded teeth. The upper leaves are alternate, the lower ones opposite. They're water-repellent, so they look as if they're covered with tiny jewels (raindrops) after it rains, accounting for the name jewelweed. If you submerge the leaves in water, their undersides will turn silvery, delighting children of all ages.

The trumpet-shaped flowers, which bloom from midsummer to early fall, are under 1 inch long, with three partially fused petals, one of which curls to form a long slipper- or sack-shaped spur. Spotted touch-me-not has orange-yellow flowers spotted with red, yellow, or white. They're usually in pairs, so the scientific name is *Impatiens biflora.* Pale touch-me-not (*Impatiens pallida*) has yellow

flowers with reddish spots. *Pallida* means "pallid." Hummingbirds, butterflies, and moths pollinate the flowers, although some are self-pollinating.

Toward the end of the growing season, this annual puts all its energy into making fruit—oblong capsules, ½ to 1 inch long. The plant is called impatiens and touch-me-not because the pods' tissues dry at different rates, creating tension that is unleashed when you touch them: They explode, ejecting a few tiny, oblong, green-and-brown seeds and curled-up pod fragments.

You can find jewelweed from early spring through fall. There are many jewelweed species growing throughout most of the United States, with no poisonous look-alikes. It grows in wet areas—along streams and ponds, in moist woodlands, in low-lying areas, along floodplains, by springs, and in swamps, favoring partial shade—the same habitat as poison ivy. But it doesn't grow near the seashore, or in dry, sandy soil (poison ivy does), so bring it along.

In early spring, you can boil jewelweed shoots measuring up to 8 inches for 20 minutes, then discard the water (or freeze the water in ice-cube trays—the water retains the medicinal properties described below), and eat them. They're not bad-tasting, but they shrink so much, they're too labor-intensive for me. **Caution:** Don't eat jewelweed shoots raw or use the cooking liquid internally. You may be poisoned by their very high selenium content.

In late summer and fall, you can surround the ripe seedpods with your hand, and grab them tightly. The seeds will pop into your hand, and you can eat them, discarding the coiled "springs." They're very tasty—walnut-flavored, but too small for more than a trail nibble. Children, who seek fun over efficiency, love learning to catch and eat jewelweed seeds.

Jewelweed contains 2 methoxy-1, 4 naphthoquinone—an anti-inflammatory and fungicide that's an active ingredient of Preparation H. If you break jewelweed's stem and repeatedly apply the juice to a fresh mosquito bite for 15 to 20 minutes, the itching stops and the bite doesn't swell. For older bites, it works only temporarily.

Jewelweed's juice also relieves bee and wasp stings, although it doesn't always cure them completely. It's good for warts, bruises, and fungal skin infections such as athlete's foot and ringworm. It's also helpful for nettle stings, minor burns, cuts, eczema, acne, sores, and any skin irritations.

If you accidentally touch poison ivy and apply jewelweed juice to the affected area before the rash appears, you probably won't get the rash. One of my best strawberry patches is also infested with poison ivy. You can't avoid touching it as you collect the irresistible fruit. I have everyone apply jewelweed to all exposed areas when we leave, and nobody ever gets a rash.

The Indians treat already-developed poison-ivy rash by rubbing jewelweed's broken stem on the rash until it draws some blood. The rash then dries out, a scab forms, and healing occurs.

There are many ways to capture jewelweed's medicinal properties: The fresh plant lasts a week in a sealed container in the refrigerator. The 1960s foraging guru and author Euell Gibbons reported that the jewelweed tincture he extracted in alcohol went moldy, but I've soaked fresh jewelweed in commercial witch hazel extract for a few weeks, and the extract of the two herbs works well and doesn't perish.

You can also make jewelweed ointment by simmering a small amount of jewelweed in light vegetable oil (any vegetable oil except olive oil, which burns) 10 to 15 minutes. Use only a small handful of jewelweed stems per quart of oil, or bubbles of jewelweed juice will form in the ointment and go moldy. Strain out the herb, add a handful of beeswax to thicken it, and heat until melted. Take out a spoonful and let it cool to test the thickness, and add more oil or beeswax as needed. Add the contents of one oil-soluble vitamin E capsule, a natural preservative, and let it cool. Refrigerated, it lasts for months.

MARSHMALLOW AND RELATIVES

(*Althaea officinalis, Malva* species)

The plant is a European import, related to cotton, hibiscus, and okra. It's a velvety, herbaceous European perennial growing 2 to 4 feet tall. It's stout, upright, hairy stem is often branched. The

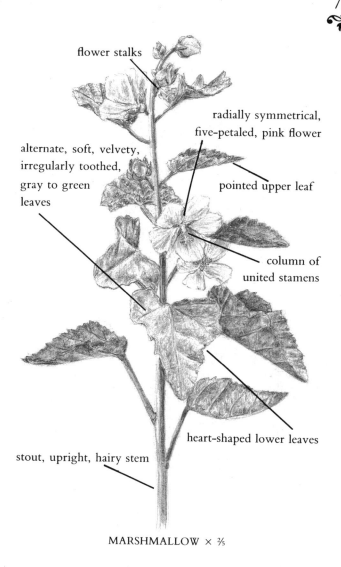

MARSHMALLOW × ⅖

alternate, soft, velvety, irregularly and coarsely toothed, gray-green leaves are 1 to 4 inches long. The upper ones are roughly oval and pointed, and the lower ones are heart-shaped or 3-lobed. The name mallow comes from *malakos*, Greek for "soft"—referring to the leaves.

The radially symmetrical, stalked, pink flowers are 1 to 1½ inches across. They have five petals, ½ to ¾ inch long, with five stamens united into a bushy column. They grow in clusters from the upper leaves' axils. The large, white taproot is mucilaginous. The dry fruits, which are hexagonal, flattened, and divided radially into fifteen to twenty segments, measure a little under 2 inches across. They resemble miniature wheels of cheese. Marshmallow leaves are in season from late spring through summer, the fruit is in season in the summer, and the roots are in season all year.

Labels on illustration: flower stalks; radially symmetrical, five-petaled, pink flower; alternate, soft, velvety, irregularly toothed, gray to green leaves; pointed upper leaf; column of united stamens; heart-shaped lower leaves; stout, upright, hairy stem

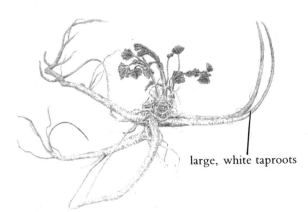

large, white taproots

MARSHMALLOW ROOT × ⅙

Unfortunately, marshmallows grow only in salt marshes and tidal rivers near the ocean in eastern North America. I've never found enough to warrant digging up the roots, although all herb stores and suppliers carry marshmallow root. However, there are other, more widespread mallows, all with similar uses.

Marshmallow leaves are too hairy to eat raw, but they make an excellent cooked vegetable, especially the younger, smaller ones, which are the most tender. I've used them as a main vegetable in quiches, in place of spinach, and they're wonderful. The flavor is very mild, the texture is soft, and the consistency slightly gooey. They're great with stronger-tasting vegetables, such as wild mustards. The fruits have the same mild, mucilaginous qualities as the leaves, but they're not hairy, so you can use them raw or cooked.

The marshmallow is as much a plant as a confection. **Caution:** Today the candy consists entirely of unhealthful artificial chemicals and refined sugar. Marshmallow candy was originally made by peeling the root and boiling the inner part in sugar water. Spoonfuls of the thick liquid cooled on a flat surface to form the candy.

You can boil peeled marshmallow roots in fruit juice, then beat with an eggbeater until fluffy, to get a substance with the consistency of beaten egg whites; fold in some fresh fruit, cool, and you have a marshmallow-root chiffon. You can also steam the peeled root, to make a very mild-tasting, gooey vegetable. Further cooking, such as sautéing with other vegetables, will reduce the stickiness. Or use the peeled roots like okra, in soups that you want to be silky in consistency.

Marshmallow root is high in minerals, especially an easily usable form of calcium. It's been used medicinally since ancient times. According to Pliny: "Whosoever shall take a spoonful of the mallows, shall that day be free of all diseases that may come to him." The generic name, *Althaea*, is Greek for "that which heals," and *officinalis* is a name for medicinal plants.

The gummy, boiled root is an emollient and demulcent: With up to 35 percent mucilage (and some asparagine and tannin), it soothes and lubricates inflamed surfaces. You can use it as a warm poultice for external sores, or apply it cold for sprains (heat increases swelling). It softens the skin, so you can also use marshmallow-root decoction, preserved in alcohol, as a skin lotion.

The tea is also internally soothing. It's good for stomach and duodenal ulcers, digestive tract irritations, urinary tract irritations, and cystitis. It's also very helpful for sore throats and irritated air passages. As an expectorant, it's good for coughs, bronchitis, and whooping cough.

A friend makes cough medicine by boiling a cup of dried root in a quart of water, then infusing a handful each of mint and coltsfoot. After a few hours, she strains out the herbs, getting as much of the thick liquid as possible. She adds half as much honey as there is tea, and simmers it down to a thick syrup.

The most common mallow species, to no one's surprise, is called common mallow (*Malva neglecta*). This inconspicuous herbaceous plant, no more than 1 foot tall, is sparsely hairy or nearly hairless. It has pleated, circular to heart-shaped alternate leaves 1 to 2 inches across, set on very long leafstalks.

Small but showy white, pink, or lilac long-stalked radially symmetrical flowers dangle from the leaf axils, blooming from spring to fall. Each flower has five notched petals, with a bushy column of stamens in the center. Less than ¾ inch across, they look like miniature marshmallow flowers.

The fruit is round, flattened, and segmented— ¼ inch across—like that of the marshmallow, only smaller. It resembles a minuscule wheel of cheese, and it's even called a cheese. Although it's com-

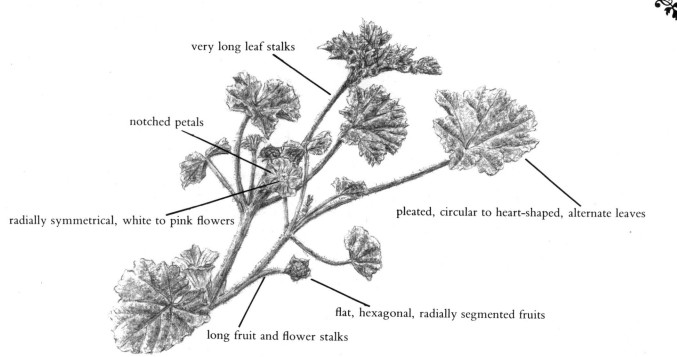

very long leaf stalks

notched petals

radially symmetrical, white to pink flowers

pleated, circular to heart-shaped, alternate leaves

flat, hexagonal, radially segmented fruits

long fruit and flower stalks

COMMON MALLOW × ⅕

pletely edible, with a pleasant sweet-gummy fla-vor-texture, strict vegetarians have been known to avoid it. You can find it from summer to fall.

Common mallow's deep root is tough and stringy, but a decoction has medicinal uses similar to marshmallow root's. Research even demon-strates activity against tuberculosis. This species is usually so common, collecting some roots won't normally endanger it.

The common mallow grows on lawns and meadows, often in cultivated areas, throughout North America. I find it from early spring to the beginning of winter in my area. It's one of the last plants to be killed by the cold, and I've collected this low-growing vegetable after hard frosts, when I desperately needed wild greens for a wild salad to feed two hundred people for my annual early winter Wild Party. In slightly warmer climates, you can collect it all year.

Another species, the swamp rose mallow (*Hibiscus palustris*), has very large, pink or white flowers, 4 to 6 inches wide and petals 2 to 3 inches long. Its long-stalked leaves are oval to three-lobed, and their undersides are whitish-downy. It grows up to 7 feet tall, and you can find it in marshes along the East Coast, and inland to the Great Lakes. The leaves are great cooked, the same

as marshmallow's. I've never found enough to dig up the roots, so I don't know whether you can use them like marshmallow's.

OTHER PLANTS OF FRESHWATER WETLANDS IN MID- TO LATE SPRING

Other Edible or Medicinal Plants

Asparagus shoots, catnip, burdock root and flower stalk, cow parsnip flower stalk, cuckoo flower, elderberry flowers, grape leaves, ground-nut tubers, jewelweed, miner's lettuce leaves, os-trich fern fiddleheads, Oswego tea, peppermint, spearmint, sweet gale leaves, thistle flower stalk, violet leaves and flowers, watercress, water mint, wild mint.

For Observation Only

Blackberry flowers, blueberry flowers, bur-dock flowers, ostrich fern fronds, grape flowers, groundnut flowers, mulberry flowers, wintercress seeds, poison ivy flowers, rose flowers, Solomon's seal flowers, false Solomon's seal flowers, water hemlock, waterleaf flowers.

PLANTS OF THE SEASHORE IN MID- TO LATE SPRING

ASPARAGUS, SPARROW GRASS

(Asparagus officinalis)

When I first saw wild asparagus in an over-grown field at summer's end, it held no resemblance whatsoever to commercial asparagus. It looked more like a Christmas decoration. Its red, shiny berries dangled perilously under the swaying, delicate, needlelike leaves. Still, there was no doubt—this was asparagus.

Since asparagus is in season in mid-spring, I returned the following year to find the young shoots we all know and love were sprouting dauntlessly from the bare ground. The previous year's bowed, dried-out stalks stood watch.

Wild asparagus stalks are thinner and more crooked than commercial asparagus. The shoots are yellow-green and spotted. Scales, which become increasingly crowded toward the tip, replace the leaves. The scaly, branched, mature plant, growing up to 6 feet tall, looks like a Christmas tree.

This relative of lily of the valley is an escape from cultivation. It probably slipped out of someone's garden late one night, when no one was looking. The plants are either male or female. Both sexes produce flowers typical of the lily family. They're small—a little over ⅛ inch long—yellow-green, six-parted, and narrowly bell-shaped. Only the female plants bear fruit. The small, globular, red berries are good to eat only if you're a bird. The three to six indigestible seeds pass through the bird's digestive tract to germinate into new plants.

Asparagus does best in salt marshes. It also grows in sandy fields and partially shaded, disturbed sites. You can find this escaped European import throughout much of North America, especially in farming areas.

Asparagus's much-branching perennial rhizomes give rise to colonies of stalks. My first find included about twenty-five sprouts. You can cut the young stalks from just below ground level, but don't harvest stalks much larger than commercial

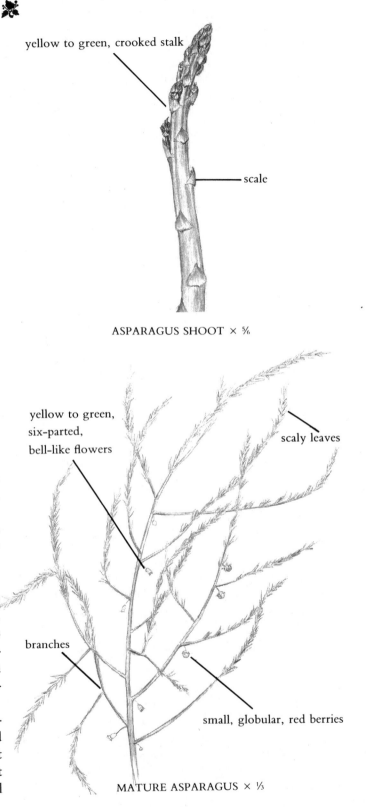

yellow to green, crooked stalk

scale

ASPARAGUS SHOOT × ⅚

yellow to green, six-parted, bell-like flowers

scaly leaves

branches

small, globular, red berries

MATURE ASPARAGUS × ⅓

asparagus, or mature plants—they're mildly toxic. Also, be sure to let some stalks mature, supplying the perennial rhizomes with nutrients for next year's growth.

I took some stalks home, tied them into a bundle (lest they try to escape again), and cooked them in enough water to cover the bottom of the pot. They have a strong flavor, superior to that of commercial asparagus. You must sometimes peel the tough coverings near the base.

Asparagus is especially rich in beta carotene and vitamin C, potassium, calcium, phosphorus, niacin, and iron. It also contains asparagine, saponins, flavonoids, volatile oil, a glucoside, gum, resin, and tannic acid.

Asparagus is a perfect diuretic for early spring. This healthful food cleanses and tones the urinary tract, and helps clear urinary obstructions. Herbalists recommend the juice for edema and kidney stones. If you eat enough, it may also act as a laxative.

Its wonderfully mild cleansing abilities also extend to the skin. Asparagus spears make an effective, stimulating facial wash, especially for oily and blemished skin (acne). Grate some raw stalks, apply them to the skin, wait ten minutes, then rinse.

BAYBERRIES

(Myrica species)

Bayberries are low, strongly aromatic shrubs with gray bark and nearly toothless, or slightly toothed, oblong leaves. Some species, especially southern ones, are often evergreen. Some are deciduous, and others are semi-evergreen (deciduous in the North, evergreen in the South). In the right habitats, these bushes are very plentiful, and their leaves are easy to collect in quantity.

Once you've identified one bayberry species, you'll simply be drawn to the other species by your nose. They're all good to use, and there are no poisonous look-alikes. Here are some common species:

cluster of inconspicuous flowers

green branches

NORTHERN BAYBERRY IN FLOWER × ⅖

The northern bayberry (*Myrica pensylvanica*) is among the most common species. This stout shrub grows 3 to 12 feet tall, with grayish branches and gray, waxy berries. The leathery, oblong, slightly toothed leaves are 1 to 4 inches long. The flowers and berries originate on the branches beneath the shrub's leafy tips. It grows in dry poor soils and marshes in areas near the Atlantic coast and the Great Lakes and in similar, dry, sunny habitats farther inland.

Common wax myrtle or southern bayberry (*Myrica cerifera*) is an evergreen shrub or small tree growing to about 26 feet tall. The leathery leaves, 1½ to 3 inches long, are wedge-shaped, with resin dots on both sides. The tiny, hard fruits are black. You'll find it in mostly sandy soil and wet places in the Southeast, from the Florida coast to southern New Jersey, west to East Texas and north to Arkansas.

Sweet gale (*Myrica gale*) is a much-branched, deciduous shrub 2 to 6 feet tall. The leaves are dull, grayish-green, slightly toothed, oblong to lance-shaped. Instead of the usual bayberries, grayish-white, waxy, conelike nutlets grow in clusters on the stiff gray bark, at the ends of the previous year's branches. You can use them as a spice, and use the leaves like commercial bay leaves. Sweet gale grows in shallow water, swamps, and wetlands

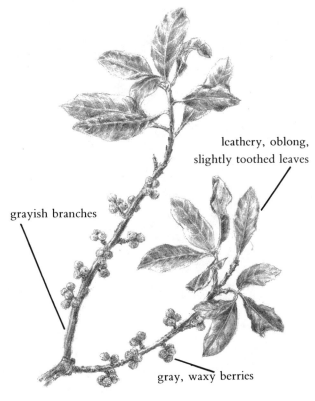

NORTHERN BAYBERRY WITH FRUIT × ⅖

leathery, oblong,
slightly toothed leaves

grayish branches

gray, waxy berries

throughout Canada and the northern United States, and on mountains to Tennessee and North Carolina. It also grows in sphagnum bogs from the Pacific Northwest north to Alaska.

The berries of all species are waxy, to prevent desiccation. They're in season from autumn until the following spring. Too resinous, hard, and waxy to eat, they're important in American history: Colonists made candles with the wax.

A friend recently tried a three-hundred-year-old recipe, spending three days collecting the fruit of the northern bayberry. Finally, he had enough to throw into boiling water. After hours of tedious labor, just enough wax rose to the surface of the boiling water to form one third of a candle! He should have used wax myrtle. A pound of the nutlets reportedly yields four ounces of wax.

Use bayberry leaves the same way as the familiar store-bought, tropical bay leaves. Your soups, stews, and sauces will all be the better for it. **Caution:** Be sure to remove the leaves after cooking. Commercial bay leaves are tough and sharp enough to cut people's stomachs, sometimes necessitating surgery. This is less likely with the wild

species, but you have nothing to lose by putting them into tea bags or tea balls, and removing them when the recipe is done.

Thanks to the resins that permeate the plants, all parts of these bushes have important medicinal uses. The bark, leaves, and roots are used as tonics, stimulants, and astringents. A decoction of bayberry bark or roots improves circulation and tones tissues, particularly in the stomach and intestines. It supposedly clears up mucus congestion of the head and stomach. A hot tea is good for colds and fevers: The constituent myricitrin has been shown to kill pathogenic bacteria, and to promote the flow of bile.

As an astringent, the decoction was used for hemorrhages (don't treat yourself for this or other life-threatening conditions), as a gargle for sore throats, and as a wash for gum disease. Externally, a compress of strong tea is used for varicose veins, swelling from insect bites, and as a douche. The leaves are the weakest, best for less severe conditions. Don't use this herb for extended periods of time. The constituent myricadiol reportedly makes you retain sodium and excrete potassium, which is unhealthful.

GLASSWORT, BEACH ASPARAGUS

(*Salicornia* species)

This is a leafless, emerald-green succulent of salt marshes that gets its name because it turns a translucent, glasslike pink in autumn. It's a salt-tolerant plant that grows where little fresh water is available. To keep from drying out, it has a thick, succulent stem with opposite branches and no leaves. The generic name, *Salicornia*, means "salt horn" in Greek, referring to the saline habitat and hornlike branches.

Tiny, inconspicuous flowers, which bloom from late summer to mid-fall, hide in groups of three to seven in joints along the stem and branches. There are several similar species, most annual, one perennial, and no look-alikes.

Common table salt, sodium chloride, is poisonous to most plants. When ancient Rome finally conquered and destroyed its longtime enemy, Car-

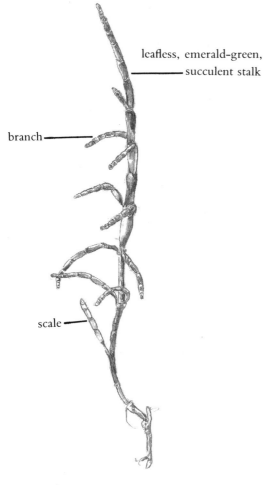

leafless, emerald-green,
succulent stalk

branch

scale

GLASSWORT × ⅕

species usually occurs in quantity. It's available from spring to fall, but the best time to collect is in late spring, when the young plant is most tender. Unless you're a fish, gather at low tide, when the habitat is most accessible.

Eating glasswort raw without other foods may irritate your throat, so take only a small nibble the first time. It makes a great garnish for cooked vegetable dishes and grains. Use it raw in spreads and salads, or cook it 10 to 15 minutes in soups, stews, or other dishes. It makes great pickles.

It's salty, so use it sparingly and omit salt from the recipe, or use it as a salt substitute. You can even dry it, grind it into a powder, and use it in place of salt. If you're on a low-sodium diet, avoid this plant altogether.

SEAWEEDS

(Algae)

Seaweeds are algae, the earliest multicellular green plants to evolve, and they still dominate the oceans. These nonflowering plants are very simple. Most have a leaflike blade, sometimes with a midrib. There's often a stem, and a holdfast. Since algae have no vascular systems (they exchange dissolved nutrients and wastes directly with the water), the holdfast simply anchors the plant.

Seaweeds are wild foods that elicit strong emotions. People either love them or hate them. Some view them as adventures in international cuisine; others refuse to admit that they're food at all. One stand-up comedian gets laughs proclaiming that he not only refuses to try seaweeds, he won't even swim where they grow.

Cultural food prejudice is an important factor in food choices. Seaweed is not yet a standard food in America. Yet people of Japan, Ireland, Iceland, Denmark, Wales, Scotland, Hawaii, Micronesia, Denmark, and China, along with their trading partners and other cultures with links to the sea, have been enjoying seaweeds for centuries. Correspondingly, most Americans enjoy corn on the cob, which we adapted from the Indians long ago. Request this dish in a fancy restaurant in Europe, where this vegetable never caught on, and the in-

thage, the Roman soldiers spread salt over the ruins so nothing would ever grow there again.

The problem with salt involves the availability of water. This vital liquid goes wherever it's the least concentrated. If a cell contains mostly water, and the region outside the cell contains salt water, the fresh-water concentration outside the cell is lower, so water will continue diffusing out through the cell membrane until the cell dies of dehydration. This is why shipwreck victims in lifeboats can't drink seawater. It would pull fresh water out of their bloodstreams and into their stomachs until they threw up, even more dehydrated than before.

Glasswort, and some other seaside plants, have a simple way to cope with their salty environment: If you can't lick 'em, join 'em. They contain enough internal salt to balance the sodium around them.

Look for glasswort on tidal flats near the beach, or on salt flats across the country. This colonial dignant waiter will tell you they don't serve pig food.

What we eat is also biologically influenced. Young children will put everything into their mouths, but as we get older, our food habits tend to become fixed. Immature chimpanzees will also try a variety of foods, avoiding only what makes them sick. But as adults, the animals will eat only what they had as youngsters. Move them to a new region, and they'll starve to death if they're not familiar with the local edibles.

You may acquire a taste for these primal vegetables if you give them a chance. They're also excellent sources of vitamins, minerals, and trace minerals. Don't let cultural conditioning and biology make a monkey out of you.

Seaweeds' one drawback is their high sodium content. Small quantities of sodium are essential for life, so we have a built-in craving for it. In larger quantities, the body must get rid of it. You get thirsty, drink lots of water, and eliminate the water and sodium, along with valuable potassium, through the urine. This also stresses the kidneys, and may lead to high blood pressure and cardiac problems.

I always wash seaweed, using the colander-and-bowl method described in the cooking section, page 274, but this still doesn't get rid of all the salt. So I eliminate other sources of salt in seaweed recipes, and enjoy these vegetables in moderation. If you're on a low-sodium diet for medical reasons, avoid seaweeds altogether.

Some species, like dulse, are delicious raw. Tougher species require cooking. You'll often find seaweeds in such large quantities, you'll want to store them. Fortunately, they dry easily. Some people dry them on clotheslines. Be sure to remove any snails and secondary seaweeds growing on your vegetables. Also, wash out any grit first, using the colander-and-bowl method.

Edible seaweeds grow in turbulent, rocky-bottom, open-sea habitats, especially in cold northern waters. They anchor themselves on rocks or man-made objects, so search rocky shores, not sandy beaches. Low tide exposes some species, while others grow in deeper regions. Unless you have scuba equipment, look for deep-growing species after storms, when many pounds may be washed ashore. Collect at low tide—the lowest occur after full and new moons. *The Old Farmers Almanac* has tide charts for the entire United States.

Avoid collecting in polluted waters. Fortunately, seaweed, at the bottom of the food chain, is the safest seafood. The fish at the top of the food chain contain all the contaminants of the smaller fish they eat. The same is true of their prey, and their prey's prey, et cetera.

Some ocean seaweeds are delicious; others are stringy, tough, and unpalatable. Only one species is possibly poisonous, but this sour-tasting, filamentous, blue-green alga doesn't resemble the leafy-looking edible ones. **Caution:** Avoid all freshwater algae. Some contain substances that displace the neurotransmitters nerve cells use to communicate. Very small amounts disrupt the nervous system, causing convulsions and death—one reason you should never drink stagnant water.

ROCKWEED, BLADDERWRACK, WRACK

(*Fucus vesicylosus*)

This very common, olive-brown seaweed grows attached to rocks in the intertidal zone—the region submerged at high tide, but exposed at low tide. It consists of flat, leathery ribbons about ½ inch wide, divided by a midrib. Y-shaped forks continually form paired branches. It has paired, oval air vessels for flotation. Kids love popping them, like bubble paper. Rockweed grows to about 3 feet long. Look for it all along the Atlantic coast. Similar edible relatives, with and without air bubbles, also grow along the Atlantic and Pacific coasts. They're in season all year.

This seaweed requires an hour of simmering, or twenty minutes in the pressure cooker, to tenderize it. It's also very strong-tasting. Use it sparingly in soups and stews, or put a small amount of precooked rockweed in the blender when you make sauces. It's a great ingredient for making stock—adding rich flavor and lots of minerals.

After much experimentation, I discovered a

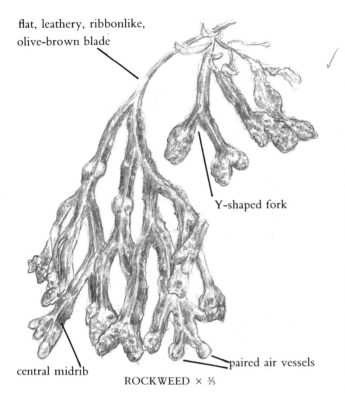

flat, leathery, ribbonlike, olive-brown blade

Y-shaped fork

central midrib

paired air vessels

ROCKWEED × ⅓

fantastic way to use this common vegetable: Dry it thoroughly, mix in seasonings and the tiniest amount of olive oil, and roast it in the oven. It becomes crisp and crunchy, somewhat like potato chips. People who shun rockweed on field walks because of its alien appearance greatly regret their decision when they taste this dish at my parties. See page 280 in the cooking section.

In the days before iodine supplements, iodized salt, and modern food transport, people living inland, with no access to seafood or other sources of iodine, often developed goiters. Herbalists treated such thyroid problems with rockweed, correcting the iodine deficiency. Research from the *American Journal of Clinical Nutrition*, the *Tokushima Journal of Experimental Medicine*, and *The Eclectic Materia Medica* also shows that rockweed works against kidney infections, bladder inflammations, cardiac degeneration, obesity, and menstrual problems.

Rockweed is also economically important. It's one of the best, most plentiful natural fertilizers. Just wash off the salt and throw it on the compost heap. In the seventeenth century, the French burned it and used the ashes to manufacture industrial soda, used to make glass and pottery.

SEA LETTUCE, GREEN LAVER

(Ulva lactuca)

This filmy seaweed, which grows up to 3 feet across, looks so much like a wet sheet of wrinkled, transparent, green, decorative plastic wrap that people are always surprised it's edible. This unmistakably distinctive annual and its similar relatives grow in east and west coast waters in the spring and summer, usually washed ashore.

Using the colander-and-bowl method described in the cooking section, page 274, rinse this vegetable over and over to get rid of all the sand. Then use it raw or cooked. Raw, it adds a nice color and texture, but little flavor, to salads and soups. If you chop it and sauté with onions and garlic, or other seasonings, it's delicious. However, it's very perishable. Use it within two days, or cook and freeze it.

Sea lettuce provides iron, protein, iodine, aluminum, manganese, and nickel. It also supplies you with some vitamins, starch, fat, and sugar.

IRISH MOSS, SEA MOSS, BLANCMANGE, JELLY MOSS

(Chondrus crispus)

This is a short, frizzy, fan-shaped perennial that's especially colorful. It may be olive green, purplish, or deep brown-red. Underwater it's sometimes iridescent-blue, and it turns white after the sun bleaches it. Each plant may reach 6 inches in length. The flattened blades are 1 to 3 inches long, arising from a short stalk. They're tough and elastic when wet, brittle when dry. They branch freely, ending in crowded, frilled-looking tips.

Irish moss grows in dense clumps, along the low tide line of the northeast coast, from New Jersey northward. A nearly identical relative, *Rhodoglossum affine*, grows along the Pacific coast. Don't confuse Irish moss with Iceland moss, a li-

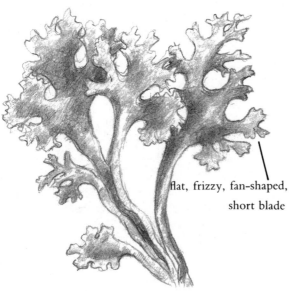

flat, frizzy, fan-shaped,
short blade

IRISH MOSS × 1

chen that doesn't grow in the sea. The two are as alike as Ireland and Iceland—only their names are similar.

Irish moss contains a strong gelling agent called carrageenan, used in many commercial products, from ice cream to hand lotion. The Irish simmer sun-blanched Irish moss (colored Irish moss discolors the dish) in milk for 20 minutes, and discard it, to make the traditional pudding, blancmange. Use 1½ cups of fresh seaweed, or ½ cup of dried seaweed (reconstituted) to thicken 3 to 4 cups of water, fruit juice, or milk. Cinnamon, lemon, and licorice are good complementary seasonings. Or use Irish moss to thicken soups or stews: Simmer the washed seaweed in the broth for 20 minutes, strain it out, and return the broth to the recipe. If you want to include the tough seaweed itself, simmer it for an hour. It all but disappears.

Irish moss is highly nutritious, and the dishes you make from it are easy for convalescents to digest. It provides iodine, potassium, calcium, phosphorus, iron, sodium, manganese, bromine, mucins, protein, and a little fat.

Carrageenan is a demulcent and emollient, coating the entire digestive tract and alleviating peptic and duodenal ulcers. It's used for thyroid problems (goiter), colon disorders, and obesity. It also relieves diarrhea and helps form bulky stools. A syrup of Irish moss with rose hips, honey, and lemon is good for a sore throat, and an Irish-moss

decoction is soothing to chapped hands. It's also supposed to be helpful for radiation contamination. Architects and children dry and spray-paint it, to use in models as "trees."

Some have suggested that Irish moss is dangerous because processed extracts are toxic. However, Irish moss in its whole form is safe. The digestive system can't release the carrageenan into the body.

DULSE

(*Palmaria palmata*)

Growing up to 1 foot high, this broad, ribless, rose-red, purple, or purple-brown seaweed has small, lobed blades resembling hands, and a small, tapering stalk. Tiny lobes often extend from the edges of the blades. Fresh dulse feels tough and leathery, and it remains pliable when dried.

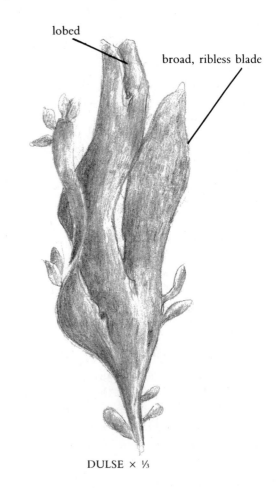

lobed

broad, ribless blade

DULSE × ⅓

Dulse grows close to or below the high-tide line, and in deeper water. Look for it along the Atlantic coast from the Arctic to Long Island Sound, and in the Puget Sound region of Washington State. Other similar, closely related species grow along the west coast from California northward. It's in season from the spring, when it's at its best, to autumn.

I love dulse's sharp, nutty flavor so much it's hard to stop eating it. Enjoy dried dulse as a snack, sauté it in oil with your choice of seasonings, roast it in the oven 10 to 20 minutes at 325°, or include it in casseroles. Add it to soups during the last ten minutes of cooking. It's great with potatoes, and wonderful baked into breads, especially sourdough. You can also make a wild gomasio (sesame salt) by grinding roasted dulse with roasted sesame seeds in the blender or spice grinder, to use as a spicy, salty seasoning.

In the spring, dulse is high in beta carotene. In autumn it's a good source of vitamin C. It also provides protein, vitamin B_6, B_{12}, calcium, potassium, magnesium, phosphorus, iron, boron, bromine, nickel, strontium, and vanadium.

LAVER, NORI

(*Porphyra* species)

Laver is a broad-to-narrow, soft, paper-thin, nearly transparent sheetlike seaweed, only one cell thick, with ruffled edges. The color ranges from purple-pink to brown. It's uncommon along the

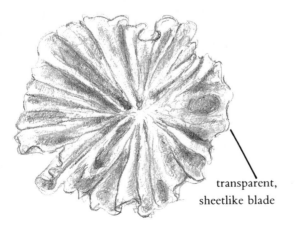

transparent, sheetlike blade

LAVER × ⅔

east coast, growing from Massachusetts to the Arctic, but it's much more plentiful along the Pacific coast. It grows in the intertidal zone, and is in season in the spring. The Japanese roast, grind, and re-form sheets of this delicious, popular seaweed to use for making sushi. It's also sold in health-food stores, at higher prices than most other seaweeds. Use it as you would use dulse, above.

EDIBLE KELP, KELP, WINGED KELP, SWEET KELP, WAKAME

(*Alaria esculenta*)

This is a long-bladed, dark-brown perennial with a distinctive, flattened midrib and thin, ruffled edges. It grows up to 10 feet long, and tends to become tattered. Butterfly-shaped sporophylls, flattened reproductive structures resembling lateral branches, extend from the long stem, just below the blade and above the holdfast.

Edible kelp is in season from spring to summer—better earlier. It grows from Long Island Sound northward. A number of closely related species also grow from the New England coast northward, and from central California northward. An Alaskan species grows up to 75 feet long.

These sea vegetables are favorites in Japan, and they're sold in health-food and ethnic stores throughout America. They have a mild, sweet, silky flavor and texture, and they thicken soups slightly. Some species require an hour's simmering to get tender; others are done in ten minutes. Cut the fast-cooking varieties into small pieces with scissors while they're still dried, add them with vegetables and a half-inch of water to a pot, cover, and simmer until done.

Edible kelp contains the sugar mannitol, especially in the tender midrib, which tastes like a salty carrot. In Wales, they sell the midrib alone, as a fresh vegetable. Kelp also provides beta carotene, vitamins B_1, B_3, B_5, B_6, B_9, B_{12}, C, E, biotin, choline, inositol, and PABA. It's the best source of minerals and trace minerals, and the best treatment for mineral deficiency. It provides iodine, potassium, calcium, iron, zinc, copper, magnesium, se-

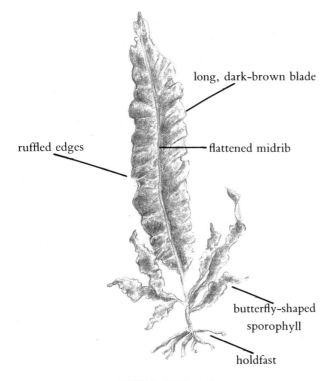

long, dark-brown blade

ruffled edges

flattened midrib

butterfly-shaped
sporophyll

holdfast

EDIBLE KELP × ⅟₂₀

lenium, sulfur, phosphorus, silicon, bromine, nickel, strontium, and vanadium.

It's reportedly very beneficial to the sensory nerves and the membranes surrounding the central nervous system. It improves thyroid function, and strengthens the nails. It contains brominated phenalic compounds, which give it antibiotic effects against both gram positive and gram negative bacteria, the two major groups of these microorganisms.

It also contains alginate, which binds bile in the intestine, so it can't be reabsorbed and reused. Instead, the liver makes new bile from cholesterol, reducing cholesterol levels and arterial plaque. Alginate also binds radioactive strontium, as well as poisonous barium, cadmium, and excessive zinc in the intestinal tract, preventing their absorption into the body.

Through unknown means, it seems to reverse prostate enlargement. There's also evidence that it inhibits the formation of breast cancer, possibly by stimulating the immune system. It's also been used for goiter, ulcers, and obesity. Experiments show that it strengthens the heart muscle.

Kelp is a traditional Asian folk treatment for uterine disorders, impotence, and infertility, as well as bladder weakness, cystitis, prostate enlargement, uterine disease, and ovarian problems. In addition to its high mineral content, it seems to be a general tonic for the male and female reproductive systems. Dried, powdered kelp is also hemostatic—it helps stop bleeding.

HORSETAIL KELP, TANGLE, KOMBU

(*Laminaria digitata*)

Horsetail kelp is a perennial that differs from edible kelp in that its broad blade is split into six to thirty straplike "fingers." The stalk is stiff and woody—you have to remove it. The holdfast is yellow-tinged.

Horsetail kelp and its close relatives grow in deep water, from New England northward along the east coast, and from California northward along the west coast. Unless you're a scuba diver,

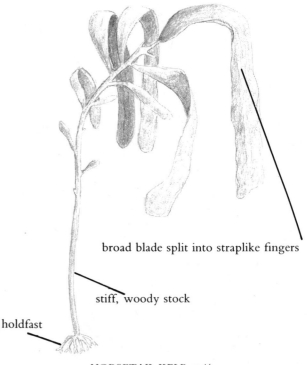

broad blade split into straplike fingers

stiff, woody stock

holdfast

HORSETAIL KELP × ¼

you'll have to wait for a storm to wash this wild food ashore. One deep-dwelling relative, long-stemmed kelp (*L. longicruris*) looks like edible kelp, but without the midrib or sporophylls. These species are very popular in Japan, and they're sold in health-food and ethnic stores in America. These require more cooking than *Alaria* species: Pressure-cook them one hour, or simmer all day. Otherwise, use them the same way as edible kelp.

Add a small strip of dried horsetail kelp to soup—it expands very greatly as it cooks. If it's still too tough at the end of cooking, discard it. The minerals and rich flavor remain behind. If tender, dice and return it to the soup.

Japanese doctors have been successfully treating high blood pressure with kombu tea for years. Research that has been documented in the *Journal of Nutrition*, as well as in the journals listed under rockweed, have shown that it also strengthens the heart. It is believed that laminine and histamine, along with possible unknown substances, account for these effects.

OTHER PLANTS OF THE SEASHORE IN MID- TO LATE SPRING

Other Edible or Medicinal Plants

Amaranth leaves, black locust flowers, sweet clover flowers, curly dock flower stalk, wild onion bulbs and leaves, field pennycress leaves, flowers, and seeds, grape leaves, groundnut tubers, lamb's-quarters leaves, orache leaves, marshmallow leaves and roots, milkweed shoots, flower buds, and flowers, mullein leaves, peppergrass leaves, flowers, and seeds, sea rocket leaves, pine needles, shoots, and pollen, seaside plantain leaves, poke-weed shoots, prickly pear pads, sheep sorrel leaves and flowers, strawberry leaves and berries, yarrow.

For Observation Only

Autumn olive flowers. Russian olive flowers, blackberry flowers, black cherry flowers, grape flowers, groundnut flowers, seaside plantain flowers, poison ivy flowers, rose flowers, sow thistle flowers, strawberry flowers, sumac flowers.

PLANTS OF MOUNTAINS IN MID- TO LATE SPRING

Edible or Medicinal Plants

Miner's lettuce leaves, pennyroyal, orpine leaves, pine pollen, shoots, and needles, roseroot leaves, sassafras leaves, flowers, and roots, sheep sorrel leaves and flowers, spicebush leaves, creeping wintergreen leaves and berries.

For Observation Only

Pin cherry flowers, chokecherry flowers, waterleaf flowers.

CHAPTER 3

EDIBLE WILD PLANTS OF SUMMER

By summer, some but not all the spring shoots, greens, and flowers go out of season, while fruits and berries come into their own. Full-grown plants pack many natural habitats, and the competition is fierce. Despite this abundance, heat and mosquitoes make foraging challenging. (Be sure to see the segment on preparing to forage.)

PLANTS OF LAWNS AND MEADOWS IN SUMMER

Edible or Medicinal Plants

Chickweed, clover flowers, dandelion flowers and roots, dayflowers, daylily flowers and tubers, wild onion seeds and bulbs, lady's thumb, common mallow leaves, flowers, and fruit, ground ivy, peppergrasses, plantain leaves and seeds, purslane leaves, stems, and seeds, sheep sorrel, wood sorrel, yarrow.

For Observation Only

Sow thistle flowers and seeds, thistle flowers, violet leaves.

PLANTS OF CULTIVATED AREAS IN SUMMER

Edible or Medicinal Plants

Apples, blueberries, burdock root, sweet cherry, sour cherry, chickweed, clover flowers, carnelian cherry, dandelion flowers and roots, dayflowers, daylily flowers and tubers, hawthorn berries, Juneberries, juniper, Kentucky coffee tree seeds, Kousa dogwood fruit, lady's thumb, lamb's-quarters leaves, linden flowers, mulberries, orpine leaves, peaches, plantain flower heads, seeds, and leaves, purslane leaves, stems, and seeds, rose hips, redbud pods, sassafras leaves and roots, sheep sorrel, wood sorrel, yucca flowers.

For Observation Only

Chestnut flowers, Hercules'-club flowers, Kousa dogwood flowers, linden seeds, orpine flowers, rose flowers, sow thistle flowers, violet leaves, wisteria pods.

PLANTS OF DISTURBED AREAS IN SUMMER

Edible or Medicinal Plants

Blackberries, brassica flowers and seeds, burdock root, carrot seeds and flowers, catnip, pin cherries, chickweed, dandelion flowers and roots, dayflowers, daylily flowers and tubers, wild onion seeds and bulbs, foxtail grass seeds, ground cherries, horseradish root, lady's thumb, lamb's-quarters leaves, povertyweed, epazote, milkweed flower buds and flowers, mugwort, mullein leaves and flowers, orpine leaves, passionflower fruit, peaches, peppergrasses, pineapple weed, plantain leaves, flowers, and seeds, purslane leaves, stems, and seeds, raspberries, rose hips, sheep sorrel, spearmint, strawberry blite leaves, sumac berries, sunflower seeds, thyme, wood sorrel.

For Observation Only

Asparagus flowers and fruits, bracken, curly dock flowers, common evening primrose flowers, garlic mustard seeds, ground cherry flowers, Japanese knotweed flowers, Jerusalem artichoke flowers, nettle seeds and flowers, orpine flowers, parsnip flowers and seeds, passionflower flowers, poison ivy berries, pokeweed flowers and berries, rose flowers, sow thistle flowers and seed heads, thistle flowers, wild lettuce flowers, wild potato vine flowers.

PLANTS OF FIELDS IN SUMMER

WILD BEAN, WILD KIDNEY BEAN

(*Phaseolus polystachios*)

This long, creeping or climbing, perennial branched vine has alternate, palmate-compound leaves, with three broadly oval, pointy-tipped leaflets, 2 to 4 inches long. The side leaflets are asymmetrical.

Long flower stalks, arising from the leaf axils, bear loose clusters of purple-red, pealike, bilaterally symmetrical flowers, about ¼ inch across. The stalked, drooping, somewhat curved, flattened pods are 1½ to 2½ inches long. They resemble commercial string beans. Within are four to six tiny beans that become chocolate-brown when mature.

The plant grows in dry woods, sandy fields, sandy streambanks, and thickets throughout most of eastern North America, especially near the seashore. Wild bean flowers and pods both occur from midsummer to the fall.

The closely related tepary bean (*Phaseolus acutifolius* and *Phaseolus metcalfei*) is very similar, but the beans are speckled. This drought-resistant species grows throughout much of the Southwest, where

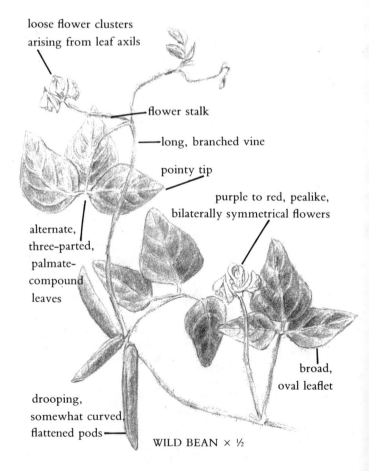

loose flower clusters arising from leaf axils

flower stalk

long, branched vine

pointy tip

purple to red, pealike, bilaterally symmetrical flowers

alternate, three-parted, palmate-compound leaves

broad, oval leaflet

drooping, somewhat curved, flattened pods

WILD BEAN × ½

the Papago Indians and others cultivate them. Like all members of the legume family, these species enrich the soil with nitrogen, thanks to a commensal relationship with nitrogen-fixing bacteria in their root nodules. Teparies are especially valuable because they can grow in sandier and more acidic soil than most other beans.

Caution: There are highly poisonous members of the legume family that can cause paralysis or death. Some resemble commercial beans so closely that, despite my sternest warnings, someone on a tour was about to risk her life for a taste. I almost had to pluck the plant from her hand en route to her mouth. While experienced foragers can identify the wild bean, beach pea (page 131), and hog peanut (page 216) with 100 percent certainty, other species in this large family are not safe. Be sure you've identified the wild bean with accuracy, and don't even think of eating any unidentified wild legume.

The youngest pods are so tender that you can pinch through them with your thumbnail. That means they're tender enough to eat—the so-called rule-of-thumb. If the pod is a little too tough to pinch through, open the pod and eat the beans inside, like peas.

I've never heard of anyone getting sick from eating a few wild beans raw. However, all beans potentially may contain some cyanide. Since cyanide is destroyed by heat, I steam or sauté wild beans lightly, or simmer them 10 minutes in soups. They're labor-intensive to collect in quantity, but delicious. However, I suspect they may interfere with memory: I keep losing tour participants whenever wild beans are in season. They forget about the rest of the tour, and seemingly try to remain collecting wild beans forever.

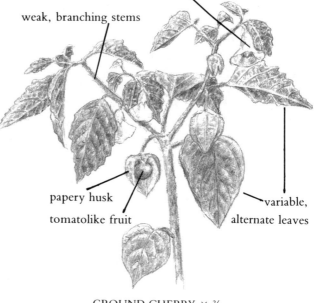

GROUND CHERRY × ⅖

GROUND CHERRY, TOMATILLO, HUSK TOMATO, JAPANESE LANTERN

(*Physalis* species)

The ground cherries are herbaceous annuals or perennials with alternate, coarsely toothed or toothless, medium-sized, triangular leaves. Depending on the species, the leaves and stems may be smooth, somewhat hairy, or very hairy. The weak stems branch. Some species are erect; others creep along the ground.

In the summer, bell- or wheel-shaped flowers arise from the leaf axils. They're often solitary, occasionally in clusters of two to five. They're usually yellow to greenish-yellow, with a purple to brown center, although the whole flower is sometimes purple.

The fruit resembles a miniature brown, yellow, red, orange, or purple tomato, ¼ to ¾ inch across and full of tiny seeds, enclosed by an inflated, ribbed, papery, lanternlike husk ½ to 1½ inches long.

This plant grows in burned areas and fields. If the fields are cultivated, it's considered a nuisance. Ground cherries will even come up through heavy mulch if the soil is moist. The fruits fall to the ground in the mid- to late summer. Gather and let them ripen in the husks for a few weeks. In some species, the ripening berry will burst through the husk.

Use ground cherries as fruits or vegetables. Mexicans boil and mash them, then add raw diced onions, chili, and cumin or coriander seeds. This makes a good sauce for Mexican burritos, tacos, and enchiladas. Ground cherries are great in relish recipes. You can also make tasty jams, pies, and puddings with them. **Caution:** Avoid berries tinged with green. They may be poisonous, as is the rest of the plant. If the fruit tastes bitter or strong, let it ripen until soft and sweet. Store it in a cool place. The husk becomes tan and papery as the fruit ripens.

This member of the nightshade family has poisonous relatives you must carefully avoid. Don't use similar-looking huskless plants, or species where the fruit is only partially enclosed. If the husk is tight instead of inflated, or not ribbed, beware! Also, avoid species with small fruits, where the husk is less than ½ inch long. All these plants are probably poisonous.

COMMON STRAWBERRY

(*Fragaria virginiana*)

This native member of the rose family grows from 2 to 6 inches tall. Its German name translates to "earthberries," because they creep along the earth; and the Anglo-Saxon word *streowberrie* indicates that it grows as though strewn around. Each

individual sprout arises from the same set of slender, horizontal runners and scaly underground rhizomes.

The three-parted, palmate-compound leaf tops a long leafstalk. The leaflets, which appear in early spring, have large, even teeth. Five-petaled, radially symmetrical white flowers ½ to 1 inch wide appear on their own stalks in mid-spring, to be replaced by the familiar berries about five weeks later.

The unmistakably familiar fruit, which is smaller than the commercial versions, makes identification easy. In addition, there are ten tiny green sepals that cup the fruit's base. Wild strawberries have a shorter season than any fruit I know— hardly more than two weeks, usually close to the summer solstice. They grow in open, sunny places, in open woods and along their margins, on moist ground, in meadows and fields, near the seashore, and on hillsides. You can find various species throughout the United States.

Wild strawberries have no poisonous lookalikes. The related wood strawberry (*Fragaria vesca*), with seeds that stand out from the surface of the fruit, is similar. It usually grows on the edges of trails in moist woods. It's edible, adding color to salads, but with absolutely no flavor. Common cinquefoil (*Potentilla simplex*) has leaves that resem-

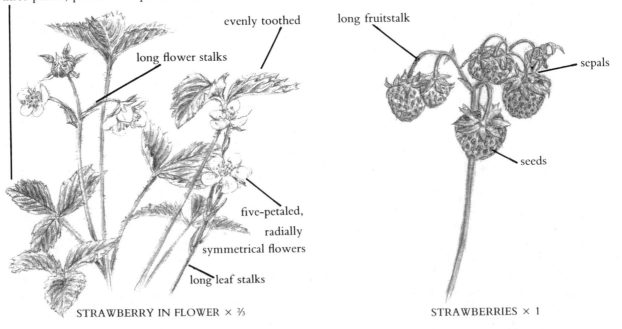

three-parted, palmate-compound leaves

evenly toothed

long flower stalks

five-petaled,
radially
symmetrical flowers

long leaf stalks

STRAWBERRY IN FLOWER × ⅔

long fruitstalk

sepals

seeds

STRAWBERRIES × 1

ble strawberry leaves, but there are five leaflets rather than three, the flowers are yellow instead of white, and there is no strawberrylike fruit.

Wild strawberries take a long time to collect, but they're worth it, with far more flavor than the ones you buy. I lay in a patch for over an hour and collected less than one cup of fruit, although I ate more than I kept. I never picked enough for a large recipe. Once I froze some with bananas, then pureed them in the blender with fruit juice, to make a sherbet. It took only a small proportion of straw-berries to get that wonderful flavor.

Strawberries provide magnesium, potassium, beta carotene, iron, and malic and citric acids. The fruit's a well-known mild laxative. It's also sup-posed to break up calcareous stones from the kid-neys, gallbladder and bladder, but as small as wild strawberries are, this is more than I can swallow.

You can collect strawberry leaves from spring to fall for tea. They're high in vitamins C and K. An infusion is diuretic and astringent, especially good for children with diarrhea or upset stomachs. It's also good for intestinal and urinary tract disor-ders. You can use a strong tea of the leaves or roots as a gargle or mouthwash, to soothe sore throats and tighten gums. The tea even makes a good rinse for acne and eczema.

Both leaves and fruit are used in cosmetics, to tone and tighten skin, and to close pores. Many commercial skin creams and lotions boast straw-berry juice, or strawberry leaf or root infusions or extracts. Strawberry juice has long been used as a facial wash, especially good for oily skin. To treat a sunburn, keep the juice on your skin for half an hour, and wash off with warm water. Strawberries supposedly remove discoloration not only from the skin but from the teeth as well. Leave the juice on for five minutes and rinse, and let me know if it works.

COMMON SUNFLOWER

(Helianthus annus)

This familiar native composite flower grows from 3 to 6 inches across, with a brown or purple

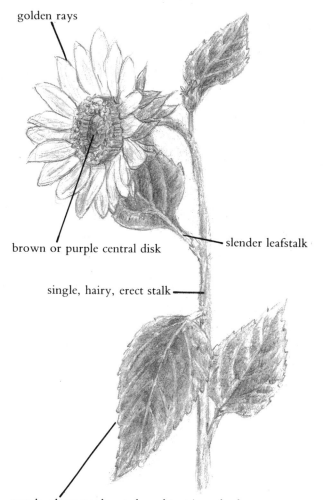

golden rays

brown or purple central disk

slender leafstalk

single, hairy, erect stalk

rough, alternate, heart-shaped to triangular leaves

SUNFLOWER × ⅓

central disk, and numerous golden rays. Each flower grows 3 to 8 feet high, atop a single erect stalk. The slender-stalked, alternate, heart-shaped to triangular leaves are 2½ inches to 1 foot long. The hairy leaves and stems are rough, like sandpa-per. Cultivated varieties are larger, up to 16 feet tall. The scientific name reflects the fact that the flower head turns to face the sun. In Greek, *helios* means "sun," and *anthos* means "flower." *Annuus* refers to the fact that this plant is an annual.

You find it in old fields, on disturbed soil, and in dry, open ground. It's most prevalent in the prairies in the Midwest (it's the state flower of Kansas). It grows scattered in the East, and its range extends to the West Coast. There are around sixty species in North and South America. All are edible, although many wild varieties are so small, their seed heads aren't worth bothering with. There are no poisonous look-alikes.

This plant has an illustrious history: The Incas of Peru worshiped this symbol of the sun. Their priestesses wore golden sunflowers on necklaces. The seeds became a big hit when they were brought to Europe. People eat them like we eat peanuts. The shells litter European streets. The Russians even cultivated giant sunflowers, which were reintroduced to American agriculture.

To harvest the seeds, cut off the flower heads in late summer, before they open fully, or the birds will eat everything. Leave enough to reproduce. Let them ripen in mesh bags at room temperature. Wild sunflower seeds are smaller than commercial ones, so the best way to enjoy them is to crack them with your teeth and eat them then and there, raw or lightly roasted.

To separate large amounts of meat from the shells, grind coarsely, then stir vigorously into water. The shells float, the seeds sink. You can also boil the ground seeds and skim off the oil that rises to the surface, to use in salad dressings or cooking. You can also roast the shells and use them as a caffeine-free coffee substitute, the way you'd use coffee beans or chicory root.

Another way to use sunflower seeds is to sprout them: Soak the mature seeds in the shells overnight to begin germination, then put them in a mesh bag or stocking and hang it from the shower head to drain. Rinse a few times every day. When the new growth is as large as the seeds, plant them in a shallow tray of rich soil. Water occasionally, and give them plenty of sunlight. In about a week, when the first leaves have opened, cut off the sprouts to use as snacks, or include them in salads, soups, or cooked vegetable dishes. They have a peppery taste and chewy texture that most adults and virtually all kids love. The clipped seeds may provide a second growth in another week.

The Indians pounded the seeds into a meal, which they put into cakes, added to gruel, or mixed with animal fat for a concentrated travel food. They used fiber from the stalks to make cloth, and extracted a yellow dye from the ray flowers. They also dried the leaves and smoked them like tobacco. The small, unopened flower buds, available in the summer, taste a little like artichoke hearts. They

may be bitter, so boil them in two changes of water.

Sunflower seeds are nutritional powerhouses. A pound has 540 nutrition-packed calories, including protein, the essential fatty acid linoleic acid, calcium, phosphorus, iron, sodium, riboflavin, iron, and vitamin C. The vitamin and protein content increases when you sprout the seeds.

Some species of sunflowers have tubers identical to the Jerusalem artichoke's (page 195), and it's hard to distinguish between Jerusalem artichokes and small wild sunflowers. *Helianthus maximiliani*, with slender, downy leaves, for example, has great tubers. Foraging guru Euell Gibbons suggested that wild sunflowers are evolving to develop tubers. You never know if there will be tubers, so check them out.

YARROW, STAUNCHWEED, SOLDIER'S WOUNDWORT

(Achillea millefolium)

This European perennial has narrow, fernlike leaves 2 to 4 inches long. They're very finely divided, hence the specific name, *millefolium*—"thousands of leaves." The basal leaves are stalked; those on the flower stalk are stalkless. The whole plant is woolly and fragrant. It reaches a height of 1 to 3 feet, arising from perennial rhizomes.

When I was a beginner, I used to think yarrow was related to wild carrot because the flower heads superficially resemble one another, although yarrow's is the smaller—2½ to 3 inches across. I was completely mistaken. The carrot flower head, Queen Anne's lace, is a loose, umbrellalike cluster of tiny flowers, and a member of the umbelliferous group. Yarrow is a composite. It takes the principle of clustering flowers one step further. The flattened flower head is a cluster of terminal clusters: If you magnify what appears to be a tiny individual flower, you'll see many petallike ray flowers surrounding numerous, tightly packed disk flowers, like a miniature daisy.

Yarrow is a common herbaceous plant of fields,

composite flowers

ray

disk

umbrellalike
flower heads

stalkless leaves

narrow, fernlike leaves

erect, slender flower stalk

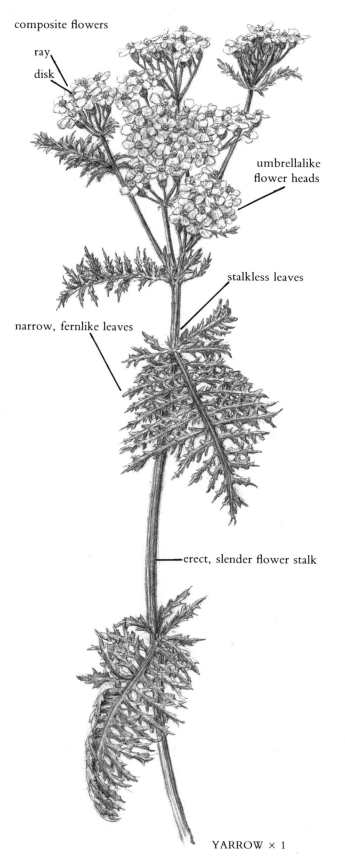

YARROW × 1

meadows, roadsides, and seashores, growing throughout most of North America. Rosettes of feathery leaves appear on the ground in early spring, and you can find the flowers from late spring to early fall. The dead flowers persist through the winter.

An infusion of the leaves, stems, and flowers makes a very popular herbal beverage. You can also stock up by drying or tincturing the herb. Yarrow is a green pharmacy. Chemical analyses have uncovered over 120 compounds.

The achilleine has been shown to stop the flow of blood and suppresses menstruation. It's balanced by coumarin, which promotes blood flow. Women used yarrow tea to relieve heavy menstruation for centuries. The tea is also good as a douche. As a styptic, dried, powdered yarrow stops bleeding. The Indians used the native species (*Achillea lanulosa*) for bruises, burns, earaches, and arrow wounds.

The tea is a tonic for the heart and circulatory system, as well as for the arterial walls. Its cyanidin affects the vagus nerve, which slows the heartbeat, while the achilleine lowers blood pressure.

A strong cup of hot tea is a diaphoretic: It helps you sweat out colds, the flu, measles, chicken pox, and fevers. Furthermore, demonstrated anti-inflammatory properties, probably due to the protein-carbohydrate complexes azulene and salicylic acid, counteract infection. **Caution:** Don't take large doses of yarrow, or use it for long periods of time. One constituent, thujone, is considered toxic.

Yarrow's generic name, *Achillea millefolium*, refers to Achilles, the nearly invulnerable legendary warrior of ancient Greece who was finally killed when an arrow struck his only vulnerable spot, his Achilles' heel. He reportedly applied yarrow compresses to treat his soldiers' wounds during the Trojan War. Yarrow promotes blood clotting. (For enemy soldiers, he deftly applied the point of his sword, which invariably increased bleeding.) Because Achilles always carried yarrow with him, it became the symbol of war.

OTHER PLANTS OF FIELDS IN SUMMER

Edible or Medicinal Plants

Apples, bayberries, blackberries, brassica flowers and seeds, burdock root, caraway seeds, carrot seeds and flowers, chicory root, chokecherries, clover flowers, cow parsnip seeds, curly dock root, wild onion seeds and bulbs, field pennycress seeds, foxtail grass seeds, groundnut pods and tubers, hawthorn berries, lamb's-quarters leaves, milkweed flowers and pods, orpine leaves, passionflower fruit, peaches, peppergrasses, pin cherries, pineapple weed, povertyweed leaves, purslane leaves, stems, and seeds, rose hips, sassafras leaves, berries, and roots, sheep sorrel, strawberry blite leaves, sumac berries, sweet cherries, yucca flowers.

For Observation Only

Autumn olive flowers, Russian olive flowers, blackberry flowers, black cherry flowers, grape flowers, groundnut flowers, orpine flowers, parsnip flowers and seeds, passionflower flowers, poison ivy flowers, pokeweed flowers and berries, rose flowers, seaside plantain flowers, sow thistle flowers, strawberry flowers, sumac flowers, thistle flowers.

PLANTS OF THICKETS IN SUMMER

BLACKBERRIES

(*Rubus* species)

The blackberry "cane" is an arching woody stem that grows up to 10 feet long. If the tip touches the ground, it takes root and spreads farther. The bark is reddish brown—covered with very visible short stiff thorns, capable of inflicting painful scratches. The leaves are palmately compound with three to seven oval, medium-sized, pointy-tipped, toothed leaflets.

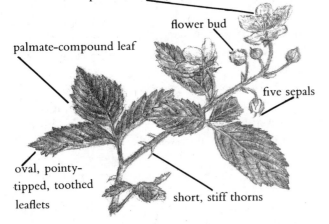

five-petaled, radially symmetrical, white flower

flower bud

palmate-compound leaf

five sepals

oval, pointy-tipped, toothed leaflets

short, stiff thorns

COMMON BLACKBERRY IN FLOWER × ⅔

The five-petaled, radially symmetrical, delicate-looking, white flowers are loosely displayed on racemes. Each flower has many ovaries. After fertilization, it produces lots of tiny, black fruits, each with its own seed. This assemblage of tiny drupes, all from one flower, is the blackberry we know and love—an aggregate fruit. The familiar, faceted, elliptical fruit grows from ½ to 1½ inches long, with a piece of the receptacle plugged into the base.

The blackberries are a very diverse group of related species that grow throughout most of North America and much of the rest of the world. This causes head-scratching and confusion among *Rubus* specialists: Like many plants, blackberries are undergoing evolutionary changes because of all the clearing, building, and planting man does. The problem is compounded by exotic varieties and hybrids, so we're left with a bunch of frustrated botanists and happy foragers. Some species are better than others, and some years are better than others. Here are a few common species:

The common blackberry (*Rubus allegheniensis*) is 2 to 8 feet high with very prickly, erect or arching, reddish or purple stems. The leaves divide into five, somtimes three, parts. The sharp-toothed, pointed leaflets are 3 to 7 inches long, and pale

COMMON BLACKBERRY × ⅔

green beneath. White flowers 1 inch or less across bloom in late spring or early summer, in elongated clusters of six to fifteen. The fruits, under 1 inch long, ripen in mid- and late summer. It grows throughout the northeast United States and southeastern Canada. Cut-leaf or evergreen blackberry (*Rubus laciniatus*) is a European species you'll see planted in city parks throughout the country. Its leaves are very deeply cut, and it's nearly evergreen. The berries are big and sweet, and sometimes the shrub is thornless. As I effortlessly gather berries from this miraculous variety, it's so easy, it seems like I'm cheating. The dewberry (*Rubus flagellaris*) has stems that creep on the ground instead of arching upward. So does the California blackberry (*Rubus ursinus*), a West Coast species of lower elevations, armed with somewhat flattened thorns. The cloudberry (*Rubus chamaemorus*) is a nonwoody, thornless species that grows in the far north and on mountaintops. Blackberry fruit has no poisonous look-alikes. The only similar fruit is the black raspberry. Its fruit is somewhat smaller

and more delicate. Most important, it's hollow after you pick it, like a thimble. Blackberries separate from their stems along with the corelike receptacle, so they're never hollow.

People sometimes confuse blackberry leaves with poison-ivy leaves. This happens only when you find blackberry brambles that have three leaves like poison ivy, or when you're so busy stuffing yourself with berries, you don't notice poison ivy growing nearby. Both plants' leaves are palmate compound, but blackberry leaves have toothed edges, while the poison ivy's leaves—sometimes irregular—are always untoothed.

I've found blackberries growing most abundantly by beaches and in dry, sunny areas inland. They also grow in old fields, along bulldozed dirt roads and roadsides, along fencerows, in pine barrens, on mountain slopes, in young woodlands and woodland margins, in wetlands, and in thickets, often cheek by jowl with poison ivy. Blackberries grow taller and more sparsely in wooded areas and on country roads. They're harder to locate in such habitats. They're easiest to locate in late spring, when they're flowering.

The berries ripen in the second half of the summer. I always know it's time for the blackberries in my area when I've finished my last cup of wild wineberries. When the blackberry season arrives, I treat it like a special holiday. The ritual begins by waiting for a relatively cool day with low humidity. I equip myself as described in the chapter on preparing to forage, and set out for my favorite blackberry stand. I usually arrange to meet a friend there, as it's almost painful to see thousands of blackberries ferment and go to waste. Blackberry "wine"-on-the-bush is not my favorite food.

Occasionally, I run into a fellow forager—usually someone who began picking as an unplanned diversion on a nature walk. This bareheaded person, wearing shorts and short sleeves, sweaty, insects alighting on the arms, soon falls by the wayside. Although a blackberry thicket on a summer day is an uncomfortable habitat for humans, it's more than worth the inconvenience.

Blackberries provide lots of potassium, magnesium, calcium, iron, phosphorus, beta carotene and vitamin C. They're delicious raw or cooked, with a powerful, earthy tang that complements their fruitiness. I've made my best-ever shakes with blackberries and yogurt or soy milk. If you include them in desserts such as pies, puddings, and cobblers, you may need additional sweetener. Blackberries' sugar content is low, so they're good for people who don't tolerate natural sugars. Apples and blackberries make a great combination in almost any dessert.

The inner bark of the stems, leaves, and especially the roots contains tannin. A decoction is a safe and effective astringent—a traditional home remedy for adult diarrhea. (For children, the milder raspberry root is favored.) Use the root for more severe cases, and save the leaves, buds, and branches for milder ones.

Blackberry tea is also used to stop bleeding and stave off dysentery, fevers, and sore throats. Since ancient times, people have chewed blackberry leaves for bleeding gums. However, the tannic acid in the leaves interferes with mineral absorption in other foods, so it's not good to use for prolonged periods. It would be better to address the cause of the bleeding.

BLUEBERRIES, HUCKLEBERRIES, AND RELATIVES

(Vaccinium species and Gaylussacia species)

This large subset of bushes within the heath family ranges from 1½ to 14 feet tall, without thorns or prickles. Most species are deciduous, but some are evergreen. The short-stalked, alternate leaves, always nearly elliptical, are smooth or slightly toothed, 1 to 2 inches long, sometimes leathery.

White flowers up to ¾ inch long, often on racemes, usually appear in the spring. They're sometimes tinged with pink, red, or green. The four to five petals fuse, forming clusters of bells.

The round, blue-black, blue, or red berries that form in the summer contain many seeds, sometimes too small to notice. All species have a five-

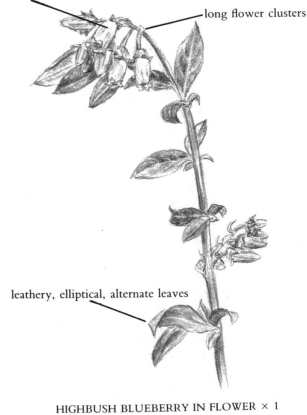

bell-shaped, five-parted, white flowers

long flower clusters

leathery, elliptical, alternate leaves

HIGHBUSH BLUEBERRY IN FLOWER × 1

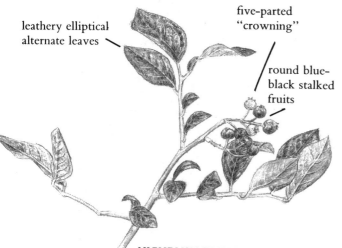

leathery elliptical
alternate leaves

five-parted
"crowning"

round blue-
black stalked
fruits

HIGHBUSH BLUEBERRY × ⅖

parted calyx or "crown" on the end of the berry, opposite its attachment to the stem. Anything fitting this description is definitely nonpoisonous (although some *Vaccinium* and *Gaylussacia* species are unpalatable). At worst, you'll eat an edible Juneberry (page 123) by mistake. All similar-looking poisonous blue-black berries lack crowns.

Blueberry and bilberry leaves are often finely toothed, sometimes woolly or leathery. Huckleberries' leaves are toothless, and splashed with distinctive, tiny, golden resin dots. In the spring and early summer, when the resin is most fluid, pressing a huckleberry leaf to your skin will leave a golden-yellow mark.

Blueberry and bilberry bark is very distinctive: The older growth is craggy and brown, and the young growth smooth and multicolored, with green and reddish hues predominant. Huckleberries' dark-colored bark is always smooth.

The fruits are round—blue to black. Some blueberries are lighter blue, powdered with a whitish bloom. The fruit is larger, with many imperceptible, soft seeds inside. Huckleberries' fruits are smaller and darker, with ten definite, hard seeds. Bilberries are Arctic *Vaccinium* species, distinguished from blueberries mainly by their range. They grow around the world in the far north. You can collect them in Canada, and on mountains in the eastern and western United States. Botanists are still fighting over other differences, but I think blueberries usually taste better.

My favorite species is the highbush blueberry (*Vaccinium corymbosum*), aptly named because it

grows over 14 feet tall. It has large berries which are easy to collect in quantity. The lowbush blueberry (*Vaccinium angustifolium*) grows only up to 1 foot tall. It forms dense thickets. The berries are small, but they're easy for children to reach. I sometimes find both species growing together.

I have found blueberries, huckleberries, and bilberries of all varieties, shapes, and flowers growing in varied habitats throughout northeastern North America, and in moist, mountainous habitats throughout the West. Different species range from the Arctic Circle to the Atlantic Ocean, and to the Gulf Coast. The environments differ, but the soil is always acidic. Bogs, mountains, woodlands, and thickets are all likely habitats. Blueberries grow best in the north, but occur as far south as Georgia. Bog bilberries grow on high mountain slopes and rocky barrens in California, as well as in New England.

The berries ripen between June and September, and the same bush can fruit one to two weeks earlier or later in different years. Highbush blueberries are usually available in early to midsummer (sometimes to late summer), while bog bilberries don't ripen until the end of the summer. The early and late lowbush blueberries' seasons are implicit.

It's not always easy to find good blueberry stands. Their energy often goes to surviving on their chronically poor, acidic soil. Bilberries, which grow under especially harsh conditions, may be especially sparse. Also, the quality of fruit varies from year to year. Even a light frost during the spring flowering season will destroy the year's crop.

Since blueberries and huckleberries are native to North America, it's not surprising that they were a great favorite of the Indians. Traditional Indians ate them raw, or dried them in the sun, or by the fire, for the winter. The Iroquois stored them in elm-bark boxes, or covered baskets, and used them in maize bread and pemmican.

I love combining blueberries with milk, soy milk, or fruit juice in the blender. The thick, rich mixture is like a sherbet. Blueberries contain pectin, a natural thickener that always makes its presence known. Jams need less agar or gelatin, and puddings and pie fillings need less thickening starches.

Blueberries, huckleberries, and bilberries have

many medicinal uses in American and European folklore. Eating blueberries relieves intestinal uneasiness. They counteract some of the bacteria that cause fermentation and inflammation. American Indians, and later European settlers, used every part of the shrub. They used a root decoction or leaf infusion to facilitate childbirth. Women began to drink it a few weeks beforehand and drank it every few minutes during labor. It stimulates and soothes the uterus, and it also has astringent properties. The tea also helps the uterus recover more quickly after miscarriage.

Blueberry or huckleberry leaf infusion is also good for diabetes. It contains neomyrtilicine, which reduces blood-sugar levels. Its astringent tannin also benefits urinary-tract inflammation. The bell-shaped blueberry flower has the strangest use: The Chippewa Indians used it to cure insanity. The inhaled fumes of the dried flowers supposedly worked wonders.

During World War II, British pilots stumbled upon an unexpected and very useful benefit simply by eating bilberry jam before leaving for night missions. Their night vision improved dramatically. Subsequent research confirmed these benefits. Besides high vitamin C levels, bilberries contain low levels of certain flavonoids called anthocyanosides. Concentrated extracts not only greatly improve night vision, but they're also useful for other vision problems, including cataracts, glaucoma, chronic eye fatigue, and day blindness. Its effect is temporary—a normal dose lasts less than a day, but the extract is completely safe.

CORNELIAN CHERRY

(Cornus mas)

This is an imported ornamental shrub or small tree, growing up to 30 feet tall. It's actually a dogwood, unrelated to cherries. Glossy, opposite leaves with pointed tips and prominent, beautifully curved veins grow on short leafstalks.

In very early spring, tiny, bright-yellow, odorless, four-petaled flowers cluster at the end of the thin green twigs. In mid- to late summer, the ob-

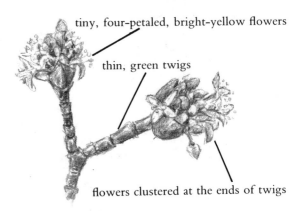

tiny, four-petaled, bright-yellow flowers

thin, green twigs

flowers clustered at the ends of twigs

CORNELIAN CHERRY IN FLOWER × 3

long, juicy, bright to dark red berries—about 1 inch long—are ready to pick. Each berry contains one large, oblong, hard seed.

There are no poisonous look-alikes. Spicebush's (page 180) similar flowers are fragrant and its leaves are alternate. The flowering dogwood is toxic, but its dry berries don't resemble the cornelian cherry's fleshy fruit. The cornelian cherry's fruit superficially resembles a small wild plum, but plum trees have alternate leaves.

Although it is still rare in the wild, landscapers across the country love it. You'll find it in sunny areas or thickets in cultivated parks, on college campuses, and in arboretums and botanical gardens throughout the United States. I expect it will continue spreading.

I identified this species long ago, but I had to forgo the berries season after season until I could determine that the fruit was safe. My first clue turned up when a contrary student popped the "forbidden" fruit into his mouth when he thought I wasn't looking. For the rest of the tour, I stood ready to catch him if he keeled over, but he lived

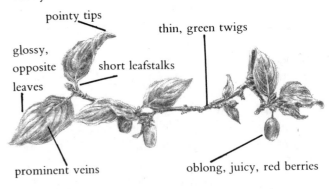

pointy tips

thin, green twigs

glossy, opposite leaves

short leafstalks

prominent veins

oblong, juicy, red berries

CORNELIAN CHERRY WITH FRUIT × ⅔

through the afternoon. I later learned that this fruit is a favorite in its country of origin, Turkey. It's even sold pickled in ethnic grocery stores in the United States.

I finally tried the fruit myself, only to be disappointed by its excessive sourness. But after another student left some in the refrigerator for a few days, she discovered that it became sweet and tender. I had tried it underripe, and it's the only berry I know that ripens off the bush. Always let cornelian cherries get very soft before you use them, and you'll love them. They taste like a combination of cherries, lemons, and plums. Enjoy them raw, or cook them with a thickener and strain out the pits to make jam, cake topping, or pie fillings.

CURRANTS AND GOOSEBERRIES

(*Ribes* species)

Gooseberries and currants are small bushes, only 2 to 5 feet tall, with arching branches. The simple, alternate, maplelike leaves vary from species to species, often with three to five lobes per leaf. Mapleleaf viburnum (page 182), which has an unpalatable but nontoxic fruit, has similar leaves that are opposite.

Currants' and gooseberries' greenish or yellowish-white to purple flowers grow on long racemes in the spring. They have five petals fused into a tubular or bell-like, five-lobed calyx tube. The fruits ripen in the summer—usually before midsummer—transforming the racemes into long, jewel-studded threads.

Gooseberries usually have thorny stems and bristly fruit, but some varieties, especially escapes from cultivation, are smooth. Currants, except for bristly currants, are thornless. The fruits are globular and stalked, not more than ¼ inch in diameter, with a tight, smooth, translucent skin. Color ranges from red to pale yellow or even black. One of the most common species is the garden red currant (*Ribes sativum*). Its branches straggle along the ground, and the red berries are light and translucent, even in the damp, dark eastern forest where I collect it.

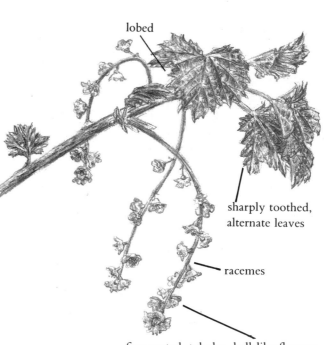

lobed

sharply toothed, alternate leaves

racemes

five-parted, tubular, bell-like flowers

CURRANT IN FLOWER × ⅕

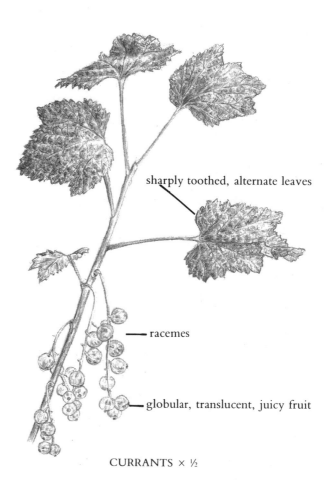

sharply toothed, alternate leaves

racemes

globular, translucent, juicy fruit

CURRANTS × ½

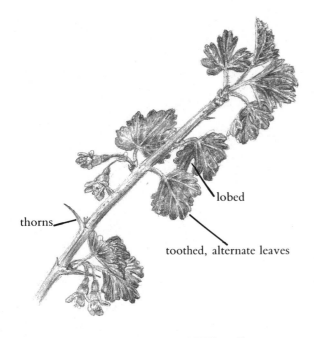

GOOSEBERRY IN FLOWER × ⅔

There are many other native and exotic species of currants and gooseberries growing wild in moist woods, prairies, hillsides, canyons, and ravines throughout the United States. They always grow where there's moisture.

Because most currants and gooseberries are grown exclusively for processing, few people recognize them, although over one hundred varieties grow in the Northern Hemisphere. Instead, we see the European gooseberry and currant jams, fillings, and liqueurs in gourmet shops, or tiny, dried raisinlike currants in supermarkets.

Wild currants have a sweet-sour flavor. Some species are good raw or cooked; others are too sour to enjoy raw. Cook all gooseberries to soften the bristles.

All these little berries are easy to use in recipes, since their seeds are so soft and tiny, you don't

have to remove them. They all contain pectin, a natural thickener, so you can use currants and gooseberries with little thickener. In Europe people use them for making juices, wines, and tarts.

The Native Americans ate some berries fresh and dried the rest for the winter, usually in the form of a cooked jam. Eastern tribes cooked currants with sweet corn. In winter, they rehydrated them to use with meats. They used animals skins to rub the bristles off gooseberries, and used them like currants.

Currants and gooseberries are high in potassium, phosphorus, calcium, and vitamins A and C. They also contain malic and citric acids, as well as pectin. The seeds of the black currant (*Ribes nigrum*) contain gamma-linolenic acid. It's expelled, put into capsules, and sold in the health-food stores for cardiovascular disease. Indians used a decoction of the root or inner bark for worms and kidney disorders.

Currant or gooseberry juice is astringent, diuretic, and diaphoretic. It's recommended for sore throats and burns, and especially for fevers. I once found a large crop of sour wild currants when I was just developing a sore throat. I ate currant after currant, and took a container home, hoping the medicinal action of the plant, along with high dosages of vitamin C–complex supplements, would help fight the cold. But the astringent currants only hurt my throat even more, and the cold ran its course. It was probably naive to hope to alleviate a systemic infection with a locally acting remedy. I don't get sick very often, but next time I have a sore throat, I'll try to use soothing herbs instead.

The young spring leaves of currants, which Europeans sometimes eat, supposedly act as a mild diuretic. Women use the leaves and roots for reproductive-system problems. In America, gooseberry root decoction was given to women with menstrual or uterine problems caused by bearing too many children.

Except for chipmunks, squirrels, raccoons, and some birds, we ignore wild *Ribes* species today. But early in the century, wild *Ribes* species were exterminated. White pines (page 217) were in great demand for lumber. The seeds were shipped to European nurseries, and the seedlings were replanted in America. However, they returned car-

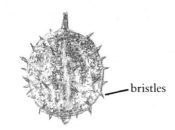

GOOSEBERRY × 2.5

rying the blister rust fungus, a European scourge of pines, which requires currants for part of its life cycle.

COMMON ELDERBERRY, AMERICAN ELDER

(Sambucus canadensis)

This member of the honeysuckle family is a shrub that grows up to 13 feet high, with smooth, gray bark. Corky bumps cover the slender branches, and there's a spongy, white pith inside the twigs and branches.

The opposite, feather-compound leaves may be over 3 feet long. The leaf is divided into five to eleven opposite, coarsely toothed, pointed, short-stalked elliptical leaflets, each 3 to 4 inches long.

In late spring or early summer, the elder bears tiny, branched, white, lacy flowers in flat-topped to slightly rounded clusters (panicles) that spread over 6 inches across. The tiny, spherical, juicy, purple-black to black, seedy berries are hardly more than ⅛ inch across. They grow in branched clusters, like the flowers, ripening from midsummer to early fall, in quantities that weigh down the branches.

The blue elder (Sambucus cerulea) has dark-blue to blackish berries, and grows in the western third of the United States. It's very similar to the common elderberry, and you can use it the same way.

Avoid elderberry species with red fruit growing in rounded, instead of flat, clusters. They may make you sick. Hercules'-club (page 57) is a shrub or small tree with feather-compound leaves that looks a little like the common elderberry. It has flat clusters of poisonous, black berries, often arranged in a ring, and a short, unbranched, thorny trunk. Elderberries are thornless.

The common elderberry often grows in large, dense stands in moist places. Look for it in marshes, along riverbanks, along roadsides, and in moist woods and thickets in eastern North America and the West Indies.

Collect the flowers by plucking off the stalk at the cluster's base. It's impossible to remove each tiny flower individually. Take a small proportion of the flowers from each bush, and collect only where they are abundant or the plant won't produce any berries. Where you find one elder bush, you usually find many more.

The flowers make wonderful food. Try elder flower (sometimes called elder blow) fritters, using your favorite tempura or pancake batter. Make a light, mild batter, so you don't overpower the delicate flowers. Try sautéing them.

Elder flowers make a pleasant-tasting tea, especially with mint. They also make a potent, fragrant wine. Steeped in vinegar they add flavor and strengthen the stomach. Taste some berries from a few bushes before you collect, so you can choose the bushes with the tastiest fruit. Gather the berries like the flowers. This is quick. The real work occurs at home: Pulling small bunches of berries from their stems, and sorting the fruit from the debris on a tray, takes time.

Avoid unripe, green berries—they'll make you sick. Even raw ripe elderberries make some people nauseated. Cooking or drying dispels the offending substance and greatly improves the flavor. Baking this fruit in muffins, cakes, and breads imbues them with a piquant crunchiness. They become the central ingredient whenever you use them in baked goods. Elderberries aren't sweet and contain no thickeners. Rely on other ingredients for these ele-

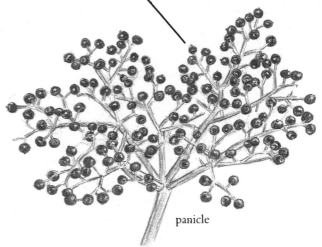

tiny, round, bluish-black berries

panicle

ELDERBERRIES × ⅕

flat-topped to slightly rounded flower cluster (panicle)

tiny, white, lacy flowers

short stalks

leading leaflet

opposite, feather-compound leaves

paired, oval, toothed leaflets

ELDERBERRY FLOWERS × ⅔

ments, especially if you're making the European favorite, elderberry jam.

The berries have few calories and lots of nutrition. They provide very large amounts of potassium and beta carotene, as well as sugar and fruit acids, calcium, phosphorus and vitamin C.

Looking at or even thinking about the elderberry bush evokes a flood of magical associations and images of the past—European ladies dousing their white skin with elder-flower water, and crystal goblets filled with elderberry wine. In European folklore, fairies and elves would appear if you sat underneath an elder bush on midsummer night. The lovely elder possessed potent magic, with the ability to drive away witches and kill serpents. Carrying the twigs in your pocket was a charm against certain diseases. One of these tales bears some truth: Sleeping under the elder supposedly produces a drugged, dream-filled sleep—the fragrance is actually mildly sedative. Perhaps the visions of fairies and elves resulted from dreaming under an elder bush.

My experience with the elder indicates that much of its charmed reputation among Europeans and Native Americans comes from its ability to heal. The flowers and fruit are medicinal. Hippocrates already recognized this in 400 B.C. (He used a smaller European species with similar properties, which doesn't grow in America.)

Due to their diuretic and detoxifying properties, people eat elderberries to lose weight. The flowers have been used in cosmetics since ancient times. Distilled elder-flower water softens, tones, and restores the skin. Elder-flower infusion cleanses the skin, lightens freckles, and soothes sunburn. Its bioflavonoids promote circulation and strengthen the capillaries.

An infusion or tincture is astringent, expectorant, and diaphoretic, great mixed with yarrow and peppermint for colds, flu, and asthma. Herbalists also use it to soothe children's upset stomachs and relieve gas. It's even applied externally for swelling, rashes, and chilblains (frostbitelike trauma to wet skin), and as an eyewash for conjunctivitis and eye inflammation. You can even steep the flowers in oil to make a soothing massage lotion that relaxes sore muscles and also soothes burns and rashes. Like the flowers, elderberry infusion is astringent and diaphoretic—good for colds,

excessive mucus, and sore throat. You can also boil the flowers in vinegar to make a black hair dye.

In 1899, an American sailor accidentally discovered that cheap port wine, which is colored with elderberries, relieved his arthritis. Other port wines didn't work. I don't recommend drinking alcohol, which causes more problems than it helps, but this result indicates elderberries' possible anti-arthritic properties. Another use for elderberry wine goes back to the movie *Arsenic and Old Lace*. Two old ladies laced it with arsenic to put lonely old men out of their misery!

Many older herb books recommend using elderberry leaves, roots, or bark medicinally, probably because Indian herbal experts used them. This doesn't guarantee safety: **Never use these parts of the elderberry.** They're poisonous. They contain a bitter alkaloid and glycoside that may change into cyanide. Children have been poisoned using elderberry-twig peashooters, and adults have been poisoned using hollowed twigs to tap maple trees. However, there is a benefit to the toxicity: People use dried, crumbled elderberry leaves in their gardens as a natural insecticide.

PASSIONFLOWER, MAYPOP, PASSION FRUIT, VINE APRICOT, WILD CUCUMBER, GRANADILLA, MARACOCK

(Passiflora incarnata)

You can find this vine up to 30 feet long, trailing along the ground, or climbing with its springlike tendrils. The alternate, lobed leaves are slightly toothed, with three to five sharply pointed lobes up to 4 inches long.

From mid-spring to early summer, large, showy flowers 1½ to 3 inches wide burst forth from the leaf axils. The inner design of three to five white sepals creates an alternating pattern with three to five white petals. This is overlaid by a fringelike crown of purple or pink threads, the corona, growing from the petals' base. The plant got its name because Spanish missionaries thought the flowers resembled Christ's crown of thorns, so the vine is supposed to symbolize faith and piety.

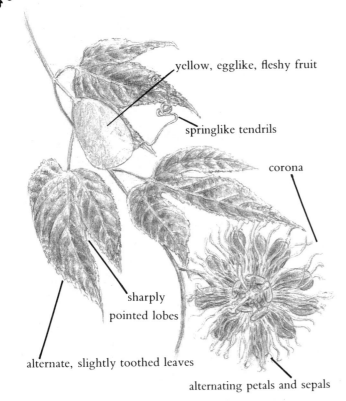

yellow, egglike, fleshy fruit

springlike tendrils

corona

sharply
pointed lobes

alternate, slightly toothed leaves

alternating petals and sepals

PASSIONFLOWER × ½

The yellow, fleshy fruits are around the size of a small hen's egg. They appear anytime from early summer to fall. There are other similar edible species in the genus, and no poisonous look-alikes.

The passionflower is a "weed" of southern states, ranging from Pennyslvania, Illinois, and Oklahoma south to Florida and west as far as Texas. It grows in sandy soil, fields, along roadsides, in disturbed habitats, and in thickets.

The fruit is called maypop because children in the South jump on it, to make it pop. Peel the ripe maypop and enjoy it raw, avoiding the many seeds. It has a sweet and cooling flavor. Maypop jams and beverages are popular in the South. The fruit is a good source of beta carotene and niacin.

Centuries ago, Peruvian Indians discovered that the fruits were a sedative, and used them for insomnia and nervousness. Europeans took this plant home, then brought it to North America.

The leaves and flowers also have antispasmodic and nervine effects. The active ingredient is a safe, nonaddictive tranquilizer called maltol, although it works properly only with its other constituents.

A tincture or infusion of the dried leaves and flowers is one of the best remedies for insomnia. Sleep encouraged by passion fruit is natural, lack-

ing the bad side effects of narcotic-incuded sleep, so you wake up refreshed. In 1979, forty-two different over-the-counter sleeping formulas available in Germany already contained passionflower extracts. The effects of this herb were scientifically verified in the 1920s, and American drug companies still show no signs of catching up.

The tea also calms frazzled nerves, and treats neuralgia, nervous headaches, high blood pressure, and certain cardiac conditions. It's also effective in small doses for hyperactive children, who are often routinely drugged with Ritalin in school, whether they really need it or not. Passionflower is much safer, without the side effects of the prescription drug. **Caution:** Huge amounts could be harmful.

PEACH AND NECTARINE

(*Prunus persica*)

The peach is a small tree with slender, curving branches. The finely toothed, narrow, curving, lance-shaped leaves are 3 to 5 inches long, tapering to a sharp point. The five-petaled, radially symmetrical, pink flowers are ½ to 2 inches across, appearing in early spring, before the leaves open.

The wild peach is like the commercial version, yellow with red blush, downy skin, and gently grooved. It ripens from mid- to late summer. Like all members of its genus, it has one large pit. The inner kernel closely resembles an almond—a close relative without the fleshy fruit. The peach has been cultivated for millennia in Europe and Asia. Peach sculptures and porcelains date back to ancient China, where the fruit was cared for under glass.

Peaches grow throughout the United States, wherever they've escaped cultivation. Soon after purchasing my first wild-food field guide, I explored an overgrown empty lot to look for asparagus. There was neither asparagus nor anything else I could identify, although subsequent explorations turned up a peach tree. Later on, I was amazed to find others in out-of-the-way places in city parks, in fields, thickets, and disturbed areas such as roadsides. People eating peaches probably threw away the seeds, which grow quite readily.

Unfortunately, the fruits I find are usually infested with insects. I had better luck with wild nectarines. I was leading a tour in Central Park late one summer, when I spotted a peach tree. After I announced my find, a student reached under the foliage and pulled out a delicious, ripe nectarine. There was plenty for everyone.

Actually, nectarines are smooth-skinned peaches, and peaches are fuzzy nectarines. That's why I was fooled. The foliage is identical. A genetic mutation, like the one that creates albinos, makes the difference. Sometimes the two fruits grow on the same tree, and you may even find fruits that are half nectarine and half peach. The popular idea that nectarines are hybrids of peaches and plums is simply erroneous.

These wild nectarines were smaller and less juicy than the peaches, but they were sweet and delicious, with no insects. This is the only nectarine tree I've ever found. It taught me that wherever you live, nature is full of surprises.

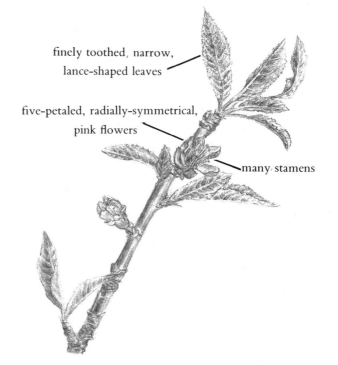

finely toothed, narrow, lance-shaped leaves

five-petaled, radially-symmetrical, pink flowers

many stamens

PEACH IN FLOWER × ⅕

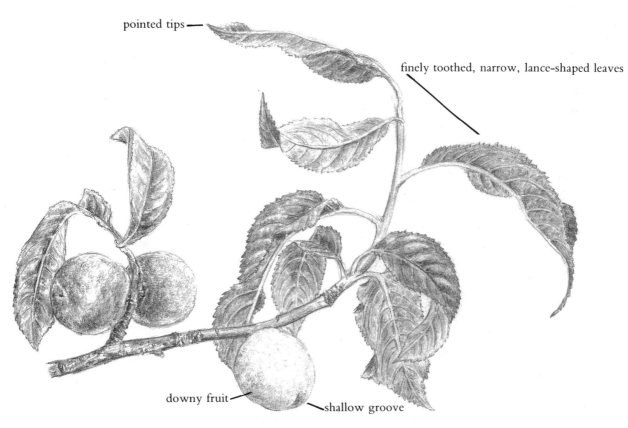

pointed tips

finely toothed, narrow, lance-shaped leaves

downy fruit

shallow groove

PEACH × ½

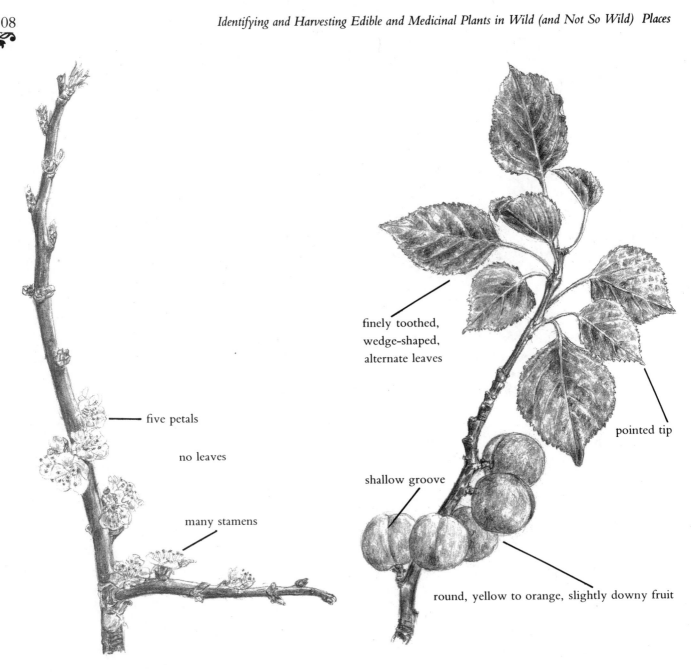

five petals

no leaves

many stamens

finely toothed,
wedge-shaped,
alternate leaves

pointed tip

shallow groove

round, yellow to orange, slightly downy fruit

APRICOT IN FLOWER × 1

APRICOTS × ½

Soon after, I found another unexpected close relative of peaches and nectarines, in an overgrown field near the seashore—a wild apricot tree (*Prunus armeniaca*). It's a small tree, with medium-sized, simple, alternate, wedge-shaped, finely toothed leaves. Apricot pits litter the ground under it all year, and the fruit is identical to commercial apricots (but much tastier), so it's easy to identify. Apricots aren't supposed to grow wild in the North (the blossoms open so early in the spring, they're vulnerable to frost), but they don't seem to realize it. I don't know how common they are, but keep your eyes open for them. You may find them escaped from cultivation in sandy fields and thickets across the country, where apricot-eating people tossed away their pits.

Peaches contain large quantities of beta carotene and potassium, as well as calcium, phosphorus, and vitamin C. Virtually everyone eats or knows how to prepare peaches, but few people know that the peach is an important medicinal plant. Peach-leaf infusion or peach-bark decoction is a demulcent—soothing to the mucous membranes. Herbalists use it for stomachaches and digestive disorders, and to clear out the intestines and the kidneys. It also has sedative, diuretic, and laxative properties.

Peach-pit decoction is also very soothing and stimulating to the stomach—a good stomach tonic to take before or during meals to stimulate enzyme activity and relax the nervous system. **Caution:** Don't eat the raw kernel inside the pit. It contains hydrocyanic acid (see Apricot, page 108), a gastric irritant (destroyed by heat) which could potentially release toxic cyanide.

PEAR

(Pyrus communis)

The pear tree, sometimes as tall as 60 feet, is closely related to the very similar-looking apple tree, but does not have the apple tree's hairy twigs and leaves with hairy undersides. I sympathize with the pear tree, since the top of my head shows the same hairless tendency.

The pear tree's early spring flowers are cream-white, 1 to 2 inches across, while the apple tree's flowers are white-pink. Pear trees have thicker, often darker twigs that are thinner and more gnarled than apples'—sometimes even spiny. (The spines are really aborted branches. They sometimes even bear minute leaf structures at their bases.) Pear's finely toothed leaves are 1½ to 3 inches long—oval to elliptical. They're rounder than apples', and the leafstalks are yellowish.

The apple tree spreads out more—it's shorter and stubbier. The pear tree looks taller and narrower, with a more upright trunk and branches. Pear bark is scaly and wrinkled, chunkier, but not as craggy-looking as the apple tree. The two species remind me of two good friends, especially when I see them growing side by side.

When wild pears ripen at summer's end or in the fall, they're long and tear-shaped—usually smaller than cultivated varieties, but clearly recognizable. They have the five-parted "crowns" on the side opposite the stem, and if you cut the fruit

creamy, white blossoms

oval to elliptical, finely toothed, alternate leaves

PEAR IN FLOWER × ⁵⁄₁₆

open, the core has a five-parted radial symmetry, like the apple. Unlike apples, pears have a gritty area just below the crown, something horticulturists are trying to breed out.

These Eurasian imports grow in thickets and woods throughout the Northeast, and locally in other regions—wherever they've escaped cultivation. Until the sixteenth century, people usually chose to stew pears into rich sauces. Soon after, lighter and juicier hybrid pears were developed. Wild pears are closer to the older varieties—heartier, grittier, and less juicy. They're best cooked.

When I collect my pears, I bring two double shopping bags, to keep ripe and unripe fruit separate. I use or refrigerate the ripe pears immediately. The rest stay in a cool place, to be re-sorted every day. Newly ripe as well as rotten fruit continually appear. Remove rotten pears before they contaminate the good ones, and separate the ripe ones before they begin to rot.

I find it difficult to do anything fancy with wild pears because I don't usually have enough ripe ones at once. My favorite recipe is in a good old-fash-

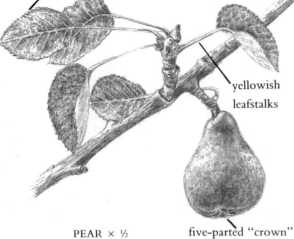

oval to elliptical, finely toothed, alternate leaves

yellowish leafstalks

PEAR × ½

five-parted "crown"

ioned pear sauce, which you prepare the same way you make applesauce. Pears contain vitamin C, and they're good sources of minerals such as potassium, calcium, and phosphorus.

Pear tree wood is valued for its hardness. It's used for sculptures and handles.

WILD PLUMS

(*Prunus* species)

Plums are in the same genus as cherries and peaches, although cherries and plums have the most in common. Both have dark bark, marked with horizontal streaks. Plum trees are generally 6 to 30 feet tall. The sharply toothed leaves are inches long—rather thin, with narrow bases and pointed tips.

The white flowers, often paired, bloom in early spring, and the ground is littered with pits. The fruit is larger than cherries, covered with a powdery bloom, and circled with two faint, longitudinal lines.

Native wild cherry (choke and black) fruits grow on a long racemes. Plums grow in radiating clusters, on separate stalks. Some plum trees are thorny, especially the wild plum, while cherry trees are thornless. The cherry pit is roundish, while the plum's is bigger and somewhat flattened, with a slight ridge between the two sides. Birds

love wild cherries but hardly touch plums, perhaps because wild cherries are smaller and easy to swallow. Foxes eat plums.

There are many native and exotic species growing in thickets, along streams, in swamps, or in cultivated areas throughout most of North America, bearing fruit from midsummer through mid-autumn.

Each tree has a slightly different flavor, so sample the plums from lots of trees before collecting. Enjoy them raw, or pit them and cook them into jams or sauces or pie fillings. They dry and freeze well, and they're easiest to pit when partially frozen.

Pitted plums are also good in breads, cakes, and muffins. Substitute them for blueberries, but use less liquid, since they're juicier. I also make plum fruit leather (see cooking section, page 275). Plums provide lots of beta carotene and potassium. They

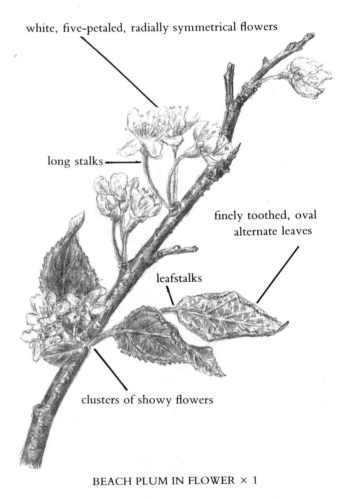

white, five-petaled, radially symmetrical flowers

long stalks

finely toothed, oval alternate leaves

leafstalks

clusters of showy flowers

BEACH PLUM IN FLOWER × 1

round, purple to black, thin-skinned fruit

finely toothed,
oval alternate leaves

BEACH PLUM × ⅔

also contain some phosphorus, magnesium, B-complex, and vitamin C.

Prunes, which are dried plums, are well-known laxatives. Eating large quantities of fresh plums has the same effect on some people. The plum tree's inner bark and twigs make an astringent decoction—a good folk remedy for mouth sores and sore throat.

Beach plums (*Prunus maritima*) are the best plums you'll ever taste. The intense, sweet and tart flavor is outstanding. The much-branched, thornless shrubs grow from 1 to 8 feet tall, usually forming large, dense stands. The finely toothed, oval, alternate, stalked leaves grow 2 to 3 inches long, and ½ to 1 inch broad.

The showy, long-stalked white flowers are

small-medium, barely over ½ inch across, with five radially symmetrical petals, growing in clusters of two to four. The round, purple-black drupe is up to 1 inch in diameter, with a thin skin covering the flesh, and a single large, flattened stone inside. Like the related cherry and peach, the stone contains high levels of hydrocyanic acid, an undesirable gastric irritant that cooking destroys. The fruit is powdered with a faint, whitish, waxy bloom. You can find large quantities of these tender, juicy plums ripening from the end of the summer to the beginning of autumn.

Amazingly, they grow in a harsh, forbidding habitat, producing abundantly in sterile sand, often within sight of the ocean. The beach plum's main range is from coastal Maine to Delaware.

Elsewhere, the American plum or wild plum (*Prunus americana*) is the most common species. The small, tart fruit is red or yellow, with only a little flesh—not quite as good as the beach plum, but still a treat. It grows in dense stands, like the beach plum, on moist soils in valleys, roadsides, low slopes, and moist edges of forests and thickets of the eastern two thirds of the United States. You can find large quantities of these tender, juicy plums ripening from late summer through fall. The Canada plum (*Prunus nigra*) is a light reddish-orange, larger and tastier species than the American plum. It grows in Canada, down the East Coast into Georgia, and west to Texas, Arkansas, and Iowa.

RASPBERRIES

(*Rubus* species)

Wild raspberries were a big hit when my mother introduced them to me as a child, on vacation in the Catskills, and I still love them today. For many foragers, the height of the fruit season occurs when raspberries ripen. They'll bear hot, muggy days, boiling sun, mosquitoes, and scratchy thorns to come home with a few quarts of this wonderful fruit. The plant doesn't make it easy, but the information in Chapter 1 on how to prepare for foraging will help.

Raspberries' woody "canes" have arching stems up to 10 feet long. They take root where the tip touches the ground, spreading more quickly than you would believe. Unlike true shrubs, they don't have a bark, and the stems live only two years. The first year they spring from perennial root systems, long and unbranched, without flowers or fruit. The second year they grow branches that produce flowers and fruits. Then they die, leaving the roots to send up new shoots the following spring. This also applies to blackberries (page 96).

Most North American species bear alternate, palmate-compound deciduous leaves with three to five leaflets. Each leaflet is roughly oval and coarsely toothed, usually broad at the base and pointed at the tip.

The small flowers grow in loose clusters. Like roses, which are relatives, they have five petals and many stamens and pistils. When a flower is fertilized, an ovary at the base of each pistil produces a tiny fruit with one seed. These fruits unite to produce the crunchy, many-seeded, globe-shaped raspberry, really a compound fruit.

Raspberries grow in a variety of habitats, such as dense thickets, on hillsides along roadsides, at the edges of woods, on disturbed soil, and near fresh water. Different species grow throughout North America, and in many parts of the world.

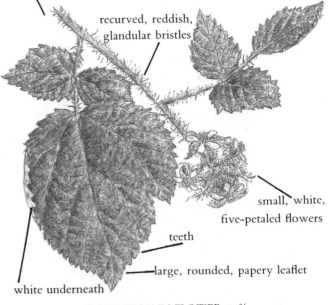

three-parted, palmate-compound leaf

recurved, reddish, glandular bristles

small, white, five-petaled flowers

teeth

large, rounded, papery leaflet

white underneath

WINEBERRY IN FLOWER × ⅔

three-parted, palmate-compound leaf

recurved, reddish, glandular bristles

white underneath

unopened fruit

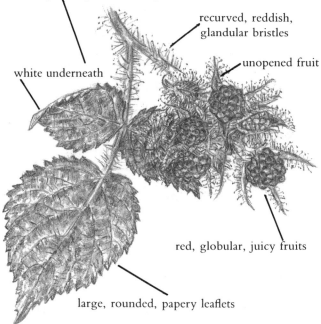

red, globular, juicy fruits

large, rounded, papery leaflets

WINEBERRIES × ⅔

white to pink flowers in flattish clusters

light underneath

curved thorns

arching cane

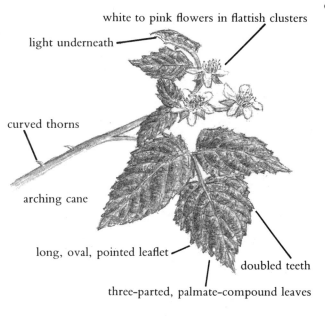

long, oval, pointed leaflet

doubled teeth

three-parted, palmate-compound leaves

BLACK RASPBERRY IN FLOWER × ⅔

The wineberry (*Rubus phoenicolasius*) comes from Asia and is common in the Northeast. It grows up to 8 feet tall; its stems, recurved, glandular bristles, and globular berries are red. The papery leaflets are roundish and large, white beneath and deep green above. The inconspicuous, white, clustered flowers are under 1 inch across, and the flower stalks are densely red and bristly. The bright red, globular, juicy fruits, around 1 inch in diameter, usually ripen in midsummer.

The black raspberry (*Rubus occidentalis*) is another alien species, with distinctive arching canes powdered by a protective greenish-blue bloom, and protected by curved prickles. The leaves are sharply double-toothed, and white-hairy beneath.

The white to pink flowers appear in mid- to late spring, in flattish clusters of three to seven individuals. The purple-black fruit, which ripens at the beginning of summer, is only ½ inch across, but it's among the tastiest. It's common from the East Coast to the Rocky Mountains, up to the northernmost section of the Great Lakes, and south to North Carolina.

The wild red raspberry (*Rubus strigosus*) is a shrub growing up to about 6 feet tall. It's similar to the black raspberry, but the stem has way more bristles and thorns, and the fruit is red.

Each leaf consists of three to seven toothed, oval leaflets. The flowers are white to greenish-white. It's sometimes erect instead of arching. It grows most abundantly north of New York, and in most of Canada, from the east to west coasts. You can find the red fruit from midsummer to late summer. This is the domestic raspberry's wild forerunner.

black, juicy fruits

sharply toothed margin

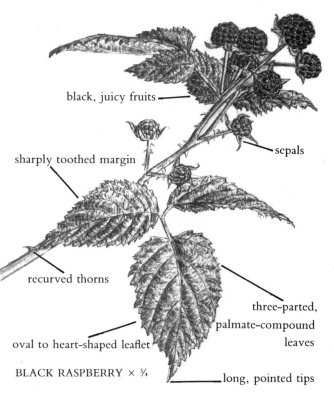

sepals

recurved thorns

three-parted, palmate-compound leaves

oval to heart-shaped leaflet

long, pointed tips

BLACK RASPBERRY × ¾

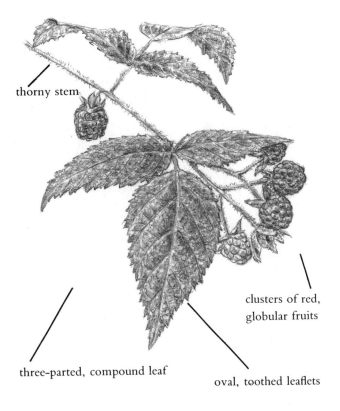

thorny stem

clusters of red,
globular fruits

three-parted, compound leaf

oval, toothed leaflets

RED RASPBERRY × ⅔

The thimbleberry (*Rubus parviflorus*) is a six-foot-high western shrub with upright to spreading branches, and glandular hairs instead of spines. The sharply toothed, simple leaves are 2 to 6 inches long, with five lobes. White to pink flowers up to 2 inches across grow in loose clusters of up to nine flowers. The bright red, sweet, juicy, globular fruit grows to nearly ¾ inch across. Local Indians ate the young shoots in the spring, as well as the berries. You can find ripe fruit in summer and fall.

The purple-flowering raspberry (*Rubus odoratus*) doesn't look like a raspberry until its fruit matures. Its showy, purple flowers are nearly 4 inches across. The simple leaves resemble large maple leaves, with three to five lobes. Erect thornless stems, covered with reddish hairs, grow 4 to 6 feet tall. The flat, red berries are drier and less abundant than other species, but quite tasty.

Many people confuse raspberries with blackberries. The most obvious difference is revealed when you pick the two fruits. Raspberries are hollow, like a thimble. A cone-shaped structure called a receptacle stays behind. When you pick a blackberry, it takes the receptacle with it, so the berry isn't hollow. Also, blackberries usually ripen later

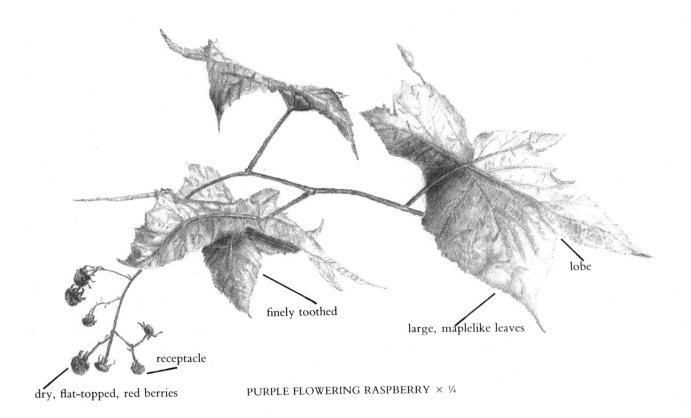

lobe

finely toothed

large, maplelike leaves

receptacle

dry, flat-topped, red berries

PURPLE FLOWERING RASPBERRY × ¼

in the summer. Blackberry canes usually have larger, more ferocious thorns than do raspberry canes. Mulberries differ even more. They grow on thornless trees, and each fruit hangs from a very thin fruit stalk rather than sitting on a cone-shaped receptacle.

Raspberry leaves somewhat resemble poison ivy, which grows in similar habitats. However, poison ivy doesn't have bristles, and the leaves aren't toothed. Goldenseal (*Hydrastis canadensis*) has crimson fruit that looks like a raspberry, but the rest of the plant is totally different. It's neither woody or bristly, and it has simple, lobed leaves. Goldenseal is poisonous in large quantities, containing dangerous alkaloids.

I eat my raspberries raw, mix them with yogurt and nuts, throw them into granola, and make jams, sauces, puddings, and pie fillings with them. Pastry filling stuffed with homemade wild raspberry jam is incredible. The fruit takes a while to collect, but there's no more processing, shelling, or pitting once you get home. The fruit is perishable. Use it within a few days. Raspberries freeze well, but they're too waterproof to dry unless you make a fruit leather (see cooking section, page 275).

Raspberries provide vitamins A, B-complex, and C, as well as calcium, phosphorus, volatile oil, sugars, citric and malic acids, pectin, and silicon. Their high iron content made them a remedy for iron-deficiency anemia for centuries. Raspberry-leaf infusion is a tasty beverage and a safe medicine. The leaves, bark, and root contain fragarine, tannin, and vitamin K. Vitamin K is necessary for proper blood clotting, and to prevent hemorrhages.

Long ago, many cultures disovered that raspberry-root decoction and raspberry-leaf infusion are extremely good tonics for the uterus and the entire female reproductive system. They strengthen the uterus wall, relieve uterine and intestinal spasms, and decrease menstrual bleeding. They balance the hormones and prepare the body for the stresses of menstruation, pregnancy, and childbirth. They hasten recovery from childbirth, and ease menopause. Outside of that, they're totally useless.

The tea is also mildly astringent. It makes a good wash or gargle for sore throat, canker sores, and loose gums. It promotes healthy bones, nails, skin, and teeth. It's also good for children's diarrhea.

Caution: Use the leaves fresh or thoroughly dried, but avoid wilted leaves, which may contain harmful substances.

SUMAC

(*Rhus* species)

The edible sumacs are a lovely group of shrubs or small trees that usually grow in dense stands. Their lemon-scented, alternate, feather-compound leaves are sometimes over 2 feet long. Each leaf is divided into numerous pointed, toothed, paired leaflets—especially noticeable in the fall, when they turn a beautiful scarlet. The stout twigs release a white, sticky sap when you break them.

Conelike flower heads, consisting of thousands of tiny greenish or greenish-yellow flowers, grow at the tops of the branches in midsummer. At summer's end, they give way to clusters of small, dry, hard, hairy, red berries. After heavy rains wash out the acids, they turn rust-colored, often persisting, brown and stiff, throughout the winter.

Edible sumacs, including regional species not covered below, can thrive almost anywhere in North America, including places inhospitable to other plants. Look for them in old fields, thickets, disturbed areas, canyons, on roadsides and on dry, sandy, or rocky soil. They do especially well at the seashore. As long as the upright berry clusters are red or red-orange, they're safe to use.

Dwarf sumac or winged sumac (*Rhus copallina*) is one of the most commonly encountered edible species of eastern and central North America. Paradoxically, it often grows taller than others, up to 30 feet. It has shiny, mostly toothless leaflets, and short-hairy fruit clusters—more open-looking than other species. The leaf's flared midrib, between the leaflets, bears flat, thin "wings."

Staghorn sumac (*Rhus typhina*) is a shrub or small tree, 4 to 15 feet tall, sometimes growing over 30 feet tall, with a similar range. *Typhina* refers to the genus *Typha*, the cattails (page 67), because the furry branches resemble mature brown cattail seed heads. The 16- to 24-inch-long leaf has eleven to thirty-one sharply toothed, lance-shaped leaflets. It's so hairy, it seems covered with velvet, like a stag's antlers. The bare shrub's irregularly forked branches add to this comparison in winter.

Male and female flowers grow in dense, cone-shaped clusters, either on the same or on separate shrubs. The tightly clustered fruiting head is so hairy that it forestalls the rain from washing away the flavor, so it has a longer season than smooth sumac. It grows at the edges of woods, along roadsides, in old fields, and along streambanks.

Smooth sumac (*Rhus glabra*) is a shrub or small tree 3 to 20 feet tall, growing in similar habitats. It has eleven to thirteen lance-shaped leaflets, 2 to 2¾ inches long. They're pointed, sharply toothed, and whitish underneath. There's a smooth, reddish midrib between the leaflets. Male and female flowers grow on dense, cone-shaped structures on separate plants. The fruit clusters are looser than staghorn sumac's, and the shrub isn't hairy. It grows throughout most of the United States.

Squawbush (*Rhus trilobata*) has a three-parted palmate-compound leaf with very coarsely

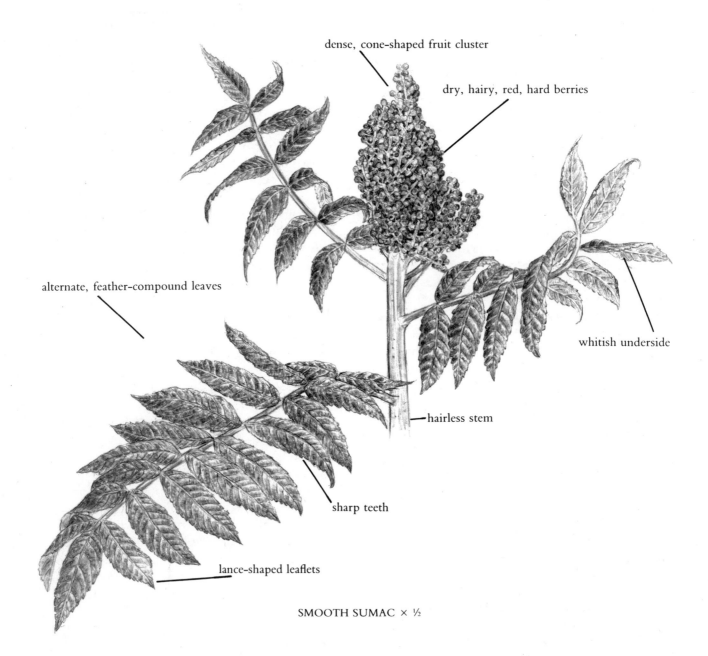

dense, cone-shaped fruit cluster

dry, hairy, red, hard berries

alternate, feather-compound leaves

whitish underside

hairless stem

sharp teeth

lance-shaped leaflets

SMOOTH SUMAC × ½

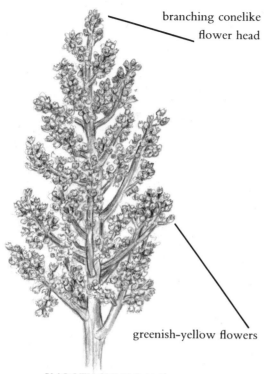

branching conelike flower head

greenish-yellow flowers

SMOOTH SUMAC × ¼

toothed, sometimes lobed leaflets. The twigs, which are densely hairy, give off a strong odor when you break them. The berry clusters are so much smaller than the eastern species, a cluster will easily fit into your hand. They begin to ripen in late summer.

Squawbush grows on slopes, plains, and canyons throughout the West, but I also found it in Central Park in Manhattan, where it must have been planted years ago. Human activity often increases wild plants' ranges.

The notorious poison sumac (*Rhus vernix*) has drooping clusters of white berries. Other species' berry and flower clusters are upright. Also, there are large spaces between poison sumac's feather-compound leaf's toothless leaflets. It would be quite difficult to confuse this species with the edible ones. It grows in bogs and partially wooded swamps with sandy soil, in eastern North America. I found it only once, in a pine barrens.

Caution: Learn to recognize this plant from a distance and avoid it. Poisoning symptoms are worse than those of its relative, poison ivy (page 38). Sensitive people may get rashes from being near it. Once you touch it, the toxin enters your bloodstream and a painful rash can last for weeks,

spreading all over your body. Symptoms may recur for months.

It's amazing how much better known some poisonous plants can be, compared to their edible relatives. To the public, the horrors of poison sumac are infamous, while the delicious taste of edible sumac lemonade is unknown. A friend brought a sample of sumac lemonade to an environmental fair a few years ago for people to try. She left some colorful red berries floating in the pitcher, to revitalize the drink as she gave out more samples. In the sunlight, the pitcher looked like rose quartz, and the fuzzy red berries like a designer moss.

Everyone wanted to try it before they knew what it was. When they learned it was sumac, most recoiled in horror, certain that death was in their cups. Yet, after a short explanation, they were ready for seconds, their eyes fixed like magnets on the container of wild pink lemonade.

Although you can see the edible species' berry clusters much of the year, they're good to use only in late summer and early autumn. Touch the bright red ripe fruit with your finger, then touch your tongue. You'll notice the strong, tart taste of ascorbic acid—vitamin C—and possibly oxalic acid. (Avoid prolonged overuse of any sources of oxalic acid—they may interfere with calcium absorption, and cause urinary-tract stones in predisposed persons.) In a drought year, the flavor persists well into late autumn, but heavy rain washes the acids into the ground.

For the best pink lemonade ever, crush the berry clusters slightly with your hands, then swish

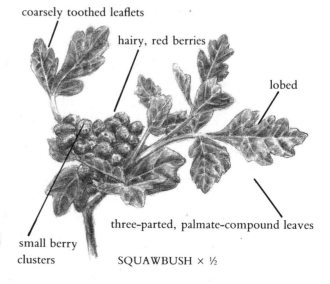

coarsely toothed leaflets

hairy, red berries

lobed

three-parted, palmate-compound leaves

small berry clusters

SQUAWBUSH × ½

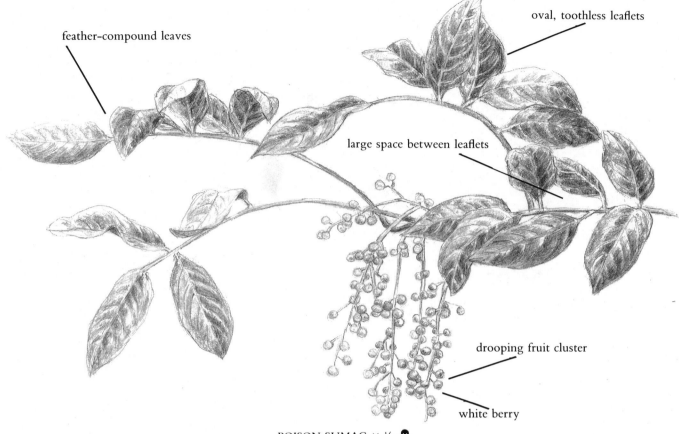

feather-compound leaves

oval, toothless leaflets

large space between leaflets

drooping fruit cluster

white berry

POISON SUMAC × ½

them in a bowl of cold water for a few minutes. (Don't use hot water, which destroys vitamin C and washes undesirable quantities of tannin into your drink.) The water will turn a pinkish-red. Strain out the fruit through a cheesecloth-lined colander, to get rid of the unpleasant little hairs. Sweeten to taste, chill, and enjoy. Swish more sumac berries in the same water for added flavor, or dilute with more water if it's too strong.

To make a sumac concentrate, repeat the process with different berries in the same water. I freeze the concentrate in ice-cube trays and store them in freezer containers. This makes a great alternative to lemon juice, and a little goes a long way. Like lemon or cranberry juice, it complements other beverages. Mix a little with fruit juice or sassafras tea for some extra flavor and zing. Some of my best salad dressings use sumac concentrate in place of lemon juice or vinegar.

Sumac-berry juice is high in vitamin C, and long before the word *vitamin* was coined, Native

Americans knew that it was good for colds, fevers, and scurvy. All parts of the bush are good for home remedies. The "berry-aid" tames fever. Leaf infusion is supposed to be stronger than "berry-aid," while inner-bark and root-bark decoctions are considered the most powerful.

Sumac is an astringent, antiseptic, and tonic. It's used for diarrhea, dysentery, asthma, urinary infections or irritations, sore throats, chronic gum problems, cold sores. The Indians would chew on the roots to ease swollen or infected gums, and to stop bed-wetting. They applied compresses to burns and cuts to stop bleeding and bring down swelling. They also mixed the dry berries half and half with tobacco, to smoke in peace pipe ceremonies. It's supposed to dilute and improve the smoke's odor. The whole plant also provides a tan to reddish-brown colorfast die.

The Indians also used the ground berries, mixed with clay, as a poultice on open sores and arrow wounds. I'd love to determine if this really works, so whenever we find sumac on my tours, I ask for a volunteer to accompany me to the archery range. Invariably, my students volunteer me.

OTHER PLANTS OF THICKETS IN SUMMER

Edible or Medicinal Plants

Amaranth leaves, black locust flowers, sweet clover flowers, curly dock flower stalk, wild onion bulbs and leaves, field pennycress leaves, flowers, and seeds, grape leaves, groundnut tubers, lamb's-quarters leaves, orache leaves, marshmallow leaves and roots, milkweed shoots, flower buds, and flowers, mullein leaves, peppergrass leaves, flowers, and seeds, sea rocket leaves, pine leaves, shoots, and pollen, seaside plantain leaves, poke-weed shoots, prickly pear pads, sheep sorrel leaves and flowers, strawberry leaves and berries, yarrow.

For Observation Only

Autumn olive flowers, Russian olive flowers, blackberry flowers, black cherry flowers, grape flowers, groundnut flowers, seaside plantain flowers, poison ivy flowers, rose flowers, sow thistle flowers, strawberry flowers, sumac flowers.

PLANTS OF FORESTS IN SUMMER

CHERRIES

(Prunus species)

The cherry tree is the only fruit tree I know where people hold the bark in as high esteem as the fruit. It's actually easier to find references to the bark in herbal literature than to other parts of the tree. The inner bark is held in such high esteem because it has important medicinal properties.

Cherry bark is unique—smooth, shiny, reddish-brown, with light-colored, horizontal stripes (lenticels). Unlike the similar black birch, it peels and becomes craggy.

Wild cherries' alternate leaves are long-oval and toothed, with narrow bases and pointed tips. In the spring, clusters of fragrant, five-petaled white flowers grow in various configurations, depending on the species.

All cherries have round fruit—drupes—with a smooth, featureless surface, and a single hard, round seed, commonly called a stone or pit. This rules out all small round fruits with crowns, like the blueberry (page 98). **Caution:** Poisonous buckthorn (*Rhamnus* species) berries resemble cherries, but contain many seeds.

Sometimes people confuse cherries with their relatives, the wild plums (page 110). Plums' larger fruit is coated with a powdery, white bloom (a wax), and the surface is marked into quarters by two vertical lines, running from the fruitstalk to the far end. Cherry colors vary from red to black—a few are even yellow. Some species' flavors are comparable to the supermarket varieties, but with more complex undertones. Others taste quite different.

One of my favorite ways to prepare wild cherries is to cook them whole in cherry juice with a thickener until soft, then strain out the pits with a

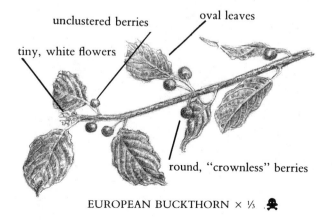

unclustered berries

oval leaves

tiny, white flowers

round, "crownless" berries

EUROPEAN BUCKTHORN × ⅓

food mill. This is especially practical for species with small fruit. Use them in pie fillings, puddings, sauces, or punch.

The sweet cherry (*Prunus avium*), a European import, is the closest to commercial varieties. Full-grown trees are quite large, up to 70 feet tall, but there are often a few good limbs to pick from within reach. The leaves are rounder than our native species', with long, narrow bases and pointed tips.

The fruits, which are larger than the native varieties, are clustered. The cherry stones, which you can find under the tree, are ⅓ inch in diameter. Each cherry grows on a relatively long stalk, and the stalks of several cherries originate from the same point on the twig.

The sweet cherry tree usually grows on the edge of woods and fields, throughout eastern North America, or under cultivation in backyards. By now, wild trees could be popping up anywhere in the country. Like many cultivated plants gone wild, we have human gardeners and hungry birds to thank.

The cherries ripen early, before summer starts in the Northeast, but you may find fruit until midsummer in other regions. They're sweeter than the native species, but not as sugary as their commercial relatives. Instead, they have a more intense cherry flavor.

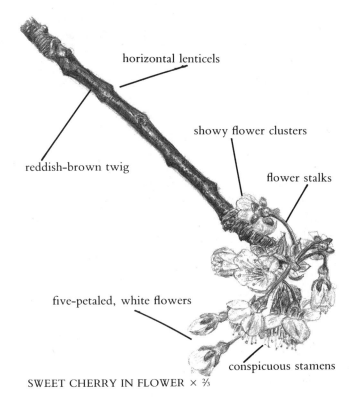

horizontal lenticels

showy flower clusters

flower stalks

reddish-brown twig

five-petaled, white flowers

conspicuous stamens

SWEET CHERRY IN FLOWER × ⅔

They're great raw. I also like them in puddings with milk or soy milk, orange juice, vanilla, and a thickener. Strain out the seeds with a food mill, or use a cherry pitter or knife. The pitted cherries are also wonderful in cereal and granola.

The sour cherry (*Prunus cerasus*) is another es-

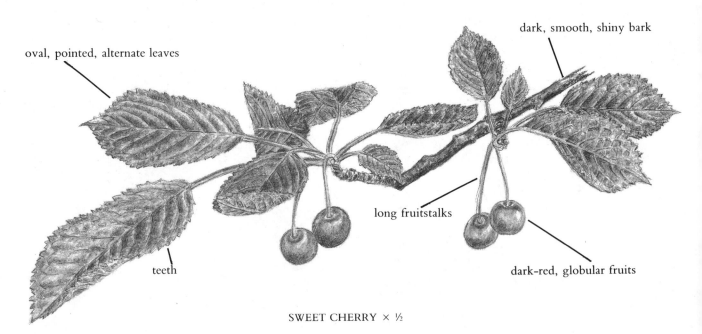

oval, pointed, alternate leaves

dark, smooth, shiny bark

long fruitstalks

teeth

dark-red, globular fruits

SWEET CHERRY × ½

cape from cultivation you may locate anywhere in North America. I usually find it in backyards. The tree gets quite large, but not as tall as the sweet cherry, and the leaves are rather broad. The fruit's soft, bright red, and especially large for a wild cherry. It also ripens close to early summer.

The flavor is quite miraculous, very tart, yet like cherry candy. With a little sweetener, the fruit makes great jams, jellies, and pie fillings. It's also good for purifying the body, with a high vitamin-C content.

The pin cherry, bird cherry, or wild cherry (*Prunus pensylvanica*) is a small tree or shrub with narrow leaves. The red-tipped twigs end with clusters of buds. the short flower clusters are umbrellalike, similar to the sweet cherry's.

It grows in northern forests in the East and West, but extends along the mountains, southward to Tennessee and Colorado. You can find it in areas cleared for lumbering or by fires (it's also called fire cherry), along steep embankments, in rocky woods and clearings, by dried-up streams, and along roadsides. Because it thrives in cleared, disturbed areas, its range is expanding.

The piquant flesh of the small, red fruit is thin and sour, but if you can collect enough, you can make a wonderful fruit sauce or jam. The Indians used to crush and pound the fruits and seed into a paste and mix it with meat and fat, sometimes with peppermint, to make a cherry pemmican.

The black cherry (*Prunus serotina*) is a medium-size native cherry tree, reaching almost 70 feet tall. I see it in various stages of growth because it's so common throughout eastern North America. The dark-green, glossy, alternate leaves are long and pointy, measuring from 2 to 6 inches in length, and from 1 to 1½ inches wide. They're lighter underneath, and if you examine the undersides with a hand lens, you'll notice the midrib is covered with rusty-colored hairs. When you scratch and sniff the twigs, you'll detect a strong aroma of "bitter almond," the same as the seed of the related apricot or peach. Unlike those of sweet or sour cherries, the flowers and fruits grow on long racemes, 4 to 6 inches long. The fruit is ⅓ inch in diameter—reddish at first, then black when ripe. The distinctive bittersweet flavor hints of grapefruit. This pioneer (or "weed") tree grows in parks,

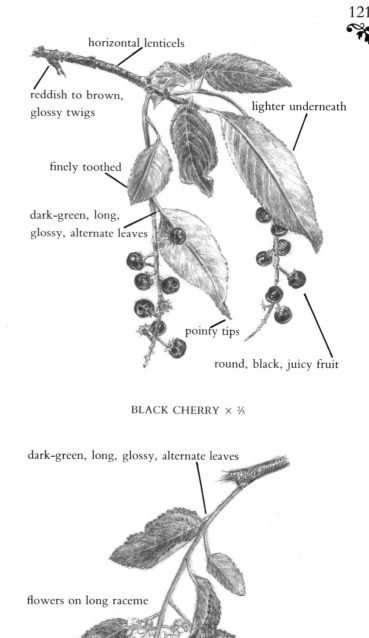

horizontal lenticels

reddish to brown, glossy twigs

lighter underneath

finely toothed

dark-green, long, glossy, alternate leaves

pointy tips

round, black, juicy fruit

BLACK CHERRY × ⅖

dark-green, long, glossy, alternate leaves

flowers on long raceme

finely toothed

radially symmetrical, five-petaled, white flowers

pointy tips

BLACK CHERRY IN FLOWER × ⅔

old fields, open woods, mixed hardwood forests, forest edges, residential streets, near the seashore, and virtually everywhere else. It flourishes in most of the eastern four fifths of the United States.

Some trees have better-tasting fruit than others, and quality and abundance vary from year to year. Many people find the flavor highly unusual and aren't immediately sure whether they like it. I didn't appreciate black cherries when I initially identified them in my first year as a beginning forager. The taste was so strange, I was afraid I had misidentified a poisonous species. Today, they're among my favorite fruits.

Still, the unusual flavor elicits strange responses: One summer I led two tours in the same park a couple of weeks apart. The first time, the first few people who tasted it immediately proclaimed their strong approval. The result: No one could eat or collect enough. I had trouble getting people to pay attention to other plants all day. They kept wandering off to the nearest black cherry trees.

The next time, the first people made no secret that they hated the fruit. Everyone else followed suit. These were the same trees, perhaps even a little riper and tastier than before. Now I always make sure a confirmed black cherry lover tries the fruit and talks about it first, and new students often tell me that learning about black cherries is a highlight of their summer.

Black cherry puree is perfect for flavoring summer punch or fruit soup, along with some cinnamon and ginger. Use it to offset sweet fruit in pies, as a cheesecake topping, or turned into a fruit leather for winter munching.

This prolific tree produces highly esteemed, beautiful lumber. We should be harvesting this renewable resource instead of destroying our old-growth forests, or importing tropical hardwoods from the vanishing rainforests.

People often confuse the black cherry tree with the similar chokecherry (*Prunus virginiana*). However, chokecherry leaves' midribs aren't hairy beneath. Their plump, dark-red or purplish red fruits, which ripen between midsummer and mid-fall, are smaller than black cherries.

Chokecherries are so astringent, they'll make your mouth pucker, accounting for the name. You have to cook and sweeten them to make them tasty.

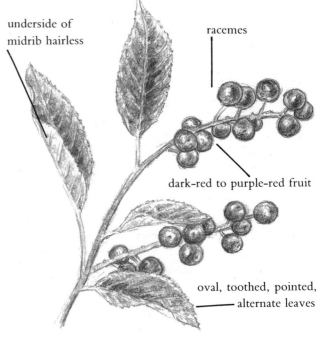

CHOKECHERRY × ¾

Although the chokecherry favors the same habitats as its cousin, the black cherry, it grows farther north, and is more common in mountainous areas. Its range includes the northern half of the United States, most of California, and the southern half of Canada.

The inner bark of the black cherry tree is a major source of medicine, originally used by the Indians. The best time to collect is in the fall, when its amygdalin level is highest. Steep it in water just off the boil. Don't boil it, or you'll lose volatile constituents. **Caution:** Amygdalin, present only in trace amounts in the fruit, is a cyanogenic glycoside that may yield dangerous levels of hydrocyanic acid: Removing the water (the *hydro* part) or hydrolyzing hydrocyanic acid can create cyanide. Red blood cells, which normally carry oxygen, prefer cyanide, so someone poisoned by this substance turns blue and asphyxiates, even though he or she is breathing.

Partially wilted cherry leaves are an excellent source of cyanide. As little as a pound can kill a large cow, and the tea can kill you within minutes, especially if you drink a large amount of water. Symptoms include difficulty breathing, twitching, spasms, coma, and death. There are reports that children have died from eating many cherry seeds, although the hydrocyanic acid doesn't normally break down into cyanide in the seeds.

On the other hand, amygdalin, under the name of laetrile, was used for terminal cancer in the late 1970s. The American Medical Association and U.S. Food and Drug Administration opposed this alternative therapy, claiming it didn't work and that there was a risk of cyanide poisoning. They took it off the market and took legal action against any doctors using it.

In an infusion of black cherry's inner bark, the hydrocyanic acid isn't converted to cyanide. It's rapidly excreted, mostly through the lungs, where it stimulates respiration, then sedates the sensory nerves that cause the cough reflex. This tea became so popular for coughs and sore throats, it became a major ingredient in the first commercial cough medicines. Drug companies still add artificial cherry flavor to cough medicines today.

Black cherry (and chokecherry) bark tea is a safe sedative, mild tonic, decongestant, expectorant, astringent, and disinfectant. It's been used for fevers and infectious diseases, asthma, ulcers, hysteria, mucus congestion, diarrhea, and inflamed gums. It's especially good as a gargle for sore throat. Various Indian tribes applied poultices of dried, powdered black cherry or chokecherry bark directly on sores, burns, or wounds. It acts as an external disinfectant and astringent.

Black cherries and chokecherries are used interchangeably for making hair rinses, and the bark's decoction improves manageability.

JUNEBERRY, SERVICEBERRY, SHADBUSH

(Amelanchier species)

These diverse shrubs or small trees have smooth to slightly furrowed, ash-gray to blackish bark, often beautifully adorned by curving, horizontal, dark-gray stripes. The alternate, oval, finely toothed, medium-sized, stalked leaves are about 2 inches long, with slightly downy, light-green undersides on some species.

The five-petaled, white flowers usually bloom

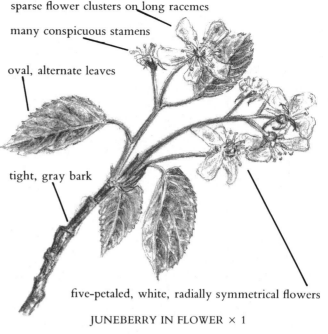

JUNEBERRY IN FLOWER × 1

early in the spring, before the leaves appear. Radially symmetrical and about ¾ inch across, they resemble apple blossoms, with many conspicuous stamens. The blossoms hang from long, sparse racemes.

The blue-black, round berries, which are red before they ripen, are about ¼ to ⅓ inch across, the size of blueberries, which they resemble. They even have the "crown"—a frilled opening on the end away from the fruitstalk. Inside are soft, almond-flavored seeds, absent in blueberries. Also, blueberries have belllike flowers, and different bark.

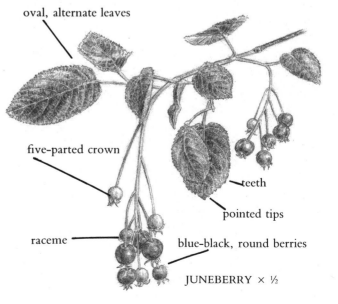

JUNEBERRY × ½

There are many native and some European species growing throughout much of the United States and Canada. Although many have delicious fruit, some have bad-tasting berries. There are no poisonous look-alikes, since no poisonous berry has a "crown," but a careless beginner could still confuse Juneberries with crownless poisonous and nonpoisonous berries. These are described under Blueberries (page 98).

Keep your eyes open for these inconspicuous woodland treasures from June through August. Most species ripen in late spring, but others come into season later. Different species favor varying habitats. Some grow on hillsides, others inhabit lowlands, while a few tolerate saltwater, and grow within yards of the sea. Look for Juneberries in moist and somewhat dry soil, in woods, along streams and lakes, on mountains, in thickets, clearings, and cultivated parks, and on the grounds of landscaped garden-apartment complexes.

Juneberries are a great surprise the first time you try them. With no similar commercial relatives, these delicious berries, related to apples, are quite unique. Although they were sold in the marketplace in the past, they're almost completely forgotten today. The fruit has a strong, sweet, and penetrating flavor, a little like pears, while the soft seeds add a nutty, almondlike flavor.

Some years there are excellent crops, but in other years, you can hardly find any berries. These shrubs are somewhat finicky about their requirements. Also, in some places, birds attack the fruit as soon as it ripens, joyously taking little bits from each berry.

It takes me a long time to gather Juneberries, especially since I can't help eating two berries for every one heading for my pail. I usually supplement them in a recipe with other fruits, to "stretch them."

Make your favorite blueberry muffin recipe using Juneberries. It will be different and fantastic. I've made my best cobblers with this fruit. They also make great jam. They contain pectin, so you don't need much thickener. The Indians, who used them like blueberries, dried them and added them to stews and pemmican.

Iroquois squaws used the fruit as a "blood remedy," to strengthen the body after the pain of childbirth. They drank a root and bark decoction to prevent miscarriage. They also used it to expel parasitic intestinal worms, as did the Chinese.

The shrub is called the Juneberry because the fruit ripens in June. It's also called "serviceberry" because it blooms in mid-April, when long-delayed religious services were held throughout nineteenth-century New England, as snow-covered roads became accessible again. But everyone was not religious and others would just as soon go fishing, especially when the first run of shad migrated upstream from the ocean, heralded by the blooming "shadbush."

MAYAPPLE, MANDRAKE

(Podophyllum peltatum)

This herbaceous perennial has a dramatic appearance. Although only 1 to 1½ feet tall, it has one or two large, spreading, lopsided, shield-shaped leaves arising from a single, long stalk. The leaves are lobed, and coarsely toothed toward the lobes' tips. In the generic name, *podos* means "foot," and *phyllum* means "leaf." The leaf outline does suggest a webfooted bird. The specific name, *peltatum*, means "shield-shaped." It looks like an exotic tropical plant, but it's a native to North America, a member of the barberry family.

The mayapple grows in dense clusters in moist, open woods, spreading by underground rhizomes. The entire colony is usually one huge, genetically homogeneous plant, connected underground.

Plants with two huge leaves are the only ones that flower. The solitary, nodding, stalked, waxy-white, fragrant flower is 2 inches wide, arising from the crotch at the base of the two leafstalks. It has 6 sepals and six to nine white, waxy petals. The stamens and pistils color its center bright yellow. It blooms in spring.

In mid- to late summer, you may find the solitary fruit, the size and shape of an egg or like a small, smooth-skinned lemon. Although the plant usually grows in partially shaded areas in the woods, only those stands that get additional sun-

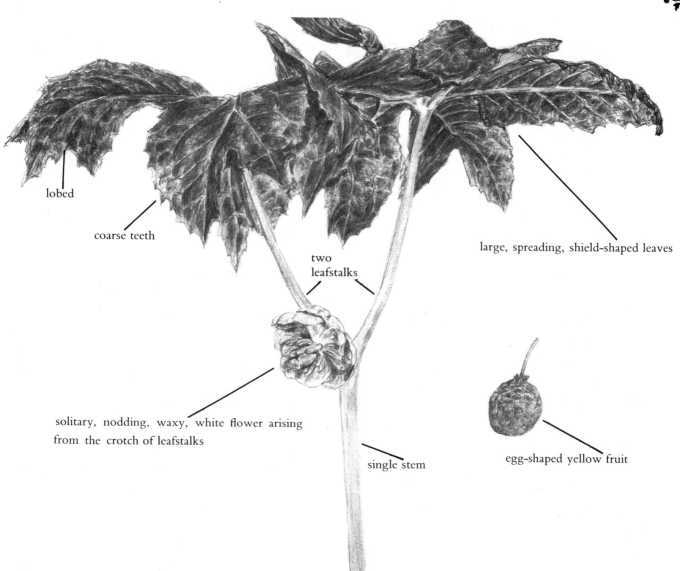

lobed

coarse teeth

two
leafstalks

large, spreading, shield-shaped leaves

solitary, nodding, waxy, white flower arising
from the crotch of leafstalks

single stem

egg-shaped yellow fruit

MAYAPPLE IN FLOWER × ½

light set fruit. The fruit usually falls to the ground unripe and green, just as the leaves are dying back. Later, it turns yellow and ripens. Hold one to your nose and enjoy the balmy, lemony fragrance.

Caution: Never eat the green, unripe fruit—it's a powerful cathartic. Let it develop its yellow ripeness at home. Peel the tough skin, and enjoy it raw or cooked. It tastes lemony. If you're preparing it in quantity, simmer the coarsely cut fruit for a few minutes to soften it, then use a food mill to separate the seeds, or remove them by hand. You'll be left with a thick, tart pulp. The best way to use it is as an unusual-tasting lemon substitute. Once

it is cooked, freeze it in ice-cube trays, store it in containers, and add the cubes to recipes as needed.

Sixteenth and seventeenth century cookbooks suggest slicing, sweetening, and stewing the whole fruit (seeds and skins) with ginger and cloves. The skin is supposed to be edible then, rather like the orange skin in orange marmalade. In the South, people sometimes use mayapple juice as lemonade. You can also blend the pulp with water or fruit juice in the blender, to make a similar beverage.

The rhizome is very toxic. It causes nausea, vomiting, and inflammation of the stomach and intestines, and has killed people. Even handling the root can create skin lesions. Getting any in your eye may cause swelling, redness, and pain. Witches used the unrelated European mandrake's (*Mandragora officinarum*) root for magic, because it would sometimes divide into five parts and resemble a human being—hence the name mandrake. According to European folklore, if you dig it up, it will emit a shriek that will cause permanent insanity. According to Native American folklore, the woman who digs up the native species will soon find herself pregnant. Clearly these roots have no place among home remedies.

Often, medical technology can make useful drugs from the most poisonous plants. Mayapple rhizomes contain a highly allergenic, resinous chemical called podophyllin, which prevents cells from growing. Doctors currently apply it topically, to treat venereal warts. Along with the associated compound, peltarine, it's being studied for testicular and bronchial cancers. The rhizome also contains etoposide, FDA-approved for use on small-cell lung cancer and testicular cancer.

People today are rightly concerned that we're losing an irreplaceable pharmacopoeia by destroying the rain forests. But we also need to appreciate how much we stand to gain in our knowledge of medicine, biochemistry, and biology by conserving our local temperate forests.

MULBERRY

(Morus species)

There are two common mulberry tree species, the native red mulberry (*Morus rubra*), and the Asian white mulberry (*Morus alba*). The red mulberry usually reaches a height of 40 to 70 feet (sometimes 105 feet), with rough, reddish-brown bark. The white mulberry grows up to only 30 to 70 feet, with rough, lighter, ocher-gray bark and spreading branches. Mulberry bark has distinctive vertical cracks or furrows, with an occasional orange-brown streak between the cracks.

Both species have roughly oval, toothed, alternate leaves 2 to 6 inches long. Sometimes they're variably lobed, sometimes they're unlobed. The

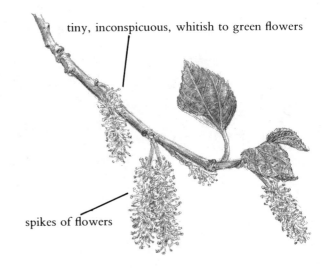

tiny, inconspicuous, whitish to green flowers

spikes of flowers

RED MULBERRY IN FLOWER × 1

red mulberry's leaves feel like sandpaper on the upper surface and smooth to somewhat hairy underneath. The white mulberry's leaves are smooth on both surfaces.

As the new leaves develop in mid-spring, tiny male and female flowers hang on separate small, slender, inconspicuous spikes. The male cluster is longer, the female rounder. When the female flowers are fertilized, an aggregate fruit results. It's globular to cylindrical, ½ to 1½ inches long, hanging from a fruitstalk.

Other, very similar, locally distributed edible species include the Texas mulberry (*Morus microphylla*), and the black mulberry (*Morus nigra*). Sassafras also has lobed and unlobed leaves, but they're fragrant and untoothed. There are no poisonous look-alikes.

Raspberry and blackberry (pages 112 and 96) fruits superficially resemble mulberries, but the fruits of these unrelated thorny canes grow upright, on receptacles, with no fruitstalks.

Ripe mulberries come in different colors: red, white, pink, and black. These colors are attributed to two different species and their hybrids. The red mulberry has red unripe berries. They darken to black, with reddish undertones, when they're ready to eat.

When I first started gathering mulberries, before reading my first field guide, I noticed that some neighborhood trees' berries never seemed to ripen, falling to the ground still white. I finally caught on that these were white mulberry trees.

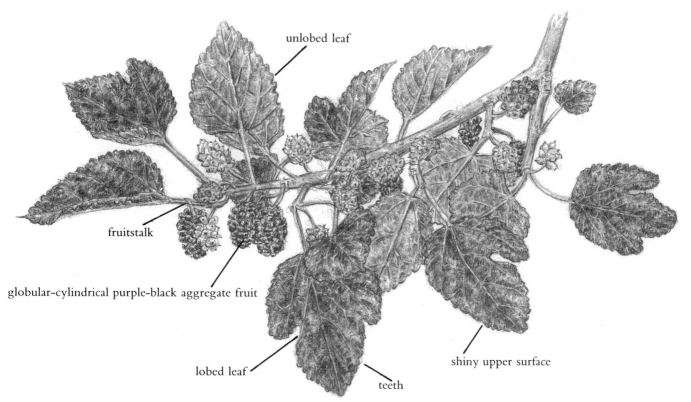

unlobed leaf

fruitstalk

globular-cylindrical purple-black aggregate fruit

lobed leaf

teeth

shiny upper surface

RED MULBERRY × ¼

Even though they're white when ripe, they're soft and juicy.

White mulberry trees were imported from Asia in the 1800s to start a silk industry. Growing silkworm caterpillars and unrolling the cocoons to make silk was too labor-intensive for us, and the effort was abandoned. But because mulberry trees are so prolific, most of America is now graced with two species plus pink hybrids.

You're most likely to find mulberries in residential neighborhoods and parks, in fields, especially along the edges, open woods, and near fresh water. They grow throughout the country, ripening in late spring and early summer.

You can spot ripe mulberries in season from a distance because the fruits make such a mess on the ground. I love taking children mulberry gathering. Everyone holds up a dropcloth, while I climb into the trees and shower the dropcloth and kids with fruit. Do this on a nice day preceded by sunny weather, because rain washes away the berries' flavor.

Use mulberries immediately. They won't last more than a couple of days in the refrigerator. They soon ferment or get moldy, probably because of their high water content and thin skins. This is why you rarely see them in stores. Eat them, cook them, dry them, freeze them—just don't let them spoil.

There are many ways to cook mulberries once you've eaten your fill of fresh fruit. Cook them in their own juice until the mixture becomes liquid, and make a sweet mulberry slurry. Add a little lemon juice and orange rind to offset the sweetness, stir in a thickener, and you have a pudding. I've made mulberry pies and mulberry muffins. You can do anything with mulberries you can do with virtually any other berry, and they dry and freeze well. Lemon or lime juice enhances their flavor, since they don't have the acidity of other fruits.

Dried mulberries are more crunchy, like (related) figs. You can grind them in a blender and mix in nut butter, sweetened to taste, to make a mulberry candy.

You can also use the young, unopened leaves in the spring. Boil them for 20 minutes and discard the water, for a mild, tasty vegetable. This water, the unripe berries, uncooked young leaves, and mature leaves are toxic and mildly hallucinogenic. While they won't kill you, they'll give you a terrific headache and an upset stomach. The primary hallucination is that you're so sick, you're going to die. However, you'll probably eventually recover.

How did red mulberries get their color? The answer lies in "Pyramus and Thisbe" the first love story ever written, compiled by Ovid from earlier Greek folklore:

Pyramus and Thisbe were neighbors who fell in love when they became adults. Their parents disapproved, but the lovers communicated secretly, through a crack in the wall separating their houses. One night, they eloped, but Thisbe was frightened away from their rendezvous point—a white mulberry tree—by a bloody-mouthed lion that had just finished a meal. She escaped and hid, but lost her cloak, which the lion mauled and bloodied.

Pyramus, seeing the bloody-mouthed lion and the cloak, imagined the worst, and impaled himself on his sword. His blood colored the mulberries red. When Thisbe found him and realized what had happened, she followed him to death on the same sword. The European mulberry species has been red ever since, colored by the lovers' blood.

In traditional European medicine, the mulberry root is a remedy for tapeworms. The tree's inner bark has been used as a laxative. The fruit's also mildly laxative.

OTHER PLANTS OF FORESTS IN SUMMER

Edible or Medicinal Plants

Apples, wild beans, black birch, blackberries, blueberries, Kentucky coffee tree, cow parsnip seeds, currants and gooseberries, dayflowers, daylily flowers and tubers, elderberry flowers and berries, wild onion seeds and bulbs, wild ginger roots, goutweed leaves and seeds, greenbrier shoots, groundnut pods and tubers, hawthorn berries, hobblebush berries, hog peanut pods, honewort seeds, horse-balm flowers and roots, jewelweed stems and seeds, linden flowers, pears, pennyroyal, pine needles, plums, raspberries, redbud pods, rose hips, salal berries, sassafras leaves, berries, and roots, spicebush leaves, strawberry leaves, wild leek bulbs, wintergreen leaves, wood sorrel.

For Observation Only

Anise root skeleton, black birch cones, bracken fronds, chestnut and chinquapin flowers, cow parsnip flowers, garlic mustard seeds, groundnut flowers, Hercules'-club flowers, hog peanut flowers, jewelweed flowers, linden seeds, nettle flowers and seeds, persimmon flowers, poison ivy flowers and berries, rose flowers, Solomon's seal berries, violet leaves, waterleaf flowers, wild leek flowers, wild lettuce flowers, wild potato vine flowers, wisteria pods.

PLANTS OF FRESHWATER WETLANDS IN SUMMER

WATER HEMLOCK, SPOTTED COWBANE

(Cicuta maculata)
Poisonous

The first time I saw this tall, beautiful plant was in a parking lot. Some friends and I had driven to a wooded park to look for wild mushrooms, but we found this plant right off the bat. The entire plant is aromatic. The smell was so sweet, I had to

resist the impulse to taste it then and there. Of course, I didn't, since I wasn't sure of its identity, and I suspected the worst. I was correct—this is the deadliest plant in North America.

Growing 2 to 9 feet tall, this native perennial has many branches. The leaves, which can be up to 1 foot long, are twice or three times feather-compound, often with some reddish coloration. The long, pointed, toothed leaflets are unlike similar plants', because the veins of the leaflets end *between* the teeth, not *in* the teeth. The smooth stem

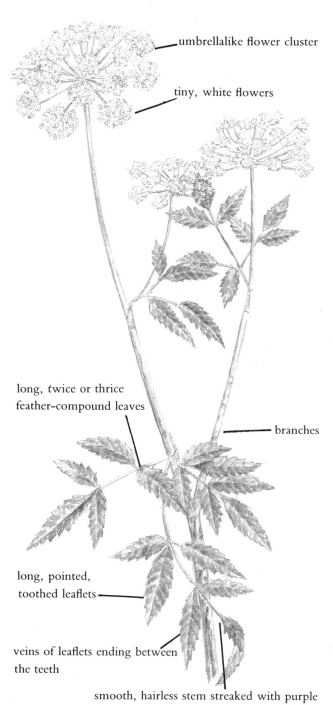

umbrellalike flower cluster

tiny, white flowers

long, twice or thrice
feather-compound leaves

branches

long, pointed,
toothed leaflets

veins of leaflets ending between
the teeth

smooth, hairless stem streaked with purple

WATER HEMLOCK × ⅓

The white root has many swollen side branches clustered together like fingers. They resemble parsnips and were reported to taste delicious, like parsnips, by those dying of water-hemlock poisoning.

Water hemlock grows in swamps, streambanks, and wet ground throughout most of North America, except for the Southwest.

It contains cicutoxin, a clear, brown resin so poisonous that one mouthful of the root can kill a healthy adult human. American Indians of both sexes used this plant to commit suicide if their spouses were unfaithful. Death was so agonizing, the surviving spouse was left with a permanent guilt trip. Symptoms include severe stomach pain, vomiting, diarrhea, pupil dilation, difficulty breathing, frothing at the mouth, violent convulsions. Death may ensue within fifteen minutes or eight hours. Victims may chew their tongues to ribbons, and asphyxiate on their vomit, unable to unclench their jaws. With prompt medical treatment, recovery may occur within twenty-four hours.

There are other plants in this family. Water parsnip (*Sium suave*) looks like poison hemlock, but its stem is strongly ridged, and its leaves are once-compound. The basal leaves, often submerged, are very finely cut, and its root is single. It also grows in swamps and wet places. Although it's edible, it resembles water hemlock so closely that I wouldn't risk eating it, and I've been foraging for decades. I strongly urge you to look for other wild foods.

Poison hemlock, fool's parsley, and other relatives are discussed under Wild Carrot (page 203). There are numerous other poisonous and nonpoisonous relatives, many of them exotics, locally distributed and not in the guidebooks, that are also beyond the scope of this book. Beginners should avoid eating edible species in the carrot family without expert supervision.

WILD RICE

(Zizania aquatica)

This tall, upright, annual grass—unrelated to rice—grows 3 to 10 feet tall. The leaf blades may

is streaked with purple and chambered inside. At the stem's base, the pith is crossed with yellow lines that eventually become more chambers.

The tiny, white flowers bloom in summer and fall. They grow in erect umbrellalike clusters, 2 to 6 inches wide, as do those of its edible relatives, wild carrots and wild parsnips.

be 40 inches long, and unusually wide for a grass. The flowers appear on drooping clusters on side branches that are 5 to 8 inches long. The spreading lower branches bear the straw-colored, purple-tinged male flowers. The upper, more vertical branches bear plumes of bright, yellow-green female flowers. The fruiting head can be over 2 feet long.

Look for wild rice in quiet waters, shallow ponds, marshes, freshwater and brackish bays. It grows throughout southern Canada, in the Pacific Northwest, and throughout the eastern United States. It does best around the Great Lakes region. Because birds eat it, hunters often plant it to attract waterfowl.

Do you ever wish that wild foods were available commercially? If you have ever bought wild rice, you've purchased a foraged food. We can't cultivate this grain efficiently with modern agricultural methods, because it grows in the water.

The grains ripen from late summer to early autumn, and fall at the slightest touch, a property called "shattering." If you're too late, the seeds will have floated away. Harvesting is labor-intensive,

which is why wild rice is so expensive. You must drift alongside the grass in a dropcloth-lined small boat or canoe, and quickly pull the plants over the boat, striking them so the grains fall aboard. Learn to apply the right amount of pressure to keep unripe seeds from falling in. Return for these another time.

Spread the grains out to dry, then parch in a 300 to 350°F oven for a few hours, stirring occasionally. Pound the grain or rub it with your hands to break off the brittle husks. Winnow them out in a natural or fan-generated breeze, tossing them from a sheet so the chaff blows away, and the grain falls back. Rinse the grain to eliminate any small bits of burned husk. Cook the grains like commercial rice: Simmer 30 to 40 minutes in two parts water to one part rice, with a dash of salt. Season if desired.

Wild rice is so tasty, it's considered a gourmet food. It's higher in carbohydrates than any of our more common grains, and it's also rich in protein. It's an excellent source of phosphorus and calcium, and it provides iron and B-complex vitamins too. A soft mush of wild rice was one of the few foods other than mother's milk Native American children under five received.

Caution: As with all grasses, there's a purplish poisonous fungus, ergot, that sometimes replaces the grains and can produce enough of an LSD-like substance to kill you. Make sure your collection isn't infected. In the Middle Ages, whole towns fell victim to tainted rye. People would start screaming about imagined tarantula attacks, throw off their clothes, and act crazy. The malady was called Saint Vitus' Dance.

OTHER PLANTS OF FRESHWATER WETLANDS IN SUMMER

Edible or Medicinal Plants

Angelica seeds, apples, bayberries, blueberries, burdock root, catnip, cattail shoots, flower heads, and pollen, coltsfoot leaves, cow parsnip seeds, curly dock root, elderberry flowers and berries, groundnut pods and tubers, jewelweed stems and seeds, Juneberries, miner's lettuce leaves, mulberries, Oswego tea, peppermint, pickerelweed leaves, plums, rose hips, salal berries, spearmint,

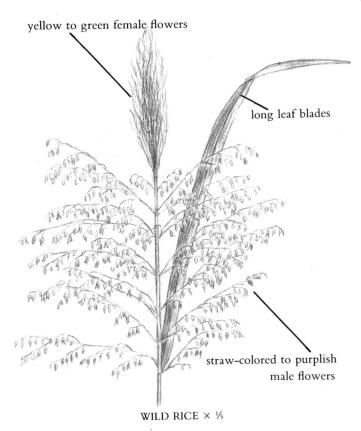

yellow to green female flowers

long leaf blades

straw-colored to purplish male flowers

WILD RICE × ⅕

watercress, water mint, wild mint, wintergreen leaves.

For Observation Only

Asparagus flowers and berries, cow parsnip flowers, cranberry flowers, curly dock flowers, false Solomon seal's berries, groundnut flowers, jewelweed flowers, parsnip flowers and seeds, pickerelweed flowers, poison ivy berries, rose flowers, Solomon's seal berries, thistle flowers, violet leaves, water hemlock, waterleaf flowers.

PLANTS OF THE SEASHORE IN SUMMER

BEACH PEA

(Lathyrus japonicus)

The beach pea is a trailing perennial herbaceous plant with pale green, smooth, somewhat stout stems that are first erect, then prostrate. It has alternate or opposite feather-compound leaves with three to twelve pairs of oval to elliptical leaflets, each 1 to 2 inches long. Curling tendrils extend from the ends of the leaves. Distinctive broad, paired, arrowhead-shaped stipules (modified leaves) clasp the base of each leafstalk.

In the late spring and early summer, a long flower stalk bears six to twelve pink to lavender, bilaterally symmetrical, pealike flowers. The smooth, stalkless, veined seedpods look like smaller versions of commercial peas, about 2½ inches long. The name *Lathyrus* is Greek for an unspecified type of legume.

The beach pea grows along the seashore on the east and west coasts of the northern United States and Canada, as well as along European and Asian coasts.

Caution: There are other members of the pea family, some in the same genus as the beach pea, that have been poisoning people since the days of ancient Greece. They can cause death or irreversible paralysis. Don't eat beach peas without expert supervision until you've become an experienced forager, and be very careful to identify the plant correctly.

Select the younger beach peas, in bright green pods, for the best flavor. If the pod is still tender enough to pinch through it, you can eat it whole or cook it like snow peas, although you may want

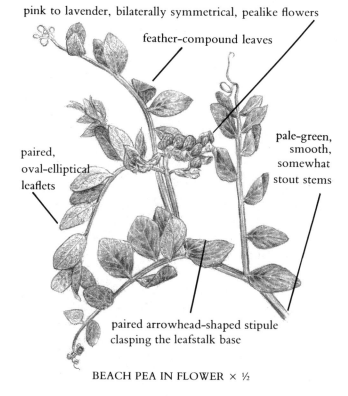

pink to lavender, bilaterally symmetrical, pealike flowers

feather-compound leaves

pale-green, smooth, somewhat stout stems

paired, oval-elliptical leaflets

paired arrowhead-shaped stipule clasping the leafstalk base

BEACH PEA IN FLOWER × ½

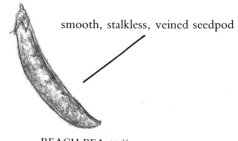

smooth, stalkless, veined seedpod

BEACH PEA × ¼

to remove the strings. Otherwise, remove and use the tiny peas, which ripen in early summer. Use them raw or cooked, the same as commercial peas. Although labor-intensive to shell, they taste great, and they contain high concentrations of B-complex vitamins, beta carotene, and protein.

OTHER PLANTS OF THE SEASHORE IN SUMMER

Edible or Medicinal Plants

Bayberries, beach plums, blackberries, carrot flowers and seeds, black cherries, curly dock root, wild onion seeds and bulbs, field pennycress leaves and seeds, foxtail grass, glasswort, groundnut pods and tubers, lamb's-quarters leaves, marshmallow leaves, fruits, and roots, milkweed flower buds, flowers, and pods, mullein leaves and flowers, peppergrasses, pine needles, prickly pear fruits, rose hips, sea rocket leaves and pods, seaside plantain, sheep sorrel, strawberry leaves, sumac berries, sweet clover flowers, wild beans, yarrow, yucca flowers.

For Observation Only

Asparagus flowers and berries, common evening primrose flowers, curly dock flowers and seeds, groundnut flowers, Japanese knotweed flowers, marshmallow flowers, parsnip flowers and seeds, poison ivy berries, pokeweed flowers and berries, prickly pear flowers, rose flowers, sow thistle flowers.

PLANTS OF THE MOUNTAINS IN SUMMER

Edible or Medicinal Plants

Cloudberry, blueberry and huckleberry, chokecherry, creeping wintergreen flowers and leaves, Juneberries, juniper, miner's lettuce, orpine leaves, pennyroyal, pin cherry, pine needles, roseroot leaves, sassafras leaves, berries, and roots, sheep sorrel, spicebush leaves.

For Observation Only

Mountain cranberry flowers, orpine flowers, roseroot flowers, waterleaf flowers.

PLANTS OF DESERTS IN SUMMER

GLANDULAR MESQUITE, MESQUITE, HONEY MESQUITE, and SCREWBEAN

(Prosopis glandulosa, Prosopis veluntina, Prosopis juliflora, and *Prosopis pubescens)*

These spreading, deciduous shrubs or small trees are legumes with edible pods. Straight, stout spines arm the branches, and occasionally the leaf bases, especially on the screwbean. The alternate, twice feather-compound leaves are typical of the pea family. Glandular mesquite has four feather-compound leaflets, each with seven to eighteen pairs of segments, which sometimes fuse. Screwbean has five to eight pairs of leaflets.

The flowers form fuzzy, willowlike spikes in spring and summer. The screwbean's are greenish-white to yellow and glandular, while the honey mequite's are yellow. The fruits of the honey and glandular mesquite are long, narrow pods 5 to 10 inches long, rounded wherever there's been a bean inside. The screwbean's pod, which grows to about 2 inches long, is dramatically twisted, accounting for its name.

These species grow throughout the Southwest. The screwbean lives along washes and streams, in dry river bottoms and gullies, often in saline soil. Glandular mesquite is common in desert valleys and on open grazing land. Honey mesquite or mesquite grow anywhere there is water.

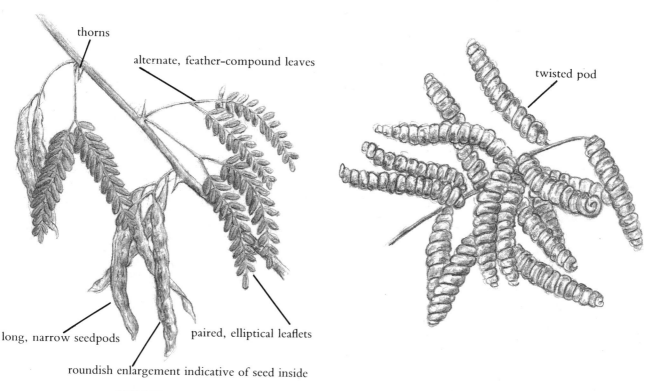

thorns

alternate, feather-compound leaves

twisted pod

long, narrow seedpods

paired, elliptical leaflets

roundish enlargement indicative of seed inside

MESQUITE × ¼

SCREWBEAN MESQUITE × ½

These plants were major foods for Indians. Archaeologists had excavated mesquite beans in 1,200-year-old storage areas. Cook the small, young seedpods like string beans, or chew on the young green pods. Sweetness varies from tree to tree. I find the screwbeans slightly sweeter than mesquite.

You can also boil the tender, young, immature sweet pods into a tasty syrup: Wash a few quarts, break them into pieces, cover with water, and simmer for a few hours. Add water if it gets too thick and threatens to burn. As the syrup cooks and the beans soften, mash everything together. Strain through a colander or sieve, put it back on the stove, add sweetener to taste, reheat, and serve.

The pods become dry and ripe in the summer, growing alongside immature pods. To make flour, toast them at a low temperature for a few hours until dry and slightly brittle. This rules out bean beetle larvae. Break the pods into manageable pieces, and grind repeatedly in a coffee or spice grinder, or a blender, discarding the harder fragments and the seeds. For the finest possible flour, regrind in a grain mill.

Replace ⅓ to ½ of the flour in any recipe with mesquite meal. The Indians simply mixed the meal with a little water to form a dry dough and baked it in the sun for a few hours. Use the leftover pod fragments, or whole pods, to make a beverage: Mash them in cold water, strain out the residue, and drink. Or just chew on the ripe pods. They're refreshing and thirst-quenching.

The sweet, gooey or dry, mealy pulp surrounding the mature pods' seeds was a favorite of many Indian tribes. They ate it straight from the pod.

Make a rich broth by boiling ripe mesquite pods, the same way you made syrup from the unripe pods. Use this sweet, aromatic liquid in beverages, baked goods, or sweet pudding, adding sweetener if necessary. You can even suck on the sweet, delicious flowers.

The tiny, hard, yellow seeds inside the pods are virtually worthless for foragers. Although they're highly nutritious—up to 60 percent protein—they're too labor-intensive to separate and grind, even with a grain mill. Most native people never bothered with them.

basal rosette

long, swordlike, evergreen leaves

YUCCA × ⅕

YUCCA, BEARGRASS

(Yucca filamentosa)

Just before I discovered wild foods, I experimented with unusual foods from ethnic stores. One such food, a root vegetable that looks like a thick tree branch, was called the yuca (often misspelled as "yucca"). However, that is a regional name of a food we call the cassava—a member of the spurge family not related to the wild yucca (a member of the agave family) at all. Common names are endlessly misleading.

Yuccas have long, swordlike evergreen leaves that grow in dense, busy clusters from the ground, like a monstrous basal rosette. In the summer, a central flower stalk grows up to 6 feet tall. It bears large terminal clusters of showy, white, six-petaled, waxy flowers. The fruit, which follows, looks like six-sided pickles.

Yucca filamentosa is a native perennial of the lily family, originating in the American desert. There are also yuccas throughout much of the rest of North America that have escaped cultivation. In the Northeast, wild yuccas grow in sandy empty lots and fields near beaches.

The fruit of a related species, the Spanish bayonet (*Yucca aloifolia*) is edible after you've scraped out the fiber and seeds and baked it 30 minutes at 350°F. This species has thick, clublike stems, often covered with downward-pointing old leaves, with a head of new leaves on top. It grows up to 15 feet tall. Unfortunately for me, it doesn't grow north of South Carolina. Some species of yucca have tough, dry, inedible fruits; others have pulpy, edible fruit. Explore your region and see what you come up with.

The flowers are edible raw in salads, or cooked. You can pickle them, or sauté them with onions and tomatoes. They're a good source of vitamin C. Use the petals and avoid their bitter green bases. There are other very similar species that you can use the same way, although some are better than others. Also, watch out for tiny black insects that live only on the seeds in the flowers. The adults pollinate the plants, and the larvae leave enough

seeds for reproduction, but they do not make an attractive addition to salads.

Some yucca species have a fleshy, edible stem at the base of the leaves. However, removing this kills the plant, and you don't want to kill any desert plants unless you're starving. They can't come back as easily as the weeds in areas with adequate rainfall.

Caution: Avoid eating yucca roots. They contain saponins, which are poisonous (Indians used them to stupefy corralled fish), although they do make a good soap. On the other hand, many arthritis preparations in health-food stores contain yucca root extracts which may be helpful.

The yucca is also one of the best plants for starting fires. The scraped stalk makes a good spindle, which you spin against a small board cut out of the plant's base. This gives you a spark for lighting your tinder, and the rest is easy. You can use a spine from the tip of a leaf as a needle. Then pound fresh leaves between stones, rinse the resulting fibers, and you have thread.

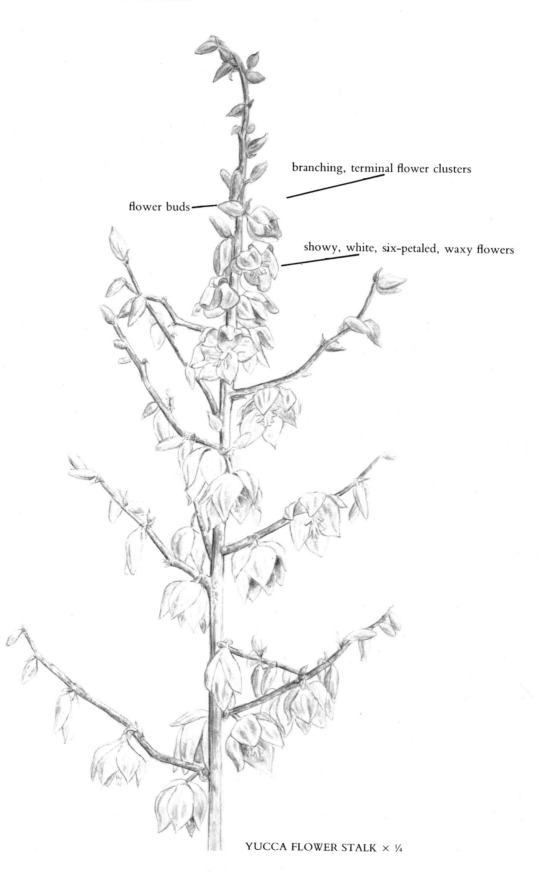

branching, terminal flower clusters

flower buds

showy, white, six-petaled, waxy flowers

YUCCA FLOWER STALK × ¼

OTHER PLANTS OF DESERTS IN SUMMER

Edible or Medicinal Plants

Digger piñon pine nuts, prickly pear fruit.

For Observation Only

Mountain cranberry flowers, orpine flowers, prickly pear flowers, roseroot flowers, waterleaf flowers.

CHAPTER 4

EDIBLE WILD PLANTS OF AUTUMN

Autumn is the most diverse time to forage. There are still plenty of persistent edible greens, and some "spring" plants even send up new growth. It's the main season for nuts and seeds, and many plants begin storing food in edible roots. Also, the procession of fruits and berries that began in summer continues right through the first frosts.

PLANTS OF LAWNS AND MEADOWS IN AUTUMN

CHICKWEED, THE HEN'S INHERITANCE

(Stellaria species and *Cerastium* species)

This delicate-looking, sprawling, European annual plant has tiny, paired oval to spade-shaped, toothless leaves growing along its stringy, flexible stem. The tiny, white flowers bloom in terminal clusters, at the ends of the plant, accompanied by some leaves. The five petals are shorter than the five sepals. The petals are so deeply divided, they look like ten. This relative of the carnation is especially beautiful under magnification.

Hundreds of species grow throughout North America, and worldwide. You'll find chickweed on lawns, in open, sunny areas, and in partially shaded spots. Look for it on lawns and disturbed areas.

Most ubiquitous is common chickweed (*Stellaria media*). Its leaves grow on short stalks, while the similar-looking star chickweed's (*Stellaria pubera*) leaves are stalkless. Both species have very little hair, while mouse-ear chickweed (*Cerastium vulgatum*) is coarsely hairy, with stalkless leaves.

Chickweed grows all year, but does best when the trees are leafless and no taller herbaceous plants grow to shade it out—from late fall to early spring,

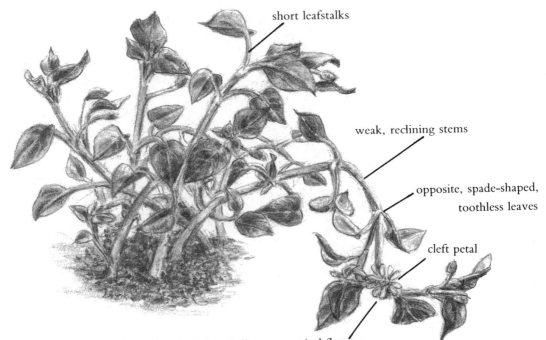

short leafstalks

weak, reclining stems

opposite, spade-shaped,
toothless leaves

cleft petal

white, five-petaled, radially symmetrical flowers

CHICKWEED × 1

and during winter thaws. This overlooked weed becomes lush, transformed by the cold into the Queen of Barren Lawns. In late winter, it's a blessing for victims of F.W.S. (Foragers' Withdrawal Syndrome), who crave a wild salad.

If there was ever a plant whose personality I would like to emulate, it's chickweed. When you look at it, it appears fragile and tender. Yet this plant also manages to be tough and hardy. It doesn't wilt under the malevolent glare of murderous gardeners. It has the vitality to fight off weed killers, stand up to frigid weather, even snow, and hold its springy shape against oblivious tramplers.

I don't collect chickweed until I find a thick mat of it. Then, as though I were giving an unruly cat a haircut, I snap off the top few inches. With a large, easily garnered bag of greens and a well-trimmed weed patch, I've done my good deed for the day.

The stem and the leaves are equally good in all recipes. Some people insist that chickweed is best raw. If they're having a wild salad, it must include chickweed, with its mild, cornlike flavor and crispy, tender texture.

Light cooking transforms the flavor into something more like spinach. You can also steam chick-

weed without water (see cooking section, page 281). Or cook it, and blend it into a cream soup, as I've done in hundreds of different ways. I've used stock, milk, soy or nut milk, and water as the liquid. Try oats, flour, potatoes, commercial yucca, corn, peas, arrowroot or sweet potato starch as thickeners. Herbs like marjoram, tarragon, and chervil complement this vegetable. Avoid using too much garlic or black pepper, and don't overcook—three to five minutes is plenty.

Chickweed is a nutritional powerhouse. It provides plenty of vitamin C, rutin, biotin, choline, inositol, PABA, vitamin B_6, B_{12}, vitamin D, and beta carotene. It's also an excellent source of the minerals magnesium, iron, calcium, potassium, zinc, phosphorus, manganese, sodium, copper, and silicon.

Chickweed also contains steroidal saponins—responsible for its ability to increase the absorptive ability of all membranes, and to eliminate congestion. As a result, the liver, kidneys, and lungs become healthier, and nutrients are more readily available to every cell. Blockages and waste are cleansed from the kidneys and gallbladder, as is congestion from irritated lungs.

Before vitamins and minerals were identified,

people used chickweed as an all-purpose health food. It's wonderfully soothing to eat—great for an upset stomach, and strengthening to the bowels. It's as safe a food as you can find anywhere, with no harsh or irritating substances. You can have as much as you feel you need. It was fed to people with debilitating conditions such as anemia, tuberculosis, and arthritis, as well as to victims of malnutrition, with good results.

Chickweed tincture is particularly good for irritated skin conditions, upset or ulcer-prone stomachs, blocked kidneys, and bladder or liver problems. A friend used it, along with plenty of chickweed salad and burdock, for a persistent bladder infection. This worked well, although if the infection had been serious, it would have been better to use antibiotics, lest the infection spread to the kidneys and do permanent damage.

The tincture is also used for arthritis, gout, joint diseases, blood diseases, eye inflammations, asthma, tuberculosis, and congested, infected lungs. Because hot chickweed tea is a diuretic, one of its most popular commercial uses is in all-natural weight-loss pills. Excreting excess water sheds pounds quickly, if only temporarily.

Chickweed is used externally to draw toxins from the skin, and as a moisturizing, soothing demulcent. It's good for all kinds of minor skin irritations and infections. Chop it up and bandage it onto cuts, rashes, burns, scratches, and wounds, or mix it with clay first to increase absorption. An oil infusion of chickweed is especially soothing for dry, hot sores and rashes. An ointment is good for hemorrhoids, and a chickweed lotion is excellent for overly dry skin.

According to folklore, you can use chickweed to predict the weather. If the flowers are blooming robustly, it won't rain for at least four hours. Otherwise, bring an umbrella.

OTHER PLANTS OF LAWNS AND MEADOWS IN AUTUMN

Edible or Medicinal Plants

Clover flowers, common mallow leaves and fruit, dandelion leaves, flowers, and roots, dayflowers, daylily tubers, wild onion leaves and bulbs, ground ivy, lady's thumb, peppergrasses, plantain leaves and seeds, purslane leaves, stems, and seeds, sheep sorrel, sow thistle leaves, storksbill, strawberry leaves, wood sorrel, yarrow.

For Observation Only

Sow thistle flowers and seeds, violet leaves and fertile flowers.

PLANTS OF CULTIVATED AREAS IN AUTUMN

GINGKO, MAIDENHAIR TREE

(Gingko biloba)

This gingko is a tall, slender tree, growing up to 90 feet, with a gray, furrowed bark. Unlike all other trees, this relic from the Age of Dinosaurs sports short, thick, bullet-shaped twigs growing at right angles to the branches. You can spot them at a glance in the winter, when no leaves obscure your view.

Even though they're gymnosperms—more closely related to conifers than to other broadleafed trees—gingkos have broad, lobed, fan-shaped leaves 2 to 3 inches across instead of needles. Broad leaves, which catch more sunlight, all originally evolved from needles. In other families, individual needles expanded over evolutionary time, but the gingko's leaves formed when many needles knit together: You can still see their outlines as veins diverging from each leaf's base.

The gingko's fruit, which ripens from mid- to

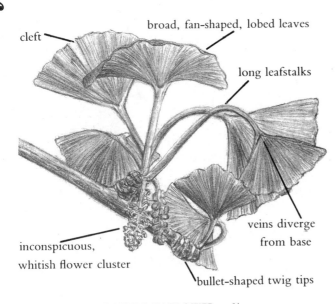

cleft

broad, fan-shaped, lobed leaves

long leafstalks

inconspicuous, whitish flower cluster

veins diverge from base

bullet-shaped twig tips

GINGKO IN FLOWER × ⅔

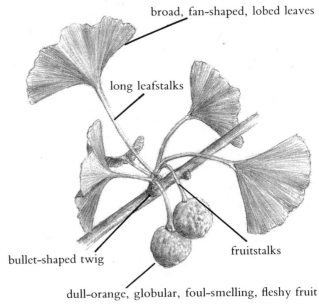

broad, fan-shaped, lobed leaves

long leafstalks

bullet-shaped twig

fruitstalks

dull-orange, globular, foul-smelling, fleshy fruit

GINGKO × ½

late fall, is a beautiful apricot color, about the size of a small fig, with an odor worse than vomit. This stench comes from butyric acid, named from its presence in rancid butter. It may have functioned to repel dinosaurs—contemporaries of this ancient tree. This theory seems difficult to prove, until you inhale and realize that it must repel all creatures, living or extinct. (In reality, butyric acid, which gives Limburger cheese its distinctive odor, is a common product of many plants and animals.)

You won't find the fruit on more than one third of the trees, because this primitive species has separate male and female trees. (Separate elongated clusters of inconspicuous, whitish male and female flowers briefly hang from the twigs in early spring.) Only the females bear fruit. Americans try to plant the males only. Fortunately for foragers, and unfortunately for building superintendents, it's very hard to determine the sex of a gingko sapling, and it's even more difficult to transplant adult trees, so there are plenty of female gingkos around. Just follow your nose. Inside the fruit is a thin beige nutshell containing an emerald-green kernel about the size of an almond.

The gingko is a unique tree, with a dramatic history extending farther back than any other living broad-leafed tree: Its habitat in China was supposedly destroyed during the Ice Age. All species in the genus and all genera in the order were believed wiped out. When the glaciers descended, North American trees sought refuge in Central America,

returning after they melted. But European and Asian trees had no retreat. Geological barriers like the Mediterranean Sea and the Himalayan Mountains cut them off. Scientists knew gingkos and their relatives only as fossils.

Then, in the late 1700s, Western explorers discovered a small number of these stately trees flourishing by a monastery in China. The monks—expert herbalists and gardeners—had saved this rare treasure by cultivating it for countless generations. When the news broke, the gingko became the greatest botanical sensation of the day. Its instant popularity was also due to its adaptability, eerie beauty, and tasty, nutritious nuts, which became an intrinsic part of Oriental cuisine.

Gingkos don't grow wild anymore, but they've been planted worldwide. I've found them in cultivated parks and on city streets across the country, except for the coldest, most northern regions. The tree has outlived all its diseases. Whatever blights or insects once attacked it became extinct when the

unopened nut   green kernel   nut shell

GINGKO NUT × 1

gingkos all but vanished, and this "living fossil" now flourishes under harsh urban conditions. Only another Ice Age could faze it. It's like a strange, hardy time traveler that likes our modern, chaotic world, and has decided to stay.

The nut is in season from mid-fall to late fall. The fruit is soft and easy to discard, but wear gloves when you handle it. One out of fifty gets a poison-ivylike rash from touching it bare-handed. Leave the vile-smelling fruit by the tree, or everyone you live with will never forgive you.

A student of mine, preparing an environmental benefit dinner, collected a huge quantity of gingkos one afternoon in Central Park. It got so late, she began stuffing whole gingkos into her bag, smelly fruit and all, but she still got caught in the rush hour. As soon as she boarded the jam-packed subway, people began turning their heads from side to side, sniffing the air. Then, a miracle occurred: She got a seat! Within a few stops, she had a whole section of the car to herself, and by the time she reached her destination, even the homeless had fled. Still, the brown-rice, gingko, and field-garlic pilaf she served at the dinner was the hit of the event.

Usually, I start collecting my year's supply after a storm, when the wind has littered the sidewalks with apricot-colored spots. The morning after the storm is best. If I start a day or so later, I find a smelly, well-trampled mess—little bits of jade-green mixed with stinking apricot paste and crushed nutshells.

Because the trees are so tall, I sometimes see ignorant people using sticks to knock down the fruit, breaking the brittle branches. This is destructive and unnecessary. All the nuts eventually fall to the ground by themselves.

In New York City, Asian people compete for the nuts. Individuals stake out a tree and become angry if someone else forages there. A cottage industry has sprung up in Central Park. On November afternoons, you'll see piles of discarded fruits under gingko trees, entire families washing the nuts in the lake and leaving the park with filled shopping carts. You rarely see gingkos sold commercially in most stores. Instead, the nuts from Central Park wind up in specialty shops in Chinatown, and on the menus of overpriced Asian restaurants.

The nuts are edible only when cooked. Wash them (still wearing gloves), bake in the shell (some people boil them, but I think they taste much better baked) at 275°F for 25 minutes, stirring occasionally. Now comes the fun part, eating gingkos. When cool, crack the shells with a light tap of a hammer or mallet, or open them with your teeth, like pistachios. The shells are thin enough to open with one gentle blow. Properly cooked gingkos are soft and rich. Undercooked nuts are hard and slightly unpleasant.

After only 15 minutes of serious cracking, you'll have enough to add color, flavor, and texture to a pilaf, soup, stuffing, or vegetable dish. Gingkos make a great snack food, full of protein and low in fat. Asian people use them as appetizers, and mix them with rice, tofu, stir-fried vegetables, and mushrooms in a variety of dishes. Don't use the cooked gingkos like the crunchy nuts you're used to. They're more like a soft, dense savory vegetable.

Gingko nuts are perishable. Don't refrigerate for more than a few days, or they'll dry out and become as hard as rocks—shelled or not. I freeze them, shelled or unshelled, within a couple of days of collecting and roasting. Briefly rebake them to defrost, and they'll be perfect.

Eating gingkos is supposed to promote digestion and diminish the effects of excessive alcohol. In Chinese folklore, the fruit is applied as a dressing for wounds, and a tea of the fruit is used for indigestion, and as an expectorant for asthma, bronchitis, and tuberculosis. But the fruit is poisonous, both internally and externally, and it's quite repulsive. I strongly advise against using it.

Modern science has found that an extract of the leaves stimulates the immune system and improves memory. A popular anti-aging preparation sold in health-food stores is based on gingko-leaf extract. It's also used for circulatory disorders, and contains an analog of platelet-activating factor, intervening in the formation of arterial plaque. It's also used for headaches, asthma, depression, toxic-shock syndrome, bleeding, bruising, and edema. It destroys free radicals, and may help prevent cancer. You can make a tincture of gingko leaves (see page 9 for tincture instructions) when they're starting to turn yellow in the fall. Before that, they contain too much tannin.

Gingko extract contains numerous valuable substances, including bioflavonoids even more effective than those in citrus fruit. This gingkolides and heterosides are antioxidants, anti-inflammatory, and vasodilators (opening blood vessels, increasing circulation, and increasing oxygenation). The extract is especially important for Alzheimer's disease—increasing circulatory efficiency, particularly in the brain. Gingko extracts are also used for tinnitus (ringing of the ears), asthma, heart and kidney disorders, and glucose utilization. The extract has a dramatic effect on blood flow. Energy production also improves from waste product elimination. Is this reason enough to call this ancient species an anti-aging miracle? Eating gingko nuts has similar benefits. Traditional Eastern herbalists advise eating the nuts, while Western herbalists give leaf extract.

Japanese folklore advises eating no more than seven nuts at a time. I eat dozens with no ill effects, but I do know people who get headaches, possibly because their blood vessels are opened too much. Some gingko overeaters experience shortness of breath, either from an allergy to gingko, or from eating too many nuts and neglecting to breathe. Eat very small amounts the first time.

KOUSA DOGWOOD

(Cornus kousa)

You can find this small Asian dogwood tree bearing fruit when it's only 8 feet tall, although it can also reach 30 feet in height. The coarsely toothed, opposite leaves are narrowly heart-shaped, about 4 inches long—not unlike many other dogwoods. But the mottled, brown and ocher bark patches are more distinctive.

The flowers, which appear from late spring to early summer, superficially resemble other dogwoods, with four showy sepals acting like petals. But in the flower's center, there's a green warty ball, smaller than a pea. In early summer, tiny, inconspicuous white flowers with four petals erupt from the warts, then disappear. The sepals fall off, the ball enlarges to the size of a table-tennis ball,

red and still warty, softening as it ripens. Now, each wart is a hard, rice-sized seed.

Thanks to landscapers' relentless efforts to fill America with foreign plants, this tree may now provide us with exotic fruit virtually anywhere in the country. I see it in cultivated areas such as parks, botanical gardens, and arboretums. I've even found it in front of private homes, where homeowners are usually astonished when I ask permission to gather the fruit. Landscapers never tell them it's edible.

This is one of the strangest-looking fruits you'll ever see, and I was reluctant to try it for years. Finally, a Japanese TV crew putting together a documentary about me gave me a Japanese foraging book as a gift, and I learned the Japanese had been eating this fruit for centuries.

The fruit's quality varies from tree to tree. Sometimes it tastes like a less-sweet combination of mango and apricot; other times it's unpleasantly bitter. Trees with larger fruit are often the best. Look for ripe fruit at the end of the summer and in the fall.

I've tried straining out the seeds and cooking the pulp, so far without success. Cooking seems to bring out bitterness and an unappetizing gray color. There must be good ways to cook it, but until I (or you) discover them, I'll be content to enjoy the fruit raw.

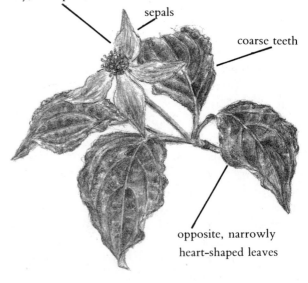

tiny, four-petaled flowers

sepals

coarse teeth

opposite, narrowly heart-shaped leaves

KOUSA DOGWOOD IN FLOWER × ⅔

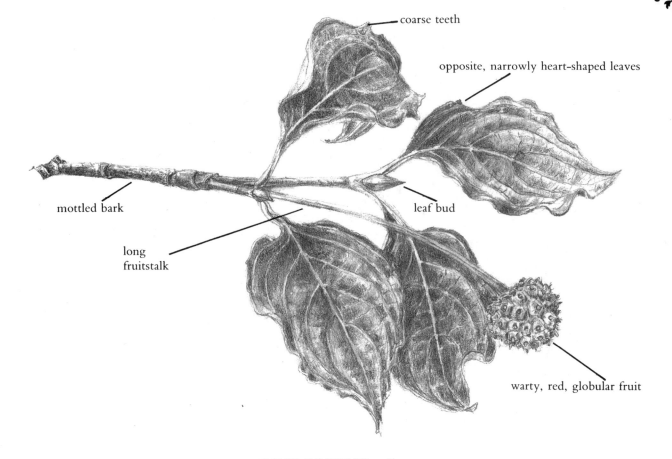

coarse teeth

opposite, narrowly heart-shaped leaves

mottled bark

leaf bud

long
fruitstalk

warty, red, globular fruit

KOUSA DOGWOOD × ⅔

YEW

(*Taxus* species)

Yews are evergreens that may grow into small trees, remain low, straggly shrubs, or get cultivated into neat hedges. They're characterized by short, flat needles under 1 inch long, different from the long, round needles of pines.

Tiny, inconspicuous, cream-colored flowers produce copious amounts of pollen in early spring. They all have the same distinctive round, red, juicy, mucilaginous berry, forming a plump cup around a single, round, hard, poisonous seed. The main difference between the species is the leaves, which vary slightly in coloring and configuration. All in all, the *Taxus* genus has evolved very similarly around the world: The berries are equally tasty, and the rest of the plant equally deadly, even

flat, short needles

small, inconspicuous, cream-colored flowers

YEW IN FLOWER × ½

round, juicy, red, cuplike fruit

hard, brown seed

YEW IN FRUIT × ⅚

though the berries are commonly held to be poisonous.

Yew species grow worldwide. The most commonly cultivated species in the United States is the Japanese yew (*Taxus cuspidata*), although another Asian species (*Taxus baccata*) is quite similar. So is our native species, the American yew or eastern yew (*Taxus canadensis*). Horticulturists often plant hybrids.

Yews are cosmopolitan—at home in pine forests, bogs, and thickets, they thrive under cultivation by urban and suburban buildings. In urban and suburban settings, they're as common as streetlights, and equally ignored. We plant them everywhere because, as evergreens, they must be flexible to shed snow in the winter. The author of *Robin Hood* had the archer make his bow from the wood of the English yew tree, and the various yew species have basked in popularity ever since. Shakespeare called the English yew doubly fatal: you can be killed by the direct action of the poison or by being shot with an arrow from a bow made of yew.

This plentiful fruit is quite tasty and a great

energizer, as long as you don't swallow the seeds. I use the fruit as a trail nibble. Separating the seeds from the fruit by hand would lead to a gooey mess, and using a food mill could be risky if parts of the seeds got mixed into the puree. The seeds are so dangerous you shouldn't eat the berries near small children, who may copy you but swallow them.

The bark, leaves, and seeds contain a heart-depressant alkaloid. If you swallow enough seeds, or decide you like the taste of shrubbery and eat fifty needles, your heart will stop beating. The Celts used the sap as an arrow poison, and the plant is the universal symbol of death.

Oddly enough, the Indians used the native species' twigs and leaves as a sedative, because minute, nonlethal dosages slow the heartbeat. They also used it in teas for urinary problems and stomach pains, and they inhaled its vapors in sweat baths, for rheumatism and paralysis. But their lives in a subsistence culture entailed greater risks than are appropriate now, when safer alternatives are available. Also, Native American shamans gained far more skill with native plants than modern herbal students. Never experiment with the yew medicinally. You easily could die.

As is very often the case, the Native Americans had discovered a plant with important medicinal properties. The western yew's (*Taxus brevifolia*) inner bark contains enough taxol (100 parts per million) for researchers to use it for uterine and breast cancer. Preliminary results are promising. According to a 1992 article in *Life* magazine, taxol can induce remission in one third of the women with terminal ovarian cancer. Subsequent research showed taxol to be less promising, but still useful. We still have much to learn from these ancient trees.

Unfortunately, you must kill several of these slow-growing, endangered trees to save a life. Also, our government and the timber industry had been cooperating to cut down the old-growth American forests, where this valuable species grows. Since it's worthless as lumber, they burned any western yews they found. Fortunately, these policies are being reevaluated.

However, taxol news is breaking all the time. The needles of many cultivated species have six times more taxol than western-yew bark, and you don't have to kill it to get the drug. Himalayan yew's (*Taxus wallichiana*) needles may have comparable quantities. Chemists are finally getting close to synthesizing taxol, and a new substance, taxotere, extracted from the European yew's needles, is also showing promise in combating cancer.

OTHER PLANTS OF CULTIVATED AREAS IN AUTUMN

Edible or Medicinal Plants

Amaranth seeds, apples, autumn olives, black walnuts, burdock roots and seeds, chickweed, chestnuts and chinquapins, clover flowers, common mallow leaves and fruits, dandelion flowers and roots, dayflowers, daylily tubers, hackberries, hawthorn berries, juniper, Kentucky coffee tree seeds, lady's thumb, lamb's-quarters leaves and seeds, orpine leaves and roots, plantain leaves and seeds, rose hips, sassafras leaves and roots, sheep sorrel, sow thistle leaves, viburnum berries, wood sorrel.

For Observation Only

Russian olive, Hercules'-club berries, sow thistle flowers and seed heads, violet leaves and fertile flowers, yucca.

PLANTS OF DISTURBED AREAS IN AUTUMN

Edible or Medicinal Plants

Amaranth, burdock roots and seeds, brassica flowers and seeds, carrot roots, seeds, and flowers, catnip, chickweed, chicory root, clover flowers, coltsfoot leaves, common evening primrose roots, curly dock roots, dandelion flowers and roots, dayflowers, daylily tubers, epazote, wild onion leaves and bulbs, foxtail grass seeds, goatsbeard, ground ivy, horseradish roots, Jerusalem artichoke roots, lady's thumb, lamb's-quarters leaves and seeds, mugwort, mullein leaves, nettles, orpine leaves and roots, parsnip roots, passionflower fruits, peppergrasses, pineapple weed, plantain leaves and seeds, povertyweed, purslane stems, leaves, and seeds, raspberry leaves, rose hips, sheep sorrel, spearmint, storksbill, strawberry blite leaves and seeds, sumac berries, thyme, wild potato vine root, wood sorrel.

For Observation Only

Common evening primrose seeds, curly dock seeds, Japanese knotweed seeds, milkweed pods, mullein seeds, poison ivy leaves and berries, pokeweed berries, sow thistle flowers and seed heads, thistles, sunflowers, wild lettuce seed heads.

PLANTS OF FIELDS IN AUTUMN

AMARANTH, PIGWEED

(Amaranthus species)

What's all the fuss about amaranth? You may have seen it displayed in health-food stores as a "wonder food." Archaeologists and historians have demonstrated that it was a staple for the Aztecs, and for earlier Indians as long ago as 6700 B.C. Nowadays, people are paying renewed respect to this granular seed, even as they unknowingly pass it by.

The Aztecs were mainly vegetarians through necessity, with no suitable game or livestock. Yet their civilization was advanced: Cortez's invading entourage was awestruck in 1519 by the grandiose beauty of the capital city, Tenochtitlán, with its shimmering palaces and temples. Aqueducts quenched the thirst of 300,000 inhabitants, while canals and paved streets followed intricately designed paths. Thousands of products were traded in marketplaces on arcaded squares. Foodstuffs the Aztecs first perfected include corn, beans, tomatoes, peanuts, and pumpkins.

Amaranth was a principal crop. With its high protein content, it was the Aztecs' key to survival. A major spiritual symbol, amaranth was linked to the deities of rain and agriculture, and because of

dense seed clusters

toothless, coarse, roughly oval to lance-shaped alternate leaves

spike arising from leaf axil

long leafstalk

COMMON AMARANTH SEED HEAD × ⅓

its amazing yields, it was associated with fertility. Cortez assured the destruction of the Aztec culture by burning all amaranth fields and making its cultivation punishable by death.

Amaranth's unobtrusive appearance may help it escape your notice. All amaranths are coarse-looking herbaceous plants with roughly oval to lance-shaped, toothless, medium-sized, alternate, long-stalked leaves, and red roots. Our wild species usually grow up to 3 to 4 feet tall, but I've seen them trailing along the ground, and reaching well over 7 feet, depending on the species and growing conditions.

Amaranth has dense flower- and seed clusters—spikes that arise from the leaf axils in late summer and fall. The flowers are green and tiny, surrounded by hairlike bracts. They'd be inconspicuous except that they're so numerous, with 100,000 to 200,000 seeds per plant.

Because of the dry, unwithering bracts that bear the seeds, the plant name is Greek for "unfading flower." Also, it's hard to destroy—just ask Cortez. Its seeds are among the most waterproof anywhere. It takes much frost and abrasion to get them to germinate. Seeds lying in fields may still sprout after forty years. Perhaps the parents of your next amaranth cereal were alive before you were born. Amaranth should be integrated into mainstream agriculture. It grows more efficiently than grains, requires less energy and water, produces greater yields, and could easily nourish the hungry in Third World countries and at home.

Inside the coarse, dry-looking bracts of wild amaranth are thousands of tiny, onyx-black granules about the size of mustard seeds. They're ripe in autumn, when they fall out with a minimum of rubbing. Commercial amaranth, domesticated since ancient times for its protein, abundance, and quick germination, has white seeds. It must be planted to grow. Another ornamental variety, popular with gardeners, has dazzlingly beet-red leaves and seeds. Lacking in nutritional content, it's favored as a dye for breads, beverages, and ceremonial offerings.

Amaranth and its relative lamb's-quarters (page 45) are both sometimes called pigweed, but lamb's-quarters leaves are more diamond-shaped, less coarse, darker green on top, and very light underneath. Fortunately, both are edible, although with lamb's-quarters, the leaves are more important than the seeds.

This genetically diverse genus grows around the world, and it's difficult to distinguish one species from another. They're all edible, but botanical historians don't know whether it was brought from Africa or Asia to Central America or vice versa, or if it grew in these places before human intervention. I often see amaranth in strange, out-of-the-way places, emerging from the cracks in sidewalks and buildings, alongside highways, and in empty lots, as well as luxuriating in gardens. Some species tolerate dry soil and environments where there is hardly any soil. However, it flourishes in rich garden soils, especially those full of nitrates. Beware of amaranth growing in overly rich, artificially fertilized, or polluted soil. If it's contaminated by nitrates, it can cause abdominal cramps and nausea.

Amaranth first appears in late spring, although new shoots sometimes emerge in autumn. Some species' young leaves are much tastier than others, flavored like strong-tasting string beans. Others are bitter, so sample before collecting. Use them raw in salads, or steam, sauté, or simmer amaranth leaves 15 to 20 minutes, or until tender.

Amaranth is related to spinach, beets, and quinoa, not to the grains, which are grasses. And its seeds' protein content and quality surpass grains'. Amaranth provides lysine, an amino acid grains lack, along with other sulfur-containing amino acids that complement one another to produce high-quality protein. Amaranth seeds contain much vitamin E and B-complex, essential for reproductive health. So the Aztecs were right to use amaranth in fertility rites. The seeds also supply mucilage and sugars, and the leaves are rich sources of vitamins C, beta carotene, niacin, riboflavin, and many important minerals, particularly calcium and iron. They contain twice the protein of most other leafy vegetables. A three-ounce serving provides 7 to 10 percent of the recommended daily allowance for protein. The leaves contain calcium, potassium, and niacin.

In late summer or fall, you can use amaranth seeds fresh, although I prefer drying and storing the seed heads in double paper shopping bags, sealed with masking tape to keep out insects. Nonporous plastic bags don't work. I often forget about my cache. Fortunately, dried amaranth keeps indefinitely. I usually resdiscover my stash later in the winter, when I hazard rummaging through my overfilled closets, searching for something else.

When you can rub the ripe fresh or dried seed heads over a tray, hundreds of seeds fall out, along with plenty of chaff. Winnow away the chaff by tossing everything into the air from a sheet, using a fan, or the wind. The light chaff blows away in the wind, while the heavy seeds will fall back onto the sheet.

You can also pass the seeds through screens available in hardware stores. When the holes are a little larger than the seeds, they pass through, and the coarse trash doesn't. When the holes are slightly smaller than the seeds, they don't pass through, but the fine trash does. How thoroughly you clean the seeds depends on your preferences.

Cook amaranth seeds in two parts water to one part seeds for 25 minutes, or until the water is absorbed. You can preroast them until light brown, if desired. Amaranth has a nutty flavor, somewhat similar to millet. You can also extend other grains with amaranth. This impoves the flavor and increases the protein quality.

The Indians mixed the lightly roasted seeds with cornmeal and salt, adding water, to form a thick batter. The Aztecs ground half their seeds, roasted the rest until they popped, and mixed the blend with honey. They molded this dough into an enormous replica of their war god, and baked it. During a ritual ceremony, the helpless deity was sacrificed and eaten in effigy by an entire city. He would be the only food they consumed for days.

FOXTAIL GRASS

(*Setaria* species)

Grasses are paradoxical. This highly successful group of flowering plants is considered primitive because the flowers aren't specialized to attract insects. They rely on inefficient wind for pollination instead. Therefore, the flowers are reduced to their bare essentials—tiny male and female parts enclosed in scaly bracts, where the seeds (grains) also

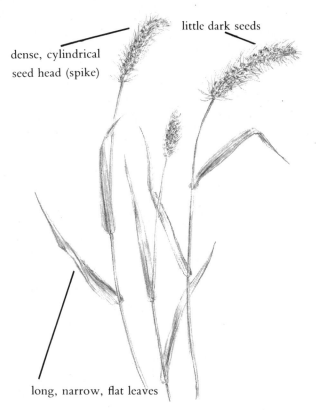

dense, cylindrical
seed head (spike)

little dark seeds

long, narrow, flat leaves

FOXTAIL GRASS × ¼

form. They flourish because they grow from the base, rather than from aboveground leaf buds. If the aboveground parts are destroyed by drought, eaten, trampled, or mowed, they quickly regenerate. Also, their fibrous roots permeate the soil, blocking other plants, and grasses can thrive in places that are too harsh or dry for many other plants.

Because of their reduced flowers, grasses are usually tough to identify, but this genus is a happy exception. The flat leaves of these annual or perennial grasses are long and narrow—common in grasses. But the dense, cylindrical seed head really does resemble a fox's tail, with its single, unbranched, bristly flower spike. This mimicry makes it especially popular with children, who also love the crunchy seeds.

Foxtail grass, which grows in fields, disturbed habitats, and near the seashore throughout the United States, ripens in late summer and fall. Collect when many of the seeds are dark-colored, and readily fall into your hand or bag when you lightly rub the seed head.

The seeds of most grasses are much smaller, and they're covered by the bracts, which you must parch in the oven, rub off, and winnow away. This is very labor-intensive, which is why few other grasses are covered in this book. (See Wild Rice, page 129.) Foxtail grass's large, huskless seeds make it one of the most forageable wild grains. One foxtail species (*Setaria italica*) is even cultivated and sold in health-food stores. It's called millet.

Gather as many of the whole seed heads as possible. Put them in a paper bag, and rub them gently, over a tray, at home. Never rub them hard, or the bristles will come off along with the ripe seeds. What happens next defies science. Physics recognizes four fundamental forces in the universe: gravity, electromagnetism, the strong force (which holds the nucleus of the atom together), and the weak force (responsible for radioactivity). If you eat foxtail grass seeds along with its bristles, you'll discover they overlooked a fifth force, more powerful than all the rest combined: It attracts the bristles to the spaces between your teeth, and holds them there to the end of time.

Collecting foxtail grass seeds is labor-intensive, but if you use the seeds sparingly along with other ingredients, they'll make their presence known. They're quite hard raw, so cook them in water with rice, barley, millet, couscous, or other grains. They add crunchiness, and a delicious, nutty flavor. You can also substitute them for poppy seeds when you're baking bread or muffins. Finally, you can grind foxtail grass seeds in a grain grinder or spice grinder, and extend the flour by mixing it with other whole-grain flowers. This will greatly improve your breads, muffins, and pastries.

HAWTHORNS

(*Crataegus* species)

Some plants have one identifying characteristic that stands out above all else—taste, smell, appearance, or even the fact that a certain animal loves it (i.e., catnip). This large group of shrubs or small trees usually has prominent, long, straight, sharp thorns, 1 to 5 inches long. Growing on the slender,

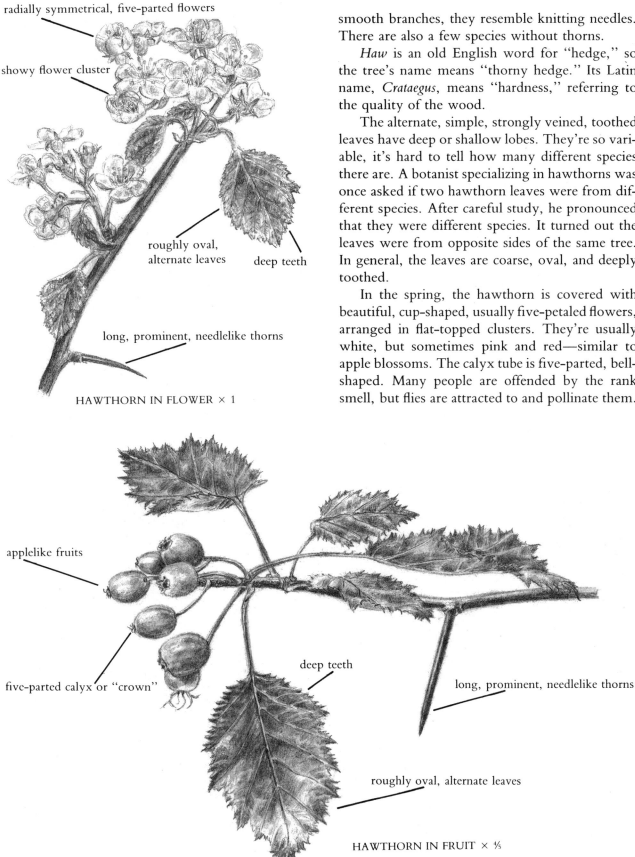

radially symmetrical, five-parted flowers

showy flower cluster

roughly oval, alternate leaves

deep teeth

long, prominent, needlelike thorns

HAWTHORN IN FLOWER × 1

smooth branches, they resemble knitting needles. There are also a few species without thorns.

Haw is an old English word for "hedge," so the tree's name means "thorny hedge." Its Latin name, *Crataegus*, means "hardness," referring to the quality of the wood.

The alternate, simple, strongly veined, toothed leaves have deep or shallow lobes. They're so variable, it's hard to tell how many different species there are. A botanist specializing in hawthorns was once asked if two hawthorn leaves were from different species. After careful study, he pronounced that they were different species. It turned out the leaves were from opposite sides of the same tree. In general, the leaves are coarse, oval, and deeply toothed.

In the spring, the hawthorn is covered with beautiful, cup-shaped, usually five-petaled flowers, arranged in flat-topped clusters. They're usually white, but sometimes pink and red—similar to apple blossoms. The calyx tube is five-parted, bell-shaped. Many people are offended by the rank smell, but flies are attracted to and pollinate them.

applelike fruits

five-parted calyx or "crown"

deep teeth

long, prominent, needlelike thorns

roughly oval, alternate leaves

HAWTHORN IN FRUIT × ⅕

The fruit's usually round to pear-shaped and red, sometimes dark purple or yellowish. They resemble tiny apples, and the genus is related to apples. If you cut one open you'll find 1 to 5 hard, rather large nutlets. Apples (page 151) have softer seeds that come in radially symmetrical sets of 5.

Hawthorns are very common—birds, squirrels, and mice eat them and spread the seeds everywhere. There are native American and European species, plus hybrids. They grow wild in old fields, orchards, woods, thickets, and along streams throughout the United States. Because of their attractive appearance, beautiful blossoms, colorful fruit, and hardiness, they're also planted in front of apartment houses, on suburban lawns, and in city parks. There are no poisonous look-alikes.

Different species ripen from late summer to late fall. Quality varies, but all varieties taste a little like mealy apples. Some are good raw, but others are better cooked, with the seeds strained out, like applesauce. They contain pectin, so the fruit sauce is self-thickening. They're also good to use with other fruit, especially those that require thickeners.

Early in the spring, you can gather hawthorn leafbuds before they unfurl. They're good cooked 10 to 20 minutes, tasting a little like lima beans. I like them in chili and soups. The flowers open when the buds unfurl. Add them to salads if you like them, or use them to make a really pungent wine.

The fruit is a good source of vitamins C, B_1, B_2, B_3, B_5, B_6, B_9, and B_{12}, choline, inositol, PABA, flavonoids and sugar, bioflavonoids, and other substances.

Reputed to have provided Christ's crown of thorns, hawthorns are supposedly imbued with special healing properties. Hawthorn's most important use is for heart disease. People use hawthorn-berry decoctions and syrups.

Studies show that hawthorn extract's flavonoids dilate the peripheral blood vessels, lowering blood pressure and reducing the burden on the heart. It even slightly dilates the coronary vessels. It increases enzyme activity in the heart muscle, making it healthier, and it increases oxygen utilization by the heart. Herbalists use it for high and low blood pressure, angina, irregular heartbeat, and Reynaud's disease (arterial spasms).

Taking effect over a period of time, it works differently from digitalis, without side effects. (Double the correct dosage of digitalis can be fatal.) The two substances even work well together, although the doctor will probably have to lower the dosage of digitalis. It's a shame more practitioners in America don't take advantage of this herb's benefits.

People used to use unripe hawthorn berries for diarrhea and hawthorn-flower tea as a safe diuretic. A decoction of the ripe berries is also used for sore throats, skin diseases, diarrhea, and abdominal distention.

Even the thorns are useful: Indians turned them into awls and fishhooks. If you're ever on a camping trip and forget a can opener, a sharp hawthorn thorn will pierce metal. Just keep them away from your eyes.

Of course, there are also dangers: A holistically oriented dietetic student who had long been studying hawthorn's medicinal properties learned to recognize it on my walks. She soon found her own tree in a park, and began eating the berries, eventually drawing attention. Soon a crowd was enjoying this fruit—all except one man. He was reluctant to explain why. Finally, he admitted that a berry he had planned to eat earlier had a worm (actually an insect larva) inside. Hearing this, someone else also checked, and found a worm in hers. Her disgusted outcry had everyone examining their fruit. There wasn't a single uninfested berry on the entire tree. The moral: Look before you eat (many hawthorns are not infested with insects, but you must inspect the fruit beforehand), and don't accept a hawthorn berry from this lady. (Fortunately for her, she isn't a vegetarian.)

❧ OTHER PLANTS OF FIELDS IN AUTUMN

Edible or Medicinal Plants

Apples, autumn olives, bayberries, brassica flowers and seeds, burdock roots and seeds, caraway roots and seeds, chicory roots, clover flowers, common evening primrose roots, cow parsnip roots, curly dock leaves and roots, wild onion leaves and roots, goatsbeard leaves and roots, groundnut pods and tubers, Jerusalem artichoke tubers, lamb's-quarters leaves and seeds, mug-wort, mullein leaves, parsnip roots, passionflower fruits, pineapple weed, pennyroyal, peppergrasses, persimmons, povertyweed leaves, purslane leaves, stems, and seeds, rose hips, sassafras leaves and roots, sheep sorrel, spearmint, strawberry blite leaves and seeds, strawberry leaves, sumac berries, thyme, wild beans, yarrow.

For Observation Only

Common evening primrose seeds, curly dock seeds, Japanese knotweed seeds, milkweed pods, mullein seeds, poison ivy leaves and berries, poke-weed berries, sow thistle flowers and seed heads, thistles, sunflowers, wild lettuce seed heads, yucca.

PLANTS OF THICKETS IN AUTUMN

APPLE

(*Malus* species)

Wild and cultivated apple trees are relatively small, somewhat gnarled trees, usually under 30 feet tall. Uncultivated individuals appear down-right bushy. The leaves are oval, slightly toothed, and woolly underneath, 2 to 3½ inches long. Even the twigs and buds look hairy under magnification. The gray bark is scaly and fissured.

The blossoms create a stunning show relatively early in the spring. The flowers are variations of pink or white—very fragrant, with five petals. Un-fortunately, their beauty is short-lived—they fall to the ground within a few days.

Many apple trees don't bear recognizable fruit. Some have fruit smaller than a cherry, others as large as Rome apples. Some are hard and green, others mushy and bright red—there are countless varieties. If you're not sure you've picked an apple, cut your specimen in half along its equator. The seeds, which are brown when the fruit is ripe, and their compartments, are always arranged radially in sets of five.

Apples have no poisonous look-alikes. The hawthorn, an edible relative (page 148), resembles a small apple, but with thorny branches, and with-out the five-part symmetry.

Wild apples often differ from cultivated ones, with rough or papery, unwaxed skins. They're compact rather than overly light and watery.

Sometimes they're very wormy. This turns many people off, but you can concentrate on naturally resistant trees, and remove any bad parts. Remember, you can't cut out the toxic pesticides in com-mercial apples, and when you peel them to discard

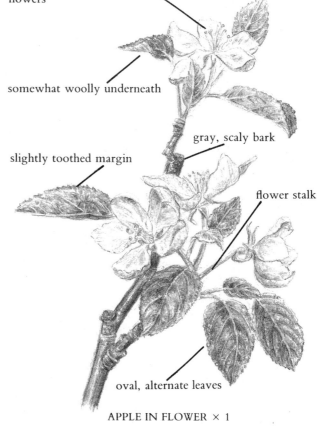

five-petaled, radially symmetrical, white-pink, showy flowers

somewhat woolly underneath

slightly toothed margin

gray, scaly bark

flower stalk

oval, alternate leaves

APPLE IN FLOWER × 1

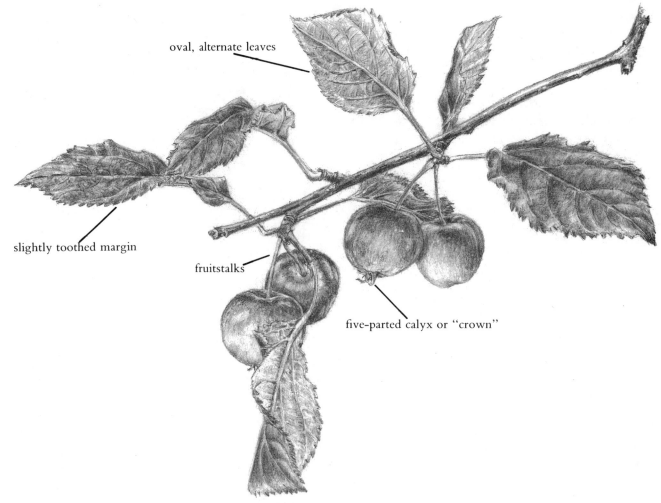

oval, alternate leaves

slightly toothed margin

fruitstalks

five-parted calyx or "crown"

CRAB APPLE × ⅕

the pesticide-laced wax coating, you lose all the nutrients concentrated just under the skin.

Wild apple trees come from seeds. Thanks to sexual reproduction, each tree is unique, though less "perfect" than cultivated ones. The diversity of size, resistance to disease and insects, flavor, ripening time, and color are left completely chance, with nature gradually selecting the fittest for subsequent reproduction.

Cultivated apple trees come from grafts of existing trees, clones of established varieties, all genetically identical. The original McIntosh tree, produced by the successful cross-fertilization of two inferior varieties, lives eternally through modern horticulture. Ornamental varieties of flowering crab apple trees are cultivated in front of civic centers and other government buildings. Politicians

seem to have them planted by force of habit—the habit of graft!?

Apple trees are very plentiful throughout most of North America, growing in forests, overgrown fields, and thickets as well as cultivated areas. Most of us come across them daily. The European common apple (*Malus sylvestris*) was planted from Pennsylvania to Illinois by Jonathan Chapman, a.k.a. Johnny Appleseed (1774–1845), who walked through the countryside for nearly fifty years, distributing seeds from cider presses to everyone he met. I've added the slightly astringent apple blossoms to salads, and steamed them lightly, but my favorite use is for making a wonderful apple-blossom wine.

The variable fruit may ripen as early as July, while some crab apples don't ripen until touched by frost in November. Prepare them according to their different qualities. Some apples are great raw. Others are better cooked into a pie or applesauce. Some make sweet juice. Crab apples are often too tart to enjoy raw or to use as a main ingredient. Combine them with other, sweet fruit to provide a lift, as you would with lemon juice.

Many crab apples seem hard and unpalatable all season. I used to give up on these until I took a chance and tasted a soft, mushy, rotten-looking one in late autumn. It didn't taste fermented at all, and the frost had ripened it enough to enjoy raw. The transformation was unbelievable.

It's difficult to remove the seeds from small apples, so I cook them and put them through a food mill (you can also use a strainer) to remove the seeds. You can also puree the seeds in the blender with some apple juice to make applesauce. Raw apple seeds are poisonous in quantity, with the potential of releasing cyanide. Cooking destroys this toxin, making applesauce with seeds safe.

Apples are very nutritious. They contain potassium, magnesium, iron and beta carotene. Malic (from *malum*, Latin for "apple") and tartaric acids produce the fruit's tartness, neutralize the products of indigestion, and assist digestion of rich foods. Because of these acids, people with uneasy stomachs, and overeaters, crave apples. This brings to mind the image of the medieval banquet, with the apple in the fatty roast pig's mouth.

Apples contain pectin, which nutritionists use to remove toxic heavy metals such as lead, mercury, arsenic, and copper from the body. This fiber prevents the intestinal damage and constipation associated with a low-fiber diet. Pectin also reduces blood cholesterol more effectively than dangerous and expensive drugs by binding with bile acids, thereby decreasing absorption of cholesterol and fat in the small intestine. Furthermore, pectin slows absorption of sugars and carbohydrates in the intestine, increasing energy efficiency and slowing the rise of blood sugar. This benefits hypoglycemics and diabetics. Researchers at Yale's Psychophysiology Center even found that the smell of spiced apples lowers blood pressure in people under stress.

Using wild apples for food and medicine is nothing new. In Europe, apples were revered since prehistoric times. Later, as civilization flourished, crab apples and other varieties were right there, entwined in the culture and traditions of Europe. Next time you find yourself taking the common apple for granted, think of how, in ancient Britain, they were considered magical, on account of their great abundance, and because cutting the apple open along its equator reveals the five-parted pentagram.

AUTUMN OLIVE

(Elaeagnus umbellata)

The autumn olive bush grows from 12 to 18 feet tall, occasionally bearing thorns, and it usually occurs in orchardlike stands. The long-oval leaves are about 2 to 4 inches long, ¾ to 1½ inches wide, toothless and leathery. Their undersides are distinctly silvery, letting you recognize the shrub from a distance. The leaves, twigs, and red, clustered berries are all speckled—especially impressive under magnification.

Crowded bunches of yellowish-white, radially symmetrical flowers appear in mid-spring, hanging from the leaf axils. They're only ½ inch long, the four petals joining at their bases to form a tube.

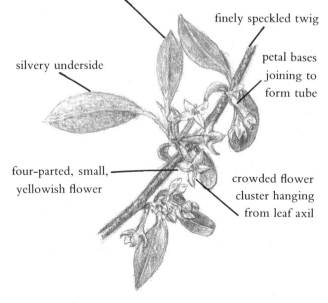

long, oval, toothless, leathery leaves

finely speckled twig

silvery underside

petal bases joining to form tube

four-parted, small, yellowish flower

crowded flower cluster hanging from leaf axil

AUTUMN OLIVE IN FLOWER × 1

long, oval, toothless, leathery leaves

silvery underneath

fruitstalk

dense clusters of scarlet, speckled berries

finely speckled twig

AUTUMN OLIVES × ½

The ripe berries are scarlet, about the size of a currant, with tiny grayish-white speckles, and small, oblong, yellowish seeds inside. Other *Elaeagnus* species with red berries and no speckles grow in Japan, but there are no poisonous look-alikes.

Like the legumes, this member of the honeysuckle family can fix nitrogen from the atmosphere and bring this vital component of all protein into the food chain. This lets it thrive in very poor soil, where there isn't much competition. Look for it near beaches, on depleted agricultural fields, in thickets, and by landscaped apartment buildings.

The autumn olive is a native shrub in Asia, where people cherish its fruits and those of its edible relatives, but it's relatively new to North America. Its hardiness makes it popular with landscapers, who often irresponsibly plant invasive exotics. I love the fruit, but this aggressive shrub displaces other species. Furthermore, birds eat the berries and spread the seeds. It's growing throughout the Northeast, and it may be spreading into other parts of North America.

Two similar relatives are common in North America: The silverberry (*Elaeagnus commutata*) is a native shrub or small tree that grows along streambanks or on hillsides in Wyoming, Utah, Colorado, and Montana. It's 6 to 15 feet high, with alternate leaves that are silvery-scaly on both sides. The twigs are mostly covered with scaly, branlike particles. Use the silvery-green berries like those of the autumn olive.

The Russian olive (*Elaeagnus angustifolia*) is another Asian escapee from cultivation that may pop up throughout North America, growing in the same habitats as the autumn olive. Like the silverberry, its narrow leaves (*angustifolia* means "narrow leaf") are silvery on both sides. However, it has sparse clusters of large, silvery-white berries. When I once sampled the fruit, it tasted great—like a sweet raisin—for the first five seconds. Then, it seemed like I had a mouthful of talcum powder. Now I stick to autumn olives and silverberries.

I knew about autumn olives long before I began eating them. I was first shown the shrub in Central Park, where the fruit was too astringent to be any good. I relegated it to the kind of "edible" plants covered in other field guides (but not this one!) that taste bad, require heroic processing, or are impractical for other reasons.

Then, one November, I was leading a tour through an abandoned seaside military base, for people who lived nearby and wanted to understand local flora. I expected a minimum of top edibles in that particular location in late autumn, but was determined to do my best.

As soon as we entered the park, we were struck by the cornucopia of relatively large, plump, juicy autumn-olive berries, a far cry from my last encounter. They turned out to be fantastic-tasting. We spent half the afternoon harvesting this choice fruit, and the rest of the time arguing over which bush was truly the best.

The fruit comes into season in late fall. It's strongly astringent at first, but it becomes sweeter as it ripens. It tastes like a wonderful combination of pomegranate, plum, currant, and raspberry—but better.

The autumn olive is delicious raw, straight from the bush, like "nectar of the gods." The berries come off the bushes in large handfuls. I pop them all into my mouth, and their sweet and sour juice bursts onto my tongue. You can spit out the small, soft seeds, although some people eat them without ill effects.

Autumn olives are also great cooked. You can simmer them over low heat, stirring constantly, until they soften and begin to stew in their own juice. Cover and simmer another 10 minutes. It's easy to strain out the seeds with a food mill, and you can also use a strainer. You're left with a wonderful-tasting if somewhat thin sauce you can use like applesauce, or spread on pancakes. If you want a smoother product, just throw it into the blender.

To transform it into a pudding or thickened sauce, reheat with your favorite thickener. I prefer kuzu and arrowroot, but you can use flour or tapioca. Agar, a vegetarian alternative to gelatin, gives you a jam (page 288). The Japanese also make a liqueur by soaking autumn olives in wine for six months.

HAZELNUTS, WILD FILBERTS

(*Corylus* species)

Hazelnuts come from spreading, smooth-barked shrubs or small trees that grow from 10 to

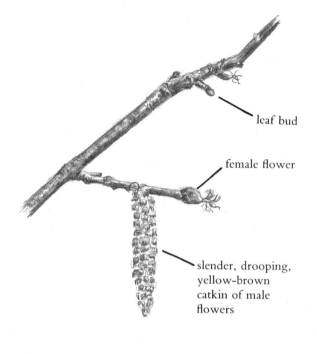

leaf bud

female flower

slender, drooping,
yellow-brown
catkin of male
flowers

BEAKED HAZELNUT IN FLOWER × 1

20 feet tall. They have alternate, simple, toothed or double-toothed, rounded to nearly heart-shaped leaves, 2 to 6 inches long.

You'll find them growing in dense stands, easiest to locate in early spring, when the slender, drooping, yellow-brown male catkins, 1 to 4 inches long, appear before the leaves unfurl. The petalless female flowers, only ¼ inch across, are harder to spot. Look for clusters of bright-red pistils projecting in all directions from the ends of their buds.

We have two species, the American hazelnut (*Corylus americana*), and the beaked hazelnut (*Corylus cornuta*). The American hazelnut's twigs and leafstalks are covered with rough hairs. Beaked hazelnuts' twigs are usually smooth. American hazelnuts' nuts have husks open at one end, while the beaked hazelnut has an extended, beaklike husk completely surrounding the nut. Both species' husks are very bristly.

The American hazelnut grows only in eastern North America, but not in the Deep South. Look

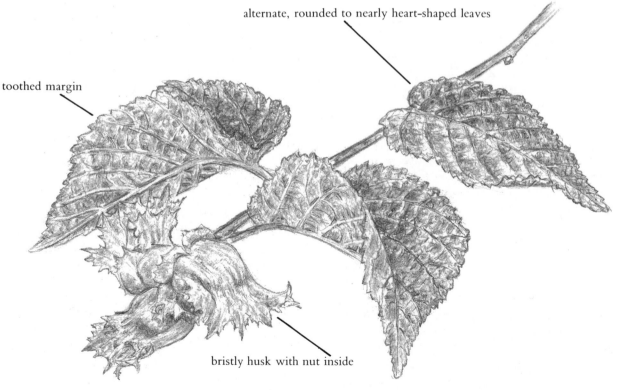

alternate, rounded to nearly heart-shaped leaves

toothed margin

bristly husk with nut inside

AMERICAN HAZELNUT × ⅗

bristly, extended, beaklike husk
completely surrounding nut

BEAKED HAZELNUTS × 1

for it in thickets and dry to moist woodlands. The beaked hazelnut grows in the northern United States and southern Canada, from coast to coast, extending south only in mountains. Look for it in dry or moist woodlands, hillsides, and thickets, and on mountain slopes.

Harvest the nuts in late summer or early fall. Collect them a little underripe, before the squirrels and birds get them, and let them ripen in storage. The nuts are ripe when they're brown and the sheath starts to dry. Handle bristly American hazelnuts with gloves, or they may irritate your hands.

Both species have thin-shelled nuts inside the husks. They taste even better than the closely related European filbert. You use them as you would any nuts, raw or cooked. Grind them in the food processor or blender. They make especially good meal that you can use with flour in pie crusts, pastries, and biscuits.

ROSES

(Rosa species)

Roses are easy to distinguish from other plants, with no poisonous look-alikes. They may be long and climbing, or trailing woody canes or shrubs. Roses boast the fiercest of thorns. They'll catch you the second you lower your guard. They're curved and barbed, and although a few species are thornless, the vast majority have plenty.

Roses have alternate, feather-compound leaves, with three to eleven small, toothed, oval leaflets on each leaf. The flower is radially symmetrical, with five petals, five sepals, and many stamens and pistils. Size and color vary: I've seen red, pink, white,

and yellow wild roses—all with the well-known, attractive fragrance.

The size of the red or orange rose hip (fruit) varies according to species. Hard and green at first, they usually get softer and fleshier in late summer or autumn, as they ripen. Some varieties are always tiny and hard—all skin and seeds—unsuitable for human consumption. Some species' fruit persists into the winter, providing a dependable trail nibble tenderized by frost.

Roses grow in fields, along the edges of trails, in thickets, fields, marshes, in openings in the forest, in disturbed sites, and at the seashore. You can find them throughout North America. Telling one wild rose species from another isn't easy. With the bees cross-pollinating the flowers, many new types are bound to occur. People also breed new varieties, confusing even botanists.

The multiflora rose is a Japanese import (*Rosa multiflora*) growing throughout eastern North America. It has white flowers with protruding stamens in the center, and small, abundantly growing rose hips with little flesh to them. It trails along the ground, forming dense cover for wildlife, and providing a favorite winter food for mockingbirds.

My favorite species is the wrinkled rose (*Rosa rugosa*). Dense bristles and thorns cover its upper branches. The dark-green leaves, composed of five to nine leaflets, are strongly wrinkled. The flowers are exceptionally large, 3 to 4 inches wide, with narrow, spoon-shaped, white or rose-purple petals. It has large, fleshy fruit, crowned with sepals,

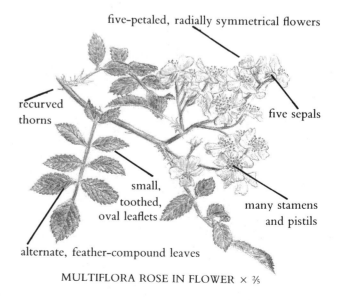

five-petaled, radially symmetrical flowers

recurved thorns

five sepals

small, toothed, oval leaflets

many stamens and pistils

alternate, feather-compound leaves

MULTIFLORA ROSE IN FLOWER × ⅖

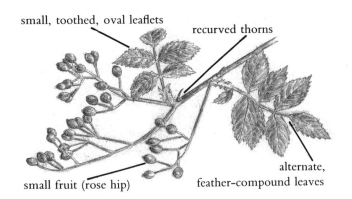

small, toothed, oval leaflets

recurved thorns

small fruit (rose hip)

alternate, feather-compound leaves

MULTIFLORA ROSE IN FRUIT × ⅖

and with a smooth, green, floral tube or hypanthium beneath.

It grows in shrublike thickets, up to 6 feet tall, on East Coast sand dunes, in the Great Lakes area, and occasionally on roadsides. It's sometimes planted in front of buildings, and in cultivated parks, throughout the United States.

The pasture rose or wild rose (*Rosa carolina*) is a low shrub, usually 2 but no more than 5 feet tall, with slender, arching green branches. The atypical thorns stick straight out, like needles. The leaves are composed of three to seven (rarely nine) coarsely toothed leaflets.

Solitary pink flowers, with lance-shaped petals, usually bloom in late spring or early summer. They give way to tiny, oval, seedy rose hips, the size of dried peas, good only for a trail nibble or tea. They ripen in late fall, often persisting through the winter. This species grows at the edges of fields and woodlands, at the seashore, throughout most of the eastern half of the United States.

The California rose (*Rosa californica*) grows over 9 feet tall, with stiff, flattened spines. The leaves have five to seven sharp-toothed leaflets. The flowers are white or pink, growing in few-flowered clusters in spring or summer. The oval or globular fruit has a distinct neck. It's a decent size, with some flesh to it, although not as large as the wrinkled rose's. The shrub grows in marshes, near rivers and streams, and in moist canyons in California and southern Oregon.

The sweetbrier rose (*Rosa eglanteria*) is over 9 feet tall with large, coarse, arching branches, covered with stiff bristles and broad, stiff spines. The hairy, glandular, double-toothed leaves have seven to nine leaflets. They smell like apples when you crush them, and their margins are glandular. The flowers, which bloom from mid-spring to mid-summer, are pink—solitary or in few-flowered clusters. The elongated fruits are fairly large. The shrub grows in fields, abandoned pastures, and woodland edges throughout the country.

To most of us, the rose was put on the earth solely for its decorative, fragrant flower. Rose flowers are indeed lovely, and receiving one from another person is a meaningful romantic symbol. But the rose's purpose goes much deeper. At one time or another, virtually all civilizations around the world have recognized the rose, in all its aspects, as an important symbol. It's been used as a healing medicine, a food, a cosmetic, a fine liquor, a potpourri, and a love potion.

The best rose hips are the largest. They taste like a combination of apricot and persimmon, with flavor more like the former, and the texture of the latter. Cut them in half and scoop out the hard, white seeds. They're not poisonous—even rich in vitamin E—but unpleasantly bitter. You can also cook whole rose hips in fruit juice, and strain out the seeds with a food mill.

If you puree the cooked, strained fruit in a blender or food processor, its pectin will thicken it to applesauce consistency. You can enjoy this wonderful food as is, or use it in puddings, cakes,

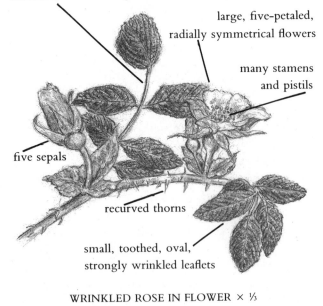

alternate, feather-compound leaves

large, five-petaled, radially symmetrical flowers

many stamens and pistils

five sepals

recurved thorns

small, toothed, oval, strongly wrinkled leaflets

WRINKLED ROSE IN FLOWER × ⅓

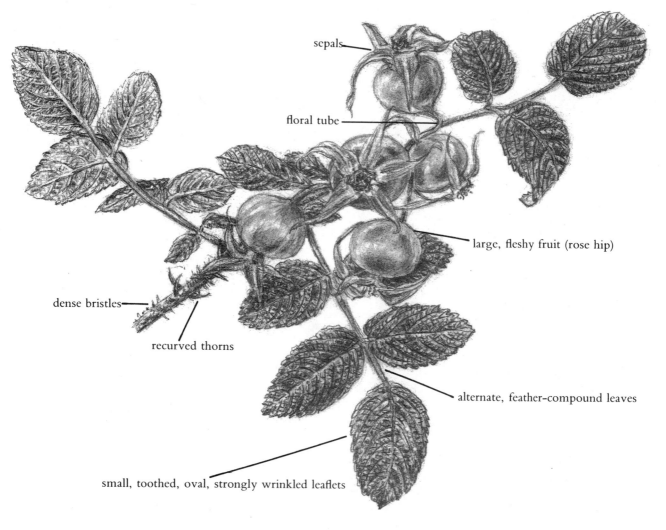

sepals

floral tube

large, fleshy fruit (rose hip)

dense bristles

recurved thorns

alternate, feather-compound leaves

small, toothed, oval, strongly wrinkled leaflets

WRINKLED ROSE IN FRUIT × ¾

cookies, and so on. I love using rose-hip sauce as a base for fruit soup. You also can extend rose-hip sauce by mixing it with applesauce or pear sauce. Sprinkle some roasted almonds on top.

Rose hips are also great raw. Eating them straight from the bush gives you a special feeling, along with an extraordinary amount of vitamin C and bioflavonoids. They also provide beta carotene, vitamins B_3, D, and E, pectin, malic and citric acids, fructose, sucrose, and zinc.

Rose hips were used both cooked and raw, as food and medicine, before the word *vitamin* was coined. It was used to treat scurvy, infections, chest congestion, and bladder problems. It also helps combat stress. Its astringency makes the fruit good for diarrhea. It's also good for the stomach: People used to eat it with fatty and fried foods, because it promotes digestion.

You can even eat rose petals. Just don't use the store-bought ones—who knows what they've been sprayed with? Different varieties of wild roses bloom from spring through most of the summer.

I've seen many elaborate recipes for rose petals—rose-petal jams, rose-petal muffins, candied rose petals, rose-petal syrup—unfortunately, all requiring lots of unhealthful sugar or honey (rose petals are somewhat bitter). Nevertheless, a few petals in a salad or sandwich adds a refreshing lift. I even garnish quiches with them. They're best as minor ingredients. Make sure to snip off the petals' bitter white base.

The Romans lavishly decorated banquets with roses, floated petals in their wine, and wore rose garlands around their heads to prevent drunkenness. Roses were associated with Venus and Cupid, goddess and god of love and romance respectively.

Heating the flowers to release the volatile fragrance, and condensing the steam into a liquid, makes rose water. To make your own, use a deep glass or enamel pot with a rounded lid. Add one quart of hot water and one pound of rose petals. Place a small glass or enamel bowl, with an ice cube inside, on a rack in the pot. Make sure the hot water isn't touching the bowl. Cover the pot with the rounded lid upside down and fill it with ice water. This is your still. Adjust the heat under the pot to medium. Water will mix with the rose petals, turn into steam, strike the cold lid, condense, flow down the lid's center, and drop into the smaller bowl. The ice cube cools it, so the volatile oils don't get lost. (When the ice cube melts, slip in a new bowl with another ice cube.) Don't take more than one cup of rose water from each batch of roses, or the flavor will get too weak.

Rose water was used in medicine and cosmetics throughout the world. It's refreshingly astringent, tightening the skin and closing the pores, while perfuming the user. People used its aroma in place of smelling salts. It's added to lotions and to cold cream, which was once called ointment of rose water.

Essential oil of the rose, sometimes used in the most expensive perfumes and cosmetics, requires thirty flowers for one drop. It's soothing and cooling: A little rubbed on the temples is good for headaches, and relaxing for an overstrained mind. When actress Rita Heyworth married playboy Aly Khan in the 1950s, Aristotle Onassis dropped four hundred roses onto a shallow reflecting pool, as a congratulatory gesture. Oil released from the petals separated and floated to the surface. Servants pulled flannel across the pool to recover this valuable substance. With enough roses and warm water, you may be able to reproduce this process without a wedding or an airplane.

Chinese herbalists use rose flower to "regulate vital energy (qi)," for poor circulation, stomachaches, liver pains, mastitis, dysentery, vaginal yeast infections, and pregnancy.

OTHER PLANTS OF THICKETS IN AUTUMN

Edible or Medicinal Plants

Bayberries, chinquapins, elderberry berries, grapes, groundnut pods and roots, hawthorn berries, hog peanut pods and roots, Jerusalem artichoke roots, mugwort, nettles, passionflower fruits, pears, plums, raspberry leaves, rose hips, sassafras leaves and roots, spiceberry leaves and berries, sumac berries, viburnum berries, wild beans, wild potato vine roots, yew berries.

For Observation Only

Hercules'-club berries, hog peanut flowers, Japanese knotweed seeds, poison ivy berries, pokeweed berries, wild lettuce seedheads.

PLANTS OF WOODLANDS IN AUTUMN

BLACK WALNUT

(Juglans nigra)

The black walnut tree is a tall, majestic denizen of the forest, 50 to 120 feet tall and 1 to 4 feet in diameter. The trunk's dark-brown bark is ropy-looking—deeply furrowed, with flattened ridges.

The elegant, feather-compound leaves are 1 to 2 feet long. Each leaf has twelve to twenty-four narrow, finely toothed, lance-shaped, unevenly paired leaflets 3½ inches long, with asymmetrical bases. The leaves are pale green above, downy beneath, and lemon-scented when crushed. Unlike the related butternut tree (page 163), there is often no individual terminal leaflet at the tip of each leaf.

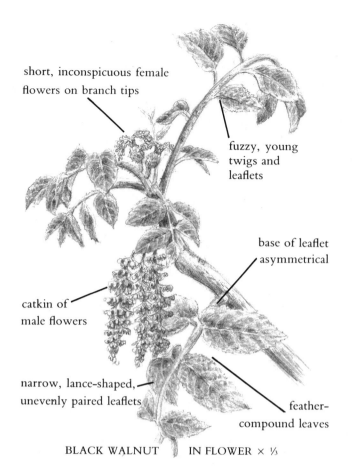

short, inconspicuous female flowers on branch tips

fuzzy, young twigs and leaflets

base of leaflet asymmetrical

catkin of male flowers

narrow, lance-shaped, unevenly paired leaflets

feather-compound leaves

BLACK WALNUT IN FLOWER × ⅓

Black walnuts are supposed to grow in rich, deciduous woods, bottomlands, and floodplains, especially in well-drained soil, but I also find them along residential streets and in cultivated parks. They grow in the eastern United States, the South, and through parts of the Great Plains. Very similar relatives, the California walnut (*Juglans californica*) and Arizona walnut (*Juglans major*), grow along streams and washes in the Southwest and southern California. Hind's walnut (*Juglans hindsii*) grows only in central California. They all bear edible walnuts.

Like all of us, black walnuts sometimes have bad years. Preparing for a nature tour, I noticed all the trees along our route weren't bearing. Yet when the group arrived, people raved about the great abundance of black walnuts. A friend and I had loaded our packs to the hilt with walnuts from a different area, and scattered them under the barren trees just before the tour began.

Unfortunately, greedy people chop down black walnut trees whenever they find them because the wood commands a high price. Each tree is worth thousands of dollars. Of course, these individuals can't be bothered to plant new trees, so black walnut trees are becoming rare wherever they're not protected.

Harvest the nuts wearing old shoes and rubber gloves. Stomp on the fruits on a concrete path, pavement, or large, flat rock to remove the fleshy, green husks. Unless you're using them for medicine or dye, the husks will weigh you down. Running over the fruits with a car also removes the husks. I have no car, but according to a friend who does, this creates a hairline fracture, and the nuts will eventually rot.

Handle the dehusked nuts with rubber gloves, or they'll stain your hands brown for weeks.

The tree flowers in spring, as the leaves are developing. The male flowers hang in slender catkins, and the short, inconspicuous female flowers grow at the tips of the branches.

The green, spherical, lemon-scented fruit, in season in autumn, is about the size of a tennis ball—up to 2½ inches in diameter. Inside is an irregularly furrowed, globular nut, similar to a commercial walnut, 1½ inches in diameter. The similar butternut has an oblong fruit, while hickory nuts are much smaller, with husks that partially divide into four segments. There are no poisonous look-alikes.

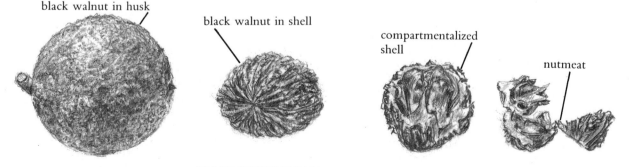

black walnut in husk

black walnut in shell

compartmentalized shell

nutmeat

BLACK WALNUT IN HUSKS × ⅕, × ⅕, × ⅕

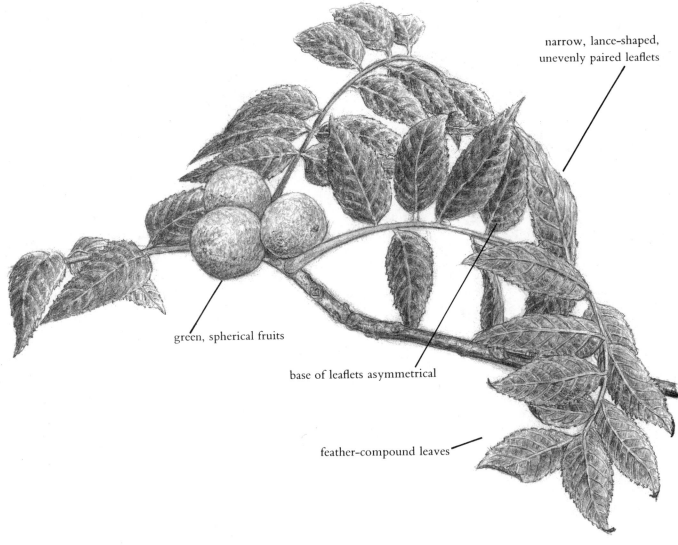

narrow, lance-shaped, unevenly paired leaflets

green, spherical fruits

base of leaflets asymmetrical

feather-compound leaves

BLACK WALNUT TREE WITH NUTS × ⅖

(Applying lemon juice as soon as possible helps remove the stain.) If you don't dehusk the nuts, insect larvae will bore into the nutmeat. Walnut husks make a well-known dye. Use them as shoe polish.

Spread the husked nuts out on newspapers in a warm, well-ventilated area, away from squirrels, to dry. Even better, place them on cookie sheets in the oven, on the lowest possible setting, with the door ajar so the moisture can escape. If you don't dry them, they become moldy. Drying helps mature the nuts and improves their quality. When dry, they can no longer discolor your hands.

You can store dry nuts in their shells in paper shopping bags, sealed with masking tape. Include lots of wild bayberry or commercial bay leaves, to repel insects. Or crack open the dry nuts at once. Use them immediately, store in the refrigerator for a few weeks, or freeze for a few months. Any longer and the oils will go rancid.

Black walnuts have very hard shells. Here are three ways to crack them:

1. Use a special black-walnut cracker, available through C E Potter, Sapula, Oklahoma 74066, (918) 224-0567. This handy device cracks the nuts evenly, without splintering the shells.

2. Use a vise, covering the nut with an old towel or clean rag so the shell slivers don't fly across the room. Insert the nut so the vise is touching the tip and base, not the sides. This isn't as good as the Potter contraption, but it works. You can purchase a vise in any hardware store.

3. Stand the nut with the pointed end up, cover it with an old towel or clean rag, and whack it with a heavy hammer. Repeat this process once or twice with the fragments, to get at the most nutmeat. This is the most labor-intensive method, but it works. The best way, of course, is to get someone else to do the work for you.

Once you've cracked open your black walnuts, use nutpicks, also called lobster picks, available in hardware stores, to remove the nutmeat from the shells.

Black walnuts taste like commercial (English) walnuts, only better. The flavor is fruitier—stronger and deeper. A little goes a long way in any recipe that calls for nuts. I usually use 1:4 parts black walnut to commercial walnuts, so the wild nuts don't overpower all the other ingredients.

Eat black walnuts right out of the shell; include them in pies, cakes, granolas, loaves, casseroles, or virtually any main course or dessert. You can also grind them into meal in the food processor or blender (in the blender, grind one-half cup at a time), and use in small amounts with whole-grain flour in baking. You can even turn black walnuts into a delicious, rich nut butter in the food processor by continuing to run the machine after you've made walnut meal.

You can tap into a black-walnut tree in early spring, and boil down the sap to make a syrup similar to maple syrup. However, this concentrated, highly processed sugar is as unhealthful as white sugar.

You can even pickle the immature nuts that fall in the summer if they're still soft enough to cut in half with a knife: Scald them by plunging briefly into boiling water, then remove the fuzz. Next, boil in numerous changes of water until the water no longer discolors and you've gotten rid of all the bitter-tasting tannin. Pack the nuts into jars with vinegar and pickling spices, let them sit a month, and they're ready to eat. Many people consider them a delicacy.

Black walnuts contain beta carotene, thiamin, riboflavin, niacin, iron, magnesium, potassium, protein, carbohydrates, and the essential fatty acids linolenic and linoleic acids. Black-walnut oil, which you can get by boiling the shelled nuts and skimming the oil from the surface of the water, is great mixed with milder oils in salad dressing, or for cooking. This oil helps alleviate skin diseases.

Black walnuts also contain alkaloids with demonstrated anticancer properties. Unfortunately, studies at the University of Missouri that led to this discovery were prematurely discontinued.

A decoction of boiled husks, with its high tannin content and juglandin, juglone, and juglandic acid, kills parasites. A compress successfully treats athlete's foot, ringworm, and other fungal infections of the skin.

People used to make beautiful buttons out of the nutshells by running the nuts through a ribbon saw, and there are still other uses for black walnuts. One day a man and woman arrived, extremely agitated. "Please help us," they said. "We got a new cat, and it has fleas! Now the fleas are all over the house—they're jumping on the carpet, they're jumping on the furniture, and now they're trying to jump on us! We heard black walnuts repel fleas. You're the only one who can help us. Please find us black walnuts." It happened to be black-walnut season, so we drove to a nearby park, and the couple grabbed every nut they could find.

I didn't hear from them again until the following autumn, when I received another desperate phone call: "Something terrible has happened! The black walnuts we scattered all over our house got rid of the fleas completely. But this year, when we left our cat in a kennel, he got fleas from the other cats. Now the fleas are jumping all over our home again! Can you help us find more black walnuts?"

BUTTERNUT

(Juglans cinerea)

This is a medium-sized tree that grows up to 80 feet tall. It has feather-compound leaves 15 to 30 inches long, with seven to nineteen paired, narrow, toothed, lance-shaped leaflets 2 to 4 inches long. Unlike its relative the black walnut (page 160), there's usually a single terminal leaflet at the leaf's tip. The gray bark has shallow grooves and comparatively broad, wavy ridges.

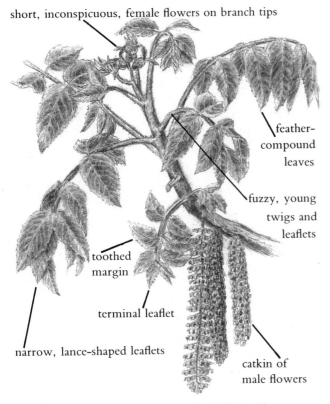

short, inconspicuous, female flowers on branch tips

feather-compound leaves

fuzzy, young twigs and leaflets

toothed margin

terminal leaflet

narrow, lance-shaped leaflets

catkin of male flowers

BUTTERNUT TREE IN FLOWER × ½

Like the black walnut, this native tree flowers in spring when the leaves are developing. Male flowers hang in slender catkins, while short, inconspicuous female flowers grow at the tips of the branches.

The fruit is oblong, green, fragrant, fuzzy, and sticky—about 3 inches long and 1¼ inches wide. The nutshell inside is deeply furrowed. It ripens

toothed margin

feather-compound leaves

narrow, lance-shaped leaflets

leading leaflet

fruitstalks

oblong, green, sticky fruit

BUTTERNUT TREE WITH NUTS × ⅙

and drops in late summer or early autumn. There are no poisonous look-alikes, and it's easy to distinguish from other nuts: Black walnuts and hickory nuts are round. Hickories are smaller, with seams that partially divide their husks into four parts.

Butternut trees grow in rich, deciduous forests throughout the eastern United States and Canada, especially in the South. They don't bear fruit every year in the North, and there's a fungus that is slowly killing many trees, but when they do bear, it's a bonanza.

After you collect butternuts, don't remove the husks. Dry the nuts in a roasting pan in the oven on the lowest possible setting, with the door ajar. The husks shrink to nothing, and the nuts ripen, becoming slightly pungent and crunchy. Unripe nuts taste like slightly bitter, limp green peas. The undried husks blacken and the nuts become insect-infested after one to two weeks. Cut undried husks will also stain your hands. Confederate soldiers were called butternuts because their uniforms' gray color came from butternut husks.

Crack the hard shell as described under black walnuts. if the butternut is too long to fit into the heavy-duty nutcracker, clip off the tip with pruning shears.

Butternuts are as nutritious as they are tasty. They provide protein, iron, magnesium, and linolenic and linoleic oils (essential oils your body requires but cannot make). They have the lowest carbohydrate content of all nuts, only 3 percent, ideal for people on a low-carbohydrate diet.

The tree gets its name because the Indians roasted the nuts, boiled the crushed nutmeat, and skimmed off the oil that rose to the surface, which they used like butter.

Butternuts taste so good, it's a wonder they were never marketed. Perhaps butternuts and black walnuts never attained mass appeal because shelling them by hand is so labor-intensive, and by the time machine-shelled nuts became available, popular tastes were already fixed.

I use butternuts for every kind of nut recipe, from pies to granola, raw or roasted. You can even grind them into meal in the food processor or blender and use it in breads, pie crusts, and pastries, along with whole-grain flour. Store butternuts the same as black walnuts.

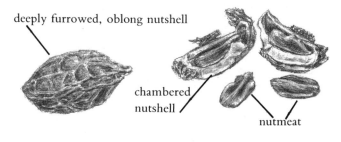

deeply furrowed, oblong nutshell

chambered nutshell

nutmeat

BUTTERNUT × ½

Native Americans used a decoction of the butternut's inner bark as a laxative and cathartic, and for liver congestion. Low doses make a safe, effective laxative that doesn't cause cramps—very popular in the nineteenth century, before it went out of style and fell into obscurity.

A compress of a strong, warm decoction of the inner bark and roots helps stop bleeding from wounds, and the decoction and the nut's oil have been used for tapeworms and fungal infections. Various parts of the tree contain juglone, which is antiseptic and herbicidal, and shows some antitumor activity.

GRAPES

(*Vitis* species)

Grapes are high-climbing vines that have been cultivated for over seven thousand years. The bark is brown and woody, with vertical strips flaking off. The leaves are maplelike, 2 to 8 inches across, partially divided into three lobes, with dense whitish to reddish felt underneath. The vine has forked tendrils, which attach to trees and fences. In the spring, there are long clusters of tiny, fragrant, inconspicuous yellow-green five-petaled flowers.

Ripe wild grapes are usually dark purple, occasionally slightly greenish or reddish, in groups of about twenty. They most closely resemble commercial Concord grapes. The fox grape (*Vitis labrusca*), the largest and tastiest wild species, is actually the Concord grape's wild ancestor. It gets its name from Aesop's fable of the fox and the grapes: When the fox couldn't reach the grapes, he

rationalized that they must be sour anyway, which is where we get the expression "sour grapes." Fox grapes are sweet and tasty, like Concord grapes, but they sometimes grow far out of reach.

There are about fifty native grape species, plus introduced varieties and naturally occurring hybrids. Most are very similar to the fox grape, although the fruit may be smaller and more acidic.

Caution: Most people don't realize that grapes have poisonous look-alikes: Canada moonseed (*Menispermum canadense*) has leaves and fruit that closely resemble grapes, with three subtle but distinct differences: 1. Each grape has many seeds, while Canada moonseed's fruit has but one flattened, crescent-shaped seed, hence its name. 2. Canada moonseed climbs by twining, without the grapevine's tendrils. 3. Canada moonseed's leafstalk attaches to the leaf beyond the leaf's base, while the grape's leafstalk attaches at the leaf's base. The berries contain a bitter toxic alkaloid that reportedly killed a child.

Virginia creeper (*Parthenocissus quinquefolia*) is a common, attractive, low-growing or high-climbing poisonous vine that has killed people. It has tendrils, but unlike the three-lobed grape leaf, this relative has a five-parted palmate-compound leaf, with toothed, elliptical leaflets up to about 6 inches long—very different from grape leaves. The tiny, globular, blue-black berries may contain unknown poisons, plus enough oxalic acid to permanently damage kidneys, although symptoms may not appear for twenty-four hours.

Porcelainberry (*Ampelopsis brevipedunculata*) is a highly invasive Asian relative of the grape that resembles the grapevine before the fruit develops. Its beautiful berries are somewhat smaller than grapes—colored blue, purple, turquoise, or white. The vine has grapelike leaves, but never develops the grape's brown, woody vine. Also, grape stems have brown pith, and porcelainberry stems' pith is white. The leaves and berries are definitely not edible, and may be poisonous.

Grapes love open woodlands. You can find them growing on lightly wooded trails, up trees and fences, along streams and riverbanks, in thickets, in wetlands, and in thickets near the seashore. They grow throughout the continent, and they're so common in eastern North America that the Vik-

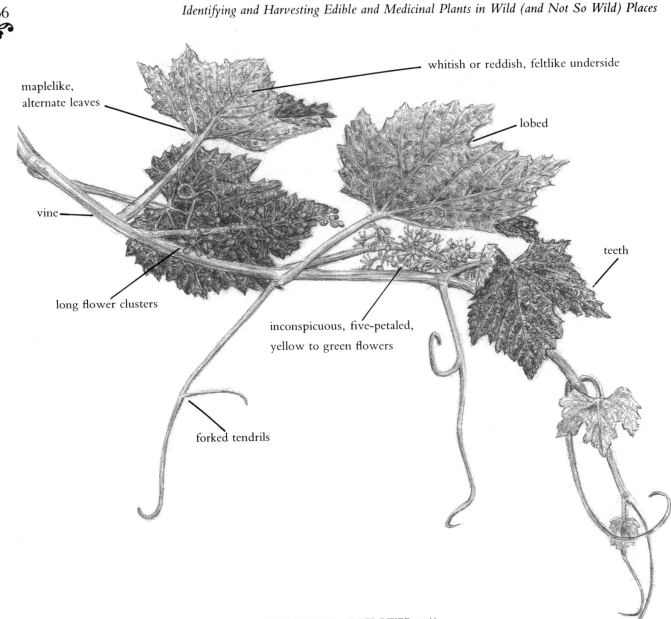

maplelike,
alternate leaves

whitish or reddish, feltlike underside

lobed

vine

teeth

long flower clusters

inconspicuous, five-petaled,
yellow to green flowers

forked tendrils

FOX GRAPE VINE IN FLOWER × ½

ing explorer, Leif Erikson, named the region Vineland.

Grapes are responsible for starting my career: One beautiful May afternoon in 1979, while bicycling past a lightly wooded area, I noticed women in the woods picking something. As a child, I had collected wild raspberries and blackberries with my mother. Now I was experimenting with ethnic cooking, so the prospect of wild fruit was enticing.

As it turned out, I had come across traditional Greek women picking grape leaves. They tried to explain what they were doing but I could barely understand them. (It was all Greek to me.) Never-

theless, they managed to convey enough for me to join them.

Commercial grape leaves are embalmed in preservatives and salt. A natural alternative was exciting. Later, I stuffed the grape leaves with a mixture of brown rice, tofu, walnuts, and raisins, seasoned with thyme and marjoram. Pressure-cooked five minutes and topped with ginger-garlic sauce, this was hardly the traditional Greek appetizer, but it tasted great. From then on, I was hooked on wild foods.

Grape leaves are good from mid-spring and early summer. Parboil them in salt water, or pickle

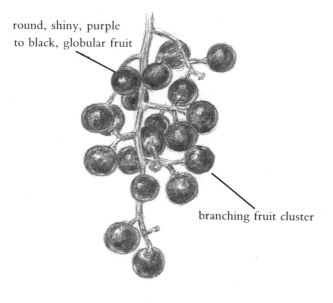

round, shiny, purple to black, globular fruit

branching fruit cluster

FOX GRAPE × ⅕

Eat the sweeter varieties raw, or put them in a blender and briefly blend on the lowest speed, to separate the seeds without breaking them. Strain out the seeds with a food mill or sieve, and use the grapes in recipes. To make a fruit sauce or jam, add grape juice for sweetening, and simmer with a thickener. Bananas, peaches, and dates also sweeten and complement grapes well.

Grapes provide potassium, beta carotene, fructose, tartaric acid, quercitrin, tannin, malic acid, gum, and potassium bitartrate. The leaves are an excellent source of beta carotene and niacin.

The fruit's a diuretic. Eating it in quantity increases the flow of urine and helps problems of water retention. Grapes are also supposed to relieve urinary-tract irritations because they decrease the urine's acidity, soothing inflamed tissues and allowing them to heal.

Resveratrol, contained in wild grapes' skin, prevents cardiovascular disease in two ways: It reduces blood platelets' tendency to stick together and form clots, and it raises high-density lipoprotein cholesterol (the good one). The wine industry is using resveratrol's benefits to increase sales, but you don't have to turn grapes into alcohol, which causes health problems, to get this substance. Japanese knotweed (page 238) also contains it.

In earlier times, Europeans squeezed juice from grape leaves to treat women for "excessive lust."

them with spices and vinegar before stuffing them. Steam medium-sized ones 15 to 20 minutes and use them as a cooked vegetable, or use them in soups and stews.

The fruit ripens in autumn, but only where there's enough sunlight. You may have to search for fruit-bearing vines. Also, some years are much better grape years than others. Wild grapes are very different from commercial varieties—less sweet, but with more flavor.

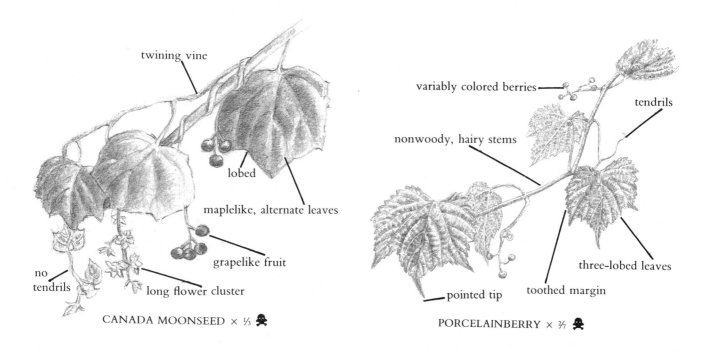

twining vine

lobed

maplelike, alternate leaves

no tendrils

grapelike fruit

long flower cluster

CANADA MOONSEED × ⅓ ☠

variably colored berries

tendrils

nonwoody, hairy stems

three-lobed leaves

pointed tip

toothed margin

PORCELAINBERRY × ⅔ ☠

Today, people in European health spas go on grape cures, where they eat nothing but grapes for days. This is supposed to detoxify the body, heal the liver, and increase the flow of bile. Grapes are very sweet, so this may not be good for diabetics or hypoglycemics. Also, don't try to detoxify using American commercial grapes. As of this writing, they're being sprayed with such dangerous insecticides that the late César Chavez's migrant workers' union is asking the public to boycott them. You're better off collecting your own wild grapes.

Cut into a grapevine for the white sap—a home remedy for weak eyes and floaters (pieces of detached tissue that interfere with vision). **Caution:** See an ophthalmologist before you put anything in your eyes.

The leaves are excellent for poultices, combined with plantain leaves (page 227), for all kinds of sores. Wrapping a bruise in fresh grape leaves is supposed to accelerate healing. Grape leaf or seed infusion is astringent, good for bleeding and diarrhea. Indians used it for hepatitis and stomachaches.

CHESTNUT

(*Castanea* species)

American chestnut bark is dark brown, with widely separated, shallow grooves. The brown, hairless twigs have a five-sided central pith. The large, spear-shaped, alternate, hairless leaves have sharp, coarse teeth that curve inward.

Fragrant, yellow-white flowers hang on long catkins in early summer. Inside each stiff, bristly husk grow two or three slightly flattened nuts that ripen in the fall.

The story of the American chestnut (*Castanea dentata*) is a sad one for nature lovers and foragers. Once upon a time, this native tree was common throughout eastern North America. It was even a street tree in New York City. Then, in the late 1800s, Asian chestnut trees were brought to this continent, along with a fungus to which they were largely resistant. The native species had no defense against it, and the blight nearly wiped them out in a few years. The trees that remained seldom grew

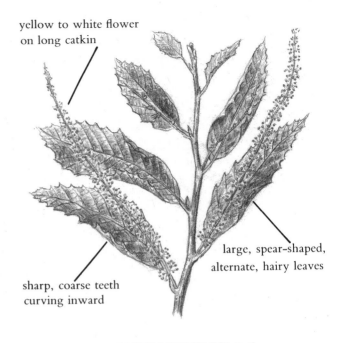

yellow to white flower on long catkin

large, spear-shaped, alternate, hairy leaves

sharp, coarse teeth curving inward

AMERICAN CHESTNUT × ½

more than 15 feet tall before the blight cut them down, and they rarely produced nuts. Now, these trees are beginning to recover. Enough healthy trees grow in the vicinity of Lehigh, New Jersey, that they're selling the nuts in the local grocery stores.

I thought I'd never taste wild chestnuts, but I was mistaken. A friend who had just moved to the mountains dropped in, with way more nuts than he could use. The nuts were smaller than commercial chestnuts, but much better-tasting. There were five foreign chestnut trees in his backyard—planted years ago and forgotten, or possibly the descen-

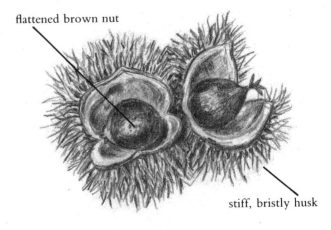

flattened brown nut

stiff, bristly husk

CHESTNUT × ⅔

dants of foreign trees. Foragers should always keep their eyes open for out-of-place wild edibles.

Abandoned cultivated chestnut trees grow in or near suburban areas, in wooded areas throughout the United States. Foraging teacher David Gould, for example, lists them growing in fifteen places within the city limits of Seattle.

The eastern chinquapin (*Castanea pumila*) is a shrub or small tree similar to the American chestnut tree, but no more than 10 to 25 feet tall. It gets its name from the Algonquin word *chincomen*, which means "large fruit." The bark is light reddish-brown, and the young twigs are densely hairy. The coarsely toothed, alternate, lance-shaped, simple leaves grow up to 5 inches long and 2 inches wide. They're whitish and hairy beneath.

Tiny, inconspicuous, beige flowers grow on long, narrow spikes in the spring. The burrlike husks are smaller than chestnuts', just under ½ inch in diameter. There's only one nut per husk, and they're not flattened.

Eastern chinquapins grow in dry woods and thickets in the Southeast, and along the East Coast north to New Jersey. They're occasionally cultivated as ornamentals throughout the country. The bush chinquapin (*Castanea sempervirens*) is a similar evergreen shrub that grows in the Northwest. Clusters of one to three nuts are each encased in a spiny burr.

Use all chestnuts and chinquapins the way you would commercial, Asian chestnuts: Remove them from their husks and cut an X in the shells with a paring knife, so they won't explode in the oven. Bake at 425°F for 15 to 25 minutes. (You can also boil them about 25 minutes.) Peel them and eat them as is, or include them in stuffing, cakes, or vegetable dishes.

Chestnuts contain lots of calories in the form of carbohydrates, plenty of potassium, moderate amounts of phosphorus and calcium, and some protein and iron. They contain less oil than any other nut, and they're easy to digest.

Caution: There are common poisonous trees planted in parks and in residential areas throughout the United States that look enough like chestnuts to tempt uninformed people to eat them. And eat them you may, but only once—you may not be around for a second helping.

The horse chestnut (*Aesculus hippocastanum*) is a large, European tree with huge, opposite, palmate-compound leaves of about seven leaflets. The large, white, showy, fragrant bilaterally symmetrical flowers grow in upright spikes in early spring, before the leaves appear. The thorny husk divides into three parts to expose a shiny, brown nut. Each nut has a large, elliptical white area that may remind you of an eye. The Ohio buckeye (*Aesculus glabra*) is a similar-looking poisonous native species, as are the sweet buckeye (*Aesculus octandra*) and the California buckeye (*Aesculus californica*). Their nuts have thornless husks.

All parts of these trees contain alkaloids, glyco-

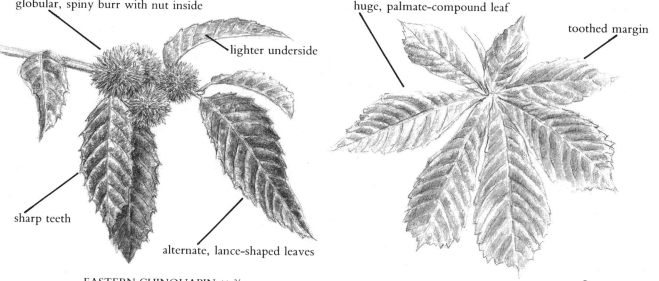

globular, spiny burr with nut inside

lighter underside

sharp teeth

alternate, lance-shaped leaves

EASTERN CHINQUAPIN × ⅖

huge, palmate-compound leaf

toothed margin

HORSE CHESTNUT LEAF × ¼

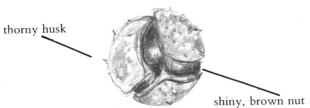

thorny husk

shiny, brown nut

HORSE CHESTNUT × ½ ☠

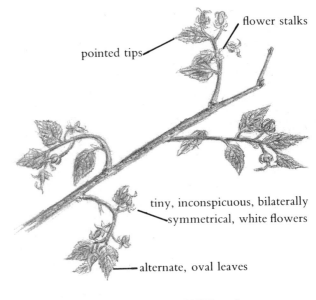

flower stalks

pointed tips

tiny, inconspicuous, bilaterally symmetrical, white flowers

alternate, oval leaves

HACKBERRY IN FLOWER × 1

sides (especially aesculin), and saponins. They've reportedly killed children in Europe. American Indians found ways to render them safe, involving days of outdoor processing—hardly worth the effort.

If you're lost and starving, and you don't know any edible wild plants, you can crush horse chestnuts and throw them into a lake, as the Indians did. The poison will stun the fish, and you'll be able to catch them with your bare hands. This unsportsmanlike form of "fishing" is unfair to the fish and completely illegal, but you'll be able to eat the fish without poisoning yourself.

Children in England have discovered the best way to use horse chestnuts and buckeyes: Drill a hole into the nut and put a string through it, tied at the end, to make a "conker." Take turns bashing it into your friends' conkers until yours is the only one with a string that hasn't broken. Then you're the "conker-er."

❧

HACKBERRY

(Celtis species)

The hackberries are medium-sized trees (although sometimes an individual can grow quite tall) with smooth, gray bark. They're the most bedraggled-looking of all American trees, with knobby insect galls in the leaves, and a nonfatal fungus infection—carried by mites and related to Dutch-elm disease—called witches'-broom. The deformed twigs messily point in all directions. Never admitting defeat, it still produces its delicious fruit.

The distinctive simple, alternate, oval, toothed leaves are 3 to 5 inches long. Their tips are long and pointed, with uneven bases. The tiny, white,

inconspicuous flowers, which appear in mid-spring, are bilaterally symmetrical. I never even noticed them until I was able to recognize the tree without fruit. Fortunately, the tree turns out to be one of the easiest to identify in winter. Some of the rotten berries stay on the twigs.

The berry varies from orange-brown to dark purple. (The desert hackberry's fruit is yellow to orange.) It's round and small, about ⅓ inch across, with a large, round seed inside. The flesh is thin, sweet, and dry. The fruit is solitary, hanging on a long slender stalk. Anything you might confuse with hackberries has clustered fruit.

There are no poisonous look-alikes. The related sugarberry (*Celtis laevigata*), which grows in the South, and the upland hackberry (*Celtis tenuifolia*) are similar to the American hackberry, but with toothless leaves. The desert hackberry's (*Celtis pallida*) evergreen leaves are toothed only from the middle to the tip, and the twigs are spiny. All have edible berries. The different varieties, which can hybridize, may not be distinct species after all.

I've found hackberry trees in city parks, in backyards, in empty lots, in the woods, and on floodplains. They're plentiful throughout North America. The desert hackberry, of course, grows in the Southwest deserts.

The berries ripen in autumn. They're a tasty trail nibble. The sweet flavor and crunchy texture reminds me of M&M's. It's too bad there's so little

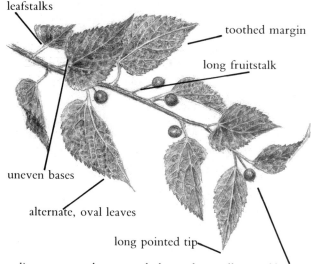

leafstalks

toothed margin

long fruitstalk

uneven bases

alternate, oval leaves

long pointed tip

solitary, orange-brown to dark purple, small, round berries

HACKBERRY TREE WITH FRUIT × ⅖

flesh on each berry. If you have the patience to collect them for an hour or so, you can cook them in fruit juice or water, strain out the seeds, sweeten to taste, and make an excellent if hard-won sweet fruit sauce or syrup.

One of the first human ancestor fossils ever unearthed was Java Man, now known to be *Homo erectus*. Piles of Asian hackberry seeds were found in his refuse heaps, indicating that our predecessor ate this fruit.

Hackberry bark or leaf infusion is supposed to make a good gargle for sore throats and a pain reliever, and the berries were used for dysentery.

HICKORIES

(*Carya* species)

Hickories are tall, common trees of rich, open woods and bottomlands. Like the related black walnut and butternut (*Juglans* species—pages 160 and 163), they have alternate, feather-compound leaves with opposite, lance-shaped, toothed leaflets. (*Carya* is "walnut" in Greek.) However, hickory leaves are usually smaller than those of *Juglans*, with only five to seven leaflets.

The trees flower in early spring, as the leaves are developing. The male flowers hang in slender catkins, and the short, inconspicuous female flowers grow at the tips of the branches.

These native trees are major components of the northeastern forests, with various species growing throughout eastern North America.

The nuts are elliptical to almost spherical, ½ to 1½ inches long, with husks that partially split into four parts when mature. Hickory trees grow slowly, and may not bear for 80 years. Fortunately, they can live for 250 years.

The best hickory species is the easy-to-recognize shagbark hickory (*Carya ovata*). The tree looks shaggy because the bark peels in striking, thin, vertical strips. The leaves—8 to 14 inches long—usually have five to seven leaflets. The thick-husked, oval nuts are delicious. Shagbark hickories grow on dry, upland slopes, and on well-drained soil in lowlands and valleys.

The mockernut hickory (*Carya tomentosa*) has tight, dark-gray, furrowed bark. The leaves are 8 to 15 inches long, with seven to nine leaflets. Their undersides and the twigs are densely woolly.

Although the mockernut's labyrinthine shell mocks you when you try to remove the tasty nutmeat, its name refers to the hard shell: *Moker noot*, means "heavy hammer nut" in Dutch.

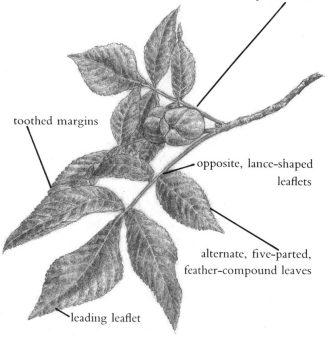

green, roundish, four-parted husk

toothed margins

opposite, lance-shaped leaflets

alternate, five-parted, feather-compound leaves

leading leaflet

SHAGBARK HICKORY TREE WITH NUTS × ¼

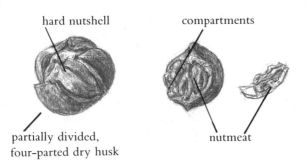

hard nutshell

compartments

partially divided,
four-parted dry husk

nutmeat

HICKORY NUT × ⅕

The pignut hickory or butternut hickory (*Carya cordiformis*) has tight, rough, grayish bark. There are five to nine leaflets per leaf. Only the bottom half of the husk splits open. The gray nutshell is heart-shaped. One taste is all you need to identify it. The nut is delicious, but only if you're a pig—it's much too bitter for people. The late Euell Gibbons, author and foraging guru of the 1960s and 1970s, wrote that you can occasionally find a pignut tree with good-tasting nuts, but I think their locations all died with him.

The well-known pecan (*Carya illinoensis*) is another hickory species. This tall tree has dark, reddish-brown bark with vertical ridges. The alternate leaves are 12 to 30 inches long with nine to seventeen lance-shaped, glossy unsymmetrical leaflets with recurved tips. The nuts usually grow in clusters of three to six. They're oblong, with smooth, thin shells, 1½ to 2 inches long, the same as commercial pecans. They grow in bottomlands, and in rich, moist soil. They're often cultivated. They grow in the Southern Mississippi Basin and the Southeast.

No hickory species is poisonous, and there are no poisonous look-alikes, but quality also varies greatly from tree to tree. The nuts are in season in early fall, and you have to get them before the squirrels and chipmunks do.

Store the nuts in paper bags until the husks dry and open, releasing mature nuts. You can store the nuts in their shells indefinitely. To crack the hard shells, use the same methods recommended for black walnuts (page 161). Once you've shelled them, the oils can react with oxygen in the atmosphere and go rancid. Store them in closed containers in the refrigerator. If you don't use them within a few weeks, you can store them in the freezer for a few months. Use them like any other nuts, raw, roasted, in pies, cakes, granola, stuffing, ground

into flour, or any way you can imagine. A friend makes seasoning by grinding the roasted nuts into a coarse meal and mixing this with dry, powdered seaweed, powdered wild greens, and salt to taste. She sprinkles it on dishes like brown rice and cooked vegetables.

Hickory nuts are very nutritious. They're very high in calories because two thirds of the shelled nut's weight is a high-quality, easy to digest oil, including the high-quality essential fatty acids shown to prevent heart disease. Hickory nuts also provide protein, carbohydrates, iron, phosphorus, potassium, trace minerals, and vitamins A and C.

School days, school days,
Dear old Golden Rule days.
Readin', and Writin', and 'Rithmetic,
Taught to the tune of a hick'ry stick . . .

If this song sounds familiar, you'll realize that the hickory apparently provided instruments of child abuse for authoritarian schoolteachers of bygone days. But the tree has other uses: You can make a maplelike syrup by tapping into the trees in early spring and boiling down the sap. The wood is prized for outdoor cooking and smoking foods, although smoked foods are carcinogenic. The leaves and green hulls make highly valued tan, brown, and blackish dyes. The wood is strong, tough, and elastic, although it decays if you repeatedly let it get wet. It's used for making tool handles. Its qualities were so admired that President Andrew Jackson's nickname was "Old Hickory." And schoolteachers no longer use it to hit kids.

oblong, smooth, thin-shelled nuts

PECANS × ⅕

inconspicuous female flowers

alternate-feather compound leaves

recurved leaf tips

toothed margins

dark red to brown bark

9 to 17 lance-shaped, glossy, unsymmetrical leaflets

PECAN IN FLOWER × ⅚

OAK

(*Quercus* species)

Oaks are common, usually large trees, with alternate, simple leaves—often lobed. In early spring, long, narrow clusters of tiny, yellow flowers hang from the twigs, on drooping catkins. They release large amounts of highly allergenic pollen into the wind while the immature leaves are still too small to impede pollination. Inconspicuous female flowers are much smaller.

All oaks bear acorns—oval, thin-shelled nuts partially surrounded by cups. Acorns have no poi-

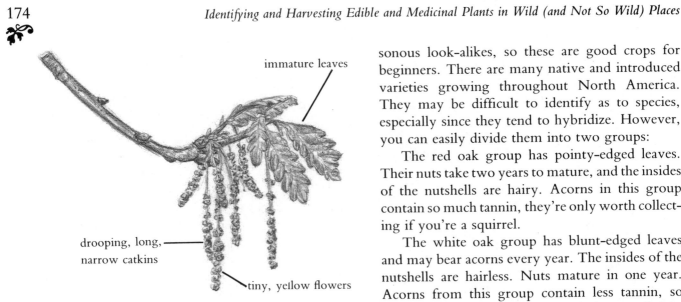

immature leaves

drooping, long,
narrow catkins

tiny, yellow flowers

WHITE OAK × ⅕

sonous look-alikes, so these are good crops for beginners. There are many native and introduced varieties growing throughout North America. They may be difficult to identify as to species, especially since they tend to hybridize. However, you can easily divide them into two groups:

The red oak group has pointy-edged leaves. Their nuts take two years to mature, and the insides of the nutshells are hairy. Acorns in this group contain so much tannin, they're only worth collecting if you're a squirrel.

The white oak group has blunt-edged leaves and may bear acorns every year. The insides of the nutshells are hairless. Nuts mature in one year. Acorns from this group contain less tannin, so they're good to use. The nuts with the most lightly

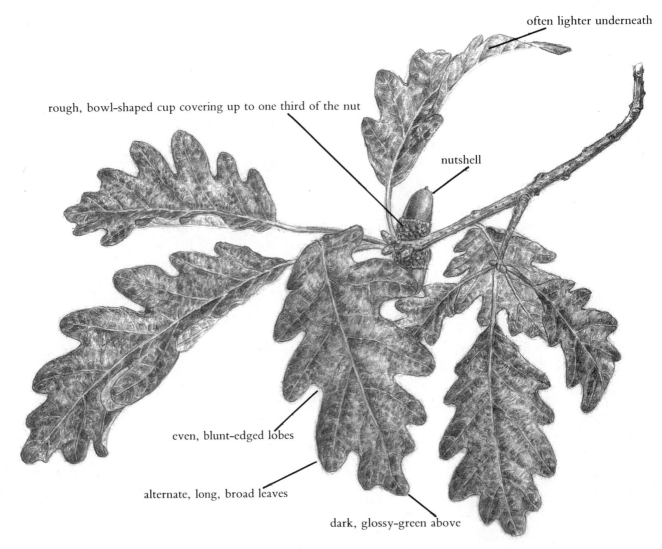

often lighter underneath

rough, bowl-shaped cup covering up to one third of the nut

nutshell

even, blunt-edged lobes

alternate, long, broad leaves

dark, glossy-green above

WHITE OAK WITH ACORNS × ½

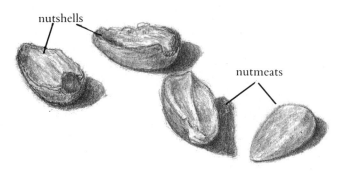

nutshells

nutmeats

WHITE OAK ACORNS × 1

colored flesh are usually the best. There are many species within this group. Here are a few:

The best-known is the white oak (*Quercus alba*). The tree grows from 60 to 120 feet tall, with light, grayish-white, somewhat furrowed and slightly scaly, flat-ridged bark. The somewhat blunt-edged leaves are 5 to 9 inches long and 2 to 4 inches wide, with seven to ten shallow or deep, even, rounded lobes. The leaves are dark glossy-green above, sometimes slightly downy underneath.

The usually stemless acorns are under 2 inches

long, light-brown when mature, with a bowl-shaped cup covering up to one third of the nut. You can find this common tree in mixed and deciduous forests in eastern North America, growing on a variety of soil types, often with other species of oak.

The chestnut oak (*Quercus prinus*) is a medium-size tree with thick, deeply furrowed, reddish-brown to dark-brown bark. The leaves are 5 to 10 inches long and 1½ to 3 inches wide with wavy margins, like chestnut leaves. The acorns are 1 to 1½ inches long, and the cup encloses half of the nut. These large, mild-tasting nuts are among the best for eating. It grows in mixed, deciduous woodlands in eastern North America, favoring dry to moist, well-drained ridges and slopes.

Gambel's oak (*Quercus gambelii*) is a shrub or small tree with grayish-brown, scaly bark. Its leaves are moderately to deeply lobed, 3 to 6 inches long and 2 to 3 inches wide. The globular acorns are up to 1 inch long, with the cup enclosing one quarter to three quarters of the nut. It has so little

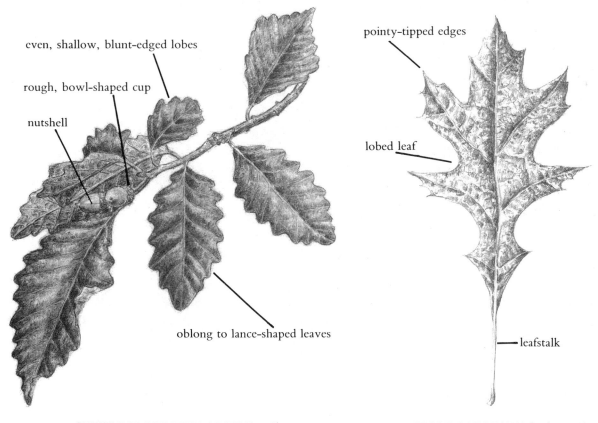

even, shallow, blunt-edged lobes

rough, bowl-shaped cup

nutshell

oblong to lance-shaped leaves

CHESTNUT OAK WITH ACORNS × ¼

pointy-tipped edges

lobed leaf

leafstalk

BLACK OAK LEAF (red oak group) × ½

tannin, some people eat the nuts unprocessed. Look for it in canyons, and on dry foothills and lower mountain slopes, in the Southwest.

All acorns contain bitter, unhealthful, water-soluble tannin, which you can leach out. Boil white-oak acorns in their shells a few minutes, to facilitate shelling. Then boil the shelled, chopped, or ground nuts in several changes of water a few minutes each time, until the water no longer darkens and the bitter taste disappears.

Red-oak acorns take days of leaching to get rid of their tannin. The Indians put them in sacks in streams for weeks. I once tried this in a clean stream. Two weeks later, the nuts I had spent a day shelling disappeared, and I was lucky to find my mesh bag blowing in the wind. Either a four- or two-legged animal must have found them.

Some years are better than others for acorns, so collect as many as possible when you can. I often freeze them in the shells in autumn and shell them in the winter. This is labor-intensive.

Eat processed acorns as is, roast them, include them in casseroles, soups, or stews. Use them immediately, or freeze or dry them. Fresh acorns get moldy within a couple of days. You can grind dry acorns into meal or flour in a blender, spice grinder, or grain mill, and store them in tightly closed jars until ready to use.

Acorn flour is great mixed fifty-fifty with whole-grain flours in muffins, breads, and pancakes. It tastes like cornmeal, only more earthy, and it's high in protein. Note: Acorn flour doesn't contain gluten, so it won't help bread rise, and it makes baked goods more crumbly. Baking low, flat loaves produces best results.

One of my favorite acorn recipes (page 279) involves flavoring processed acorns with lemon juice, garlic, and curry powder. I got it published in a major London newspaper as "Park Nuts" after the chef at the British embassy to the United Nations attended my tours and used wild foods for a special dinner. I sent press releases to all the London papers, headlined FRENCH AND BRITISH AMBASSADORS DINE ON WEEDS FROM CENTRAL PARK. The British press ate it up.

White-oak bark contains vitamin B_{12}, calcium, potassium, phosphorus, sulfur, cobalt, and sodium. All oaks contain the biologically active compound tannin, as well as gallic acid and ellagi-

tannin. Tannin is astringent, antiseptic, and antiviral, with both antitumor and anticarcinogenic activity. Although it is present in commercial Chinese tea, long-term intake is unhealthful—contributing to constipation and interfering with calcium absorption. Prolonged high doses can cause kidney damage.

Indians and Europeans made a bitter decoction from the white oak's inner bark. They used it as a gargle for sore throat, for diarrhea, and as a douche for yeast infections. It's a good wash for skin ulcers, ringworm, and other fungal skin infections, and a cold compress is good for cuts and burns. You can even powder the inner bark and inhale it like snuff for nosebleeds.

The Celtic word for "oak" is *duir*, which also means both "protection" and "Druid." The Druids worshiped oaks, and in English, *duir* became "door," which was protective and made of oak.

PAWPAW, CUSTARD APPLE, FALSE BANANA

(Asimina triloba)

This fruit-bearing species is the only North American representative of the tropical custard-apple family. The pawpaw is a small shrubby tree, 10 to 40 feet tall. The dark, smooth bark, spotted with gray, is sometimes broken in places. The

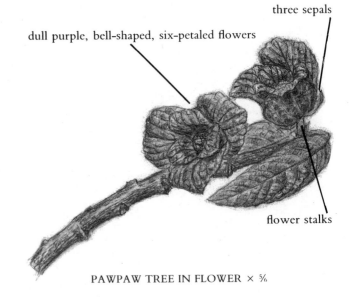

three sepals

dull purple, bell-shaped, six-petaled flowers

flower stalks

PAWPAW TREE IN FLOWER × ⅚

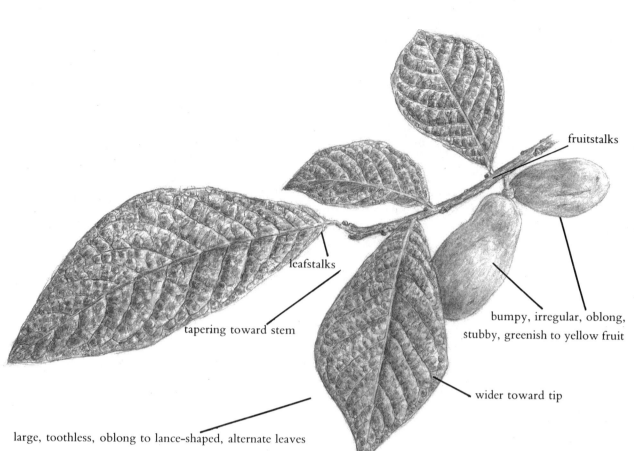

fruitstalks

leafstalks

tapering toward stem

bumpy, irregular, oblong,
stubby, greenish to yellow fruit

wider toward tip

large, toothless, oblong to lance-shaped, alternate leaves

PAWPAW TREE WITH FRUIT × ⅓

large, toothless, oblong to lance-shaped, alternate leaves grow from 4 inches to 1 foot long, and 2 to 6 inches broad— light green above and pale green below. Thin and membranous, they're wider toward the tip and taper toward the stem.

In springtime, just before or when the leaves develop, dull purple, bell-shaped, stalked flowers, 1 to 1½ inches across, emerge from the leaf axils of the previous year's shoots. They're unusual, with three sepals and six thick, nearly round, inch-long petals—three spreading and three standing upright.

The pawpaw's clumsy-looking, bumpy, irregularly oblong, aromatic fruits are 3 to 7 inches long and 1 to 2 inches thick. Unripe, they look like clusters of stubby, greenish-yellow bananas. They ripen in late summer or autumn, turning yellow, dappled with brown spots and often powdered by a whitish, powdery bloom. Peel the skin, and a sweet, aromatic pulp comes oozing out. Rows of large, flat, brown seeds lie inside.

Pawpaws grow from east Texas and Florida to New Jersey and the Great Lakes, and west to Iowa. They prefer deep, rich, moist soil, growing along streambanks, in forests, and in river valleys and bottomlands. You can also find them neatly labeled in many botanical gardens.

Harvest before raccoons or opossums get them. Gently shake the tree to loosen the ripest fruits. Put a dropcloth under the tree if possible—falling pawpaws can break and ooze their pulp, making a gooey mess. Or collect unripe, full-size fruits. Store them in a cool place, away from animals, and they'll ripen off the tree.

The pulp has a banana's texture, with overtones of pear, strawberry, and mango. Different people prefer various stages of ripeness. Children love them dead ripe, sweet as honey. Some like them when they're just turning brown, and still a little firm. Avoid overripe fruit, which turns black and produces a heavy, unpleasant flavor.

Many people eat ripe pawpaws from the tree,

spitting out the seeds, which are poisonous. You can also remove the seeds by hand and mash the pulp for a bowl of natural instant pudding. Or use the pulp in your favorite banana-bread recipe. Make pawpaw frosting by mixing it with cream cheese or tofu, and lemon juice. Pawpaws are also great in yogurt puddings, with nuts, raisins, and vanilla.

According to the U.S. Department of Agriculture, pawpaws are high in calories and carbohydrates, but without other nutrients. High-calorie wild foods are few and far between, and the pawpaw is one of the best sources. In our overfed society, we shun high-calorie foods. But in survival situations, warmth is often more important than food or water, hypothermia being the most common cause of death and high-calorie foods are essential.

Pawpaw fruits may also be slightly laxative and sedative. Some people get drowsy after eating them. The whole plant is full of many poorly understood, biologically active substances. We know it's safe only because people have been eating them for hundreds of years.

Caution: The unripe fruit is emetic, and not everyone can tolerate the ripe fruit. This is the only wild fruit that ever disagreed with me. The first time I found it, it caused no ill effects, but the following year, one serving made me nauseated. Six months later, a small serving made me throw up, and I learned the meaning of "forbidden fruit." Be extra sure to eat very little of this fruit the first few times.

The fresh or tinctured seeds are emetic and produce a stupor, but they're deadly if you're an insect. People used to apply them, powdered, to children's heads, to kill lice. Unknown factors in the bark make it too dangerous to use, although people used to use a decoction as a tonic and stimulant. However, scientists are studying compounds in the unripe fruit, bark, and seeds for anticancer activity.

PERSIMMON, DATE PLUM

(*Diospyros virginiana*)

'Possum up the 'simmon tree,
Raccoon on the ground.

Raccoon says to the old 'possum,
"Won't you throw those 'simmons down?"
—Lyric to Southern folksong

Diospyros means "Zeus' fruit" in Greek. The top god had good taste.

The most noticeable characteristic of this tree, which is from 15 to 100 feet tall, is its thick, gray-black bark. Divided into square blocks of about 1½ inches per side, the trunk resembles a mosaic. This member of the ebony family has hard, dense wood, like its relatives. The alternate, toothless, short-stalked leaves are elongated ovals, up to 6 inches long and 3 inches broad, with the tips and bases coming to points. The upper surface is dark-green and glossy; the underside is lighter.

I had a problem trying to sketch the early summer flower. Fortunately, as I stood under the tree gazing into the distant branches, along came an acquaintance who had scaled limbless coconut trees in the tropics. He scrambled up the tall tree with only the irregularities of the bark for support and tossed down a beautiful, pale-yellow, fragrant, bell-shaped persimmon flower. It was only ¾ inch long, with four petals and eight stamens. This was the fertile flower (with male and female parts), and

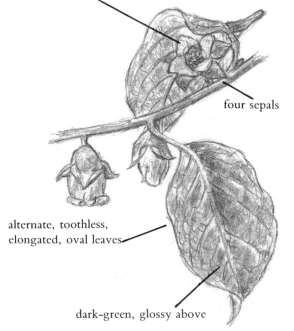

pale-yellow, four-petaled, radially symmetrical flower

four sepals

alternate, toothless, elongated, oval leaves

dark-green, glossy above

PERSIMMON TREE IN FLOWER × 1

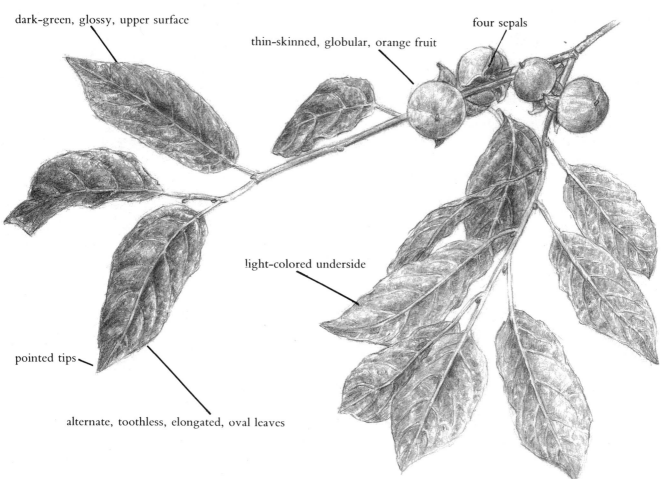

dark-green, glossy, upper surface

thin-skinned, globular, orange fruit

four sepals

light-colored underside

pointed tips

alternate, toothless, elongated, oval leaves

PERSIMMON TREE WITH FRUIT × ½

there are also sterile (male only) flowers that produce pollen but make no seeds, growing in clusters of two or three. They're similar, but just ⅜ inch long, with sixteen long stamens.

The persimmon's globular fruit is orange throughout, 1 to 1½ inches across, covered by a thin, leathery skin that discolors yellowish-brown. There are one to six large, brown, flat seeds inside.

The persimmon has no poisonous look-alikes. A closely related edible species, the Texas persimmon (*Diospyros texana*) is similar, but with smooth, gray bark and black fruit.

Look for persimmons in dry woods, on rocky hillsides, along roads, in rich bottomlands, in or along the edges of old fields, and along roadsides. They're often cultivated in botanical gardens and urban parks throughout the country.

American persimmon trees thrive in the Southeast. Although they normally don't seed themselves north of New Jersey, you can find planted persimmons into New England. Persimmons also range westward into Texas and the Central States. The Texas persimmon grows throughout the Southwest.

Harvest time is highly variable. I found ripe fruit as early as September in Virginia, disproving the myth that all wild persimmons must be touched by frost to ripen. In the Northeast, ripe persimmons litter the ground from mid- to late fall. Some years they ripen after the first frost; other times they're still falling onto the snow, one by one, in the winter—a feast from heaven for foxes, deer, raccoons, and opossums. Ripening time is less dependent on cold than on nights with long periods of darkness.

Often, the fruit is hopelessly out of reach. Some authorities claim that you can shake the fruit from the tree, and they're correct—but only with a crane. With the most strenuous effort, I may sometimes dislodge five or six fruits, although half of them usually vanish into the brush. I've tried shaking trees in unison with a friend or two, but it's like my clumsy attempts to learn to dance: My partners and I are always at cross-purposes.

Ripe persimmons are very soft to the touch, with a gooey, datelike center ready to pour out through the thin skin. Their fragility and small size account for their absence in the marketplace. Also, the ripe fruit looks unattractive, even rotten, turning a dull brownish-orange on its bed of fallen leaves. Sometimes the impact of the fall breaks the skin, so the pulp gets mixed with twigs and leaves, but the fruit usually lands softly and lies plumply, just begging to be eaten. It has an unforgettably sweet taste and rich texture, much more intense than that of commercial Oriental persimmons.

Not all the persimmons that fall to the ground are ripe. Any fruit even slightly hard to the touch is unfit to eat. They're very astringent and can make your whole mouth pucker and go dry. The best cure is to eat a ripe persimmon.

I visit persimmon trees over a period of weeks, bending and squatting around each tree until I've gathered my fill. Separate ripe and unripe persimmons, using shallow plastic food containers, so the soft, sticky fruit doesn't get mangled with the debris. Go through your supply every two days. Refrigerate unripe fruit in a container with unwaxed apples, which emit ethylene gas that hastens ripening. Discard any hard, mold-hardened fruit. Avoid using unripe fruit. I've ruined recipes with slightly unripe persimmons. I sometimes have so many persimmons, I have no choice but to freeze them, or prepare a king-sized recipe. You can freeze them raw or cooked.

Persimmons are great in puddings, supplying all the sweetness and thickening you need. Oranges make a perfect complement, so I often include their juice or rind. I often make persimmon-nut bread, persimmon-orange pancakes, and persimmon ice cream. Still, there's nothing like eating persimmons raw. A few drops of lemon juice add a contrast of tartness that brings out their sweet flavor even more.

Persimmons are one of the most caloric, filling fruits. They're a great source of potassium and vitamin C, and provide a lots of calcium and phosphorus. Persimmon-leaf infusion is very high in vitamin C, and tasty too.

American Indians used unripe persimmons for burns. Their astringency counteracts the burns' tendency to ooze, and persimmons are slightly antiseptic. They also help treat wounds and stop bleeding.

Like the papaya and pineapple, they contain the digestive enzymes papain and bromelain—included in digestive enzyme preparations. These enzymes are so effective, they actually work from a distance: I just have to see a persimmon tree in season, and my mouth starts watering.

COMMON SPICEBUSH, WILD ALLSPICE, FEVERBUSH

(Lindera benzoin)

This 5- to 20-foot-tall, spreading bush is a native member of the laurel family. The bushes are usually colonial, spreading by the roots. Crush or scratch the thin, brittle twigs, or any part of spicebush to release its lemony-spicy fragrance.

The bright-green, alternate, toothless, pointy-tipped, stalked leaves are elliptical, 2 to 6 inches long. In the early spring, before the leaves appear, dense clusters of tiny, yellow flowers in the axils scent the air, attracting early-season insects. The spiciest parts are the hard, oval, stalked, scarlet berries, each with one large seed. They grow in clusters, from the leaf axils of the female bushes, in autumn.

Look for spicebushes in damp, partially shaded, rich woodlands, on mountains' lower slopes, in thickets, and along streambanks, throughout the eastern United States, except the nothernmost regions. Pioneers knew that this was good land for farms, with moist, fertile soil.

The berries, which taste a little like allspice, are an irreplaceable seasoning for me. Rinse them, pat them dry, and chop them in a blender or spice grinder. If you have neither, put them under a towel and crush them with a hammer. Some people remove the seeds, but I crush them along with the rest of the berries.

Since spiceberries are ripe in apple season, they often find themselves in the same pot. I love compotes with sliced apples, walnuts, orange rind, and

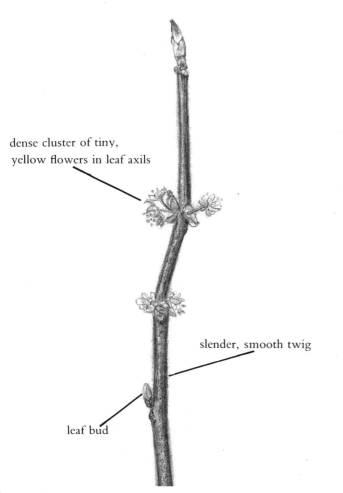

dense cluster of tiny, yellow flowers in leaf axils

slender, smooth twig

leaf bud

SPICEBUSH IN FLOWER × 2

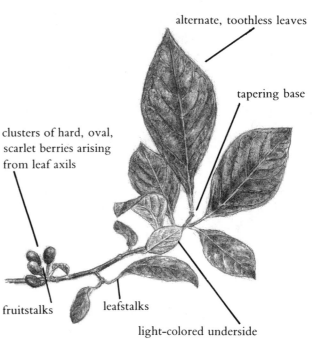

alternate, toothless leaves

tapering base

clusters of hard, oval, scarlet berries arising from leaf axils

fruitstalks

leafstalks

light-colored underside

SPICEBUSH WITH BERRIES × ½

spiceberries, simmered about 15 minutes. Spiceberries don't go quite so well with some other later autumn fruits, such as autumn olives and persimmons. Wild raisins, on the other hand, get a much-needed zing from spiceberries. The seasoning is also wonderful for main courses and in pastries, like commercial allspice.

To store long-range, don't dry the berries. They're too oily, and may go rancid at room temperature. Spread the chopped berries out on a plate or cookie sheet and freeze them, then pack into a freezer container. This way, you can remove small amounts of herb as needed, and your seasoning doesn't stick together. I think half a teaspoon is plenty for a recipe that serves six, but it depends on your personal preference.

Collect the twigs year-round for teas, or use the leaves from mid-spring to fall. In one cup of water, steep either one-half cup of fresh leaves

(dried leaves lose their flavor) or twigs, or two tablespoons of chopped berries.

Pioneers called this plant feverbush because a strong bark decoction makes you sweat, activating the immune system and expelling toxins. They used it for typhoid and other fevers, and to expel worms. I use a tincture of the leaves, along with wild ginger and field garlic, plus vitamin C and zinc lozenges, at the first sign of a cold or sore throat, and it sometimes works.

The Indians used a spiceberry infusion for coughs, colds, delayed menstruation, croup, and measles. They used the oil from the berries, externally, for chronic arthritis. It's also good for flatulence and colic. Spicebush leaf, bark, or berry tea compresses are also good for mild skin irritations, such as rashes, itching, and bruises.

VIBURNUMS

(*Viburnum* species)

These members of the honeysuckle family are all shrubs or small trees with opposite, simple, toothed leaves. In the spring, they have large, showy, flat umbrellalike clusters of 5-petaled white flowers. Similar clusters of berries occur in late summer or autumn. Each berry contains one flat seed.

That's where the similarity between species ends. Leaf shapes vary. Some species have blue-black, oval berries, others have bright-red, round berries. Edible viburnum fruits are pulpy and sweet. Inedible species' fruits are hard and dry—all skin and seed, with no flesh. No viburnum is poisonous. The wild raisin and nannyberry, described below, are two delicious edible viburnums with black, oval, pulpy berries.

Many species are native to eastern North America, but these and some European species have been planted throughout much of the continent.

You can divide the viburnums into four groups, based on leaf shapes: First is the maple-leaf group, with three lobes per leaf. The maple-leaf viburnum (*Viburnum acerifolium*), very common in eastern North American woods, has unpalatable, inedible berries. This shade-tolerant shrub emerges from the ground over a wide area, snagging fallen leaves that shelter the soil from the impact of raindrops. A class of schoolchildren can collect hundreds of berries, poke holes into the soil with pencils, and plant enough of this maintenance-free native shrub to solve serious erosion problems on shady hillsides.

The native highbush cranberry (*Viburnum trilobum*), not related to true cranberries, also has maplelike leaves. This elegant, upright native shrub grows up to 16 feet tall, with red berries that taste somewhat like its ground-creeping namesake.

The leaves are 2½ to 4½ inches long, and nearly as wide. They're coarsely toothed with pointed tips, and partially divided into three lobes, similar to some inedible viburnums. There are two kinds of flowers in dense, round-topped clusters—an outer circle of large, sterile flowers attracts insects to a central cluster of smaller, fertile flowers.

The shrub bears spectacular clusters of brilliant red, soft, juicy, translucent, oval to round berries, ½ inch long, growing in generous clusters. They look especially festive against the greens, oranges, and yellows of fall, and they persist throughout the winter, brightening the snowy landscape.

A cultivated European look-alike, the cramp

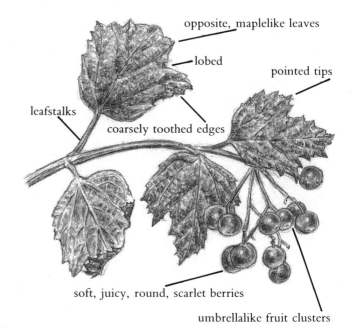

opposite, maplelike leaves

lobed

pointed tips

leafstalks

coarsely toothed edges

soft, juicy, round, scarlet berries

umbrellalike fruit clusters

HIGHBUSH CRANBERRY × ⅖

bark or guelder rose (*Viburnum opulus*), has unpleasantly bitter fruit, although a decoction of its inner bark is supposed to be good for cramps. It was given to women with menstrual cramps, or for cramps in the limbs during pregnancy. This shrub has tiny, concave glands on the leafstalk, near the base of the leaf. The same glands on the highbush cranberry are convex or dome-shaped. You need a magnifying lens to detect this distinction. Unfortunately, no matter how good a hand lens I use, every prospective highbush cranberry I've ever examined always turned out to be a guelder rose.

Highbush cranberries grow from coast to coast in the northernmost states and southern Canada, along streambanks, in moist thickets, and in woodlands or at their edges. The similar cranberry tree (*Viburnum pauciflorum*) has leaves with hairy under-surfaces. It grows in the Northwest, and you use it like the highbush cranberry. Guelder roses grow in parks and cultivated areas throughout America.

Highbush cranberries are very tart and a little bitter, the better to perk you up in the cold weather. They're particularly refreshing to the palate, getting milder as winter progresses. The best way to prepare them is by cooking, straining out the seeds and sweetening to taste. Add a pinch of orange rind for a great "cranberry" sauce. You can also freeze the strained puree and add a little to other fruit desserts. A few tablespoons mixed into an apple-pie filling or stirred into a simple vanilla pudding add a touch of distinction.

The highbush cranberry is a good source of vitamin C, and a good detoxifier. Its astringent quality made it popular for treating swollen glands.

Another group of viburnums is the wayfaring tree group, which includes the edible hobblebush or wayfaring tree, also called witches' hobble (*Viburnum alnifolium*). This northeastern species grows up to 10 feet tall. It has very large, rough, heart-shaped, finely double-toothed leaves 4 to 9 inches long.

Specializing in rocky or mountainous woods, it gets its name from the way it hobbles along the ground: It spreads when the stems droop over and take root in spots of soil between rocks. If you trip on these branches, you'll hobble too.

Its flowers are like the highbush cranberry's, with large, sterile flowers surrounding small, fertile ones. The fruits are on the small side, ¼ inch or less in diameter, purplish-black, with typical, flat viburnum seeds. The berries taste good, but they're too small for more than a trail nibble, and they're only abundant on shrubs growing in open, sunny areas.

The arrowwood viburnum group has very coarsely toothed, oval leaves, with pointy tips and

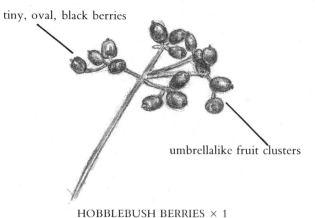

tiny, oval, black berries

umbrellalike fruit clusters

HOBBLEBUSH BERRIES × 1

egg-shaped bases. The berries are inedible—bitter and dry—although chipmunks and some birds eat them. The Indians used the tough, durable wood to make arrow shafts, and they smoked the leaves.

The last group, and foragers' favorite, is the black haw group. Fortunately these edible viburnums are not hard to find. They seem to grow in a very wide region, all over woodland areas in eastern North America. Furthermore, landscapers have planted them in cultivated areas across North America, especially around garden-apartment complexes.

The most common varieties are northern wild raisins (*Viburnum cassinoides*), nannyberries (*Viburnum lentago*), and black haw (*Viburnum prunifolium*). All members of this group have narrow, oval leaves with either finely toothed or smooth edges. The dark, shiny berries are covered with a powdery bloom. They're sometimes roundish, often oval, and sometimes more flat.

The black haw is the best. This large shrub, 6 to 30 feet tall, is very bush—densely covered with twigs. The dull, ovate to elliptical leaves are quite small, about 2 inches long—somewhat leathery and blunt-tipped rather than long and pointy. The oval berries are about ½ inch long. They start out green, turn bright red in late summer, then turn blue-black when they ripen in autumn. They taste like dates, with a texture like prunes. Look for them in bogs and moist woods, throughout the eastern four fifths of the United States.

The nannyberry, which can grow over 30 feet tall, also has delicious fruit. It's quite similar to the black haw, but the long, pointed, finely toothed leaves are larger—2 to 4 inches long. The leafstalks are winged, with a thin-ridged margin. The fruits

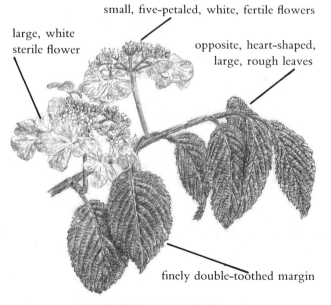

small, five-petaled, white, fertile flowers

large, white sterile flower

opposite, heart-shaped, large, rough leaves

finely double-toothed margin

HOBBLEBUSH IN FLOWER × ¼

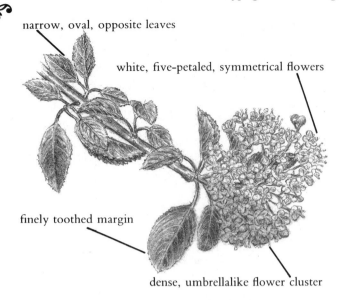

narrow, oval, opposite leaves

white, five-petaled, symmetrical flowers

finely toothed margin

dense, umbrellalike flower cluster

BLACK HAW IN FLOWER × ⅔

are ½ inch long and egg-shaped. The fruits go straight from green to a ripe black.

The northern wild raisin's young shoots are covered with rusty scales. The leaves are dull above, not leathery like the black haw's, and usually very finely toothed or wavy edged. The wild raisin has very small fruits, with only a little pulp and a relatively large seed.

Although these viburnums flower in midspring, the fruits don't ripen until mid-fall. I find them as late as November, and they sometimes persist well into the winter. The black haw meets my expectations for autumn fruit—heavy, warming, and full of carbohydrates, very different from the light, juicy, cooling fruits of summer.

I've often wondered why some of the viburnums have not been grown for the market. They're nonperishable: Black haws can keep in the refrigerator for over a month. They taste great, and like all fruits, I'm sure they can be bred for more pulp.

Black haws and their relatives are great raw or cooked. I find them more nourishing and energizing than any other fruit. Simmer about 20 minutes, almost covered with fruit juice. Tart juice is best to offset their heavy sweetness—I often use a combination of orange and lemon juice. Strain out the seeds with a food mill. This can be labor-intensive unless you get someone else to do it.

I've used the resulting fruit puree in many recipes. I once added lemon and orange rind, cinnamon and wild spiceberries for seasoning. It was very

strongly flavored, so I stretched it with some pureed tofu (or substitute ricotta cheese), sunflower-seed oil, and soy milk (or substitute milk). I used this black, rich filling for strudels and pastry. It was also great as a pancake topping, and as a filling for muffins.

The inner bark of the trunk and roots of all viburnums is a traditional medicine. References dating back to the sixteenth century refer to all species as antispasmodic, although the black haw is the best. It relaxes the nervous system and tones the uterus. In former times, drying and selling viburnum barks was a small business in this country, and you could buy them dried and powdered in neighborhood pharmacies. They were used for painful menstruation, cramps, amenorrhea, threatened abortion, bleeding, and pregnancy, as well as asthma, charley horses, epilepsy, and convulsions. Women who tended to miscarry would take a tincture or decoction several times a day.

The antispasmodic properties of viburnums were independently discovered by Europeans, Native Americans, and Asians. One active principle is a bitter glucoside called viburnin. Black haw,

narrow, oval, opposite leaves

finely toothed margin

dark, shiny, bluish-black, long to oval berries in umbrellalike clusters

BLACK HAW WITH FRUITS × ⅔

once called "false cramp bark," has three times the viburnin of cramp bark or guelder rose. Highbush cranberry bark has the least viburnin.

Collect the bark in the spring, when the sap is rising and it's easier to peel. Cut off a few stems, preferably in places where the shrub looks like it can use some pruning. Separate the inner and outer bark from the hard woody center with a sharp pocket knife. It's best to make a tincture, since the dried bark loses strength with time.

OTHER PLANTS OF WOODLANDS IN AUTUMN

Edible or Medicinal Plants

Aniseroot leaves and roots, black birch, Kentucky coffee tree pods and seeds, cow parsnip roots, dayflowers, daylily tubers, elderberry berries, wild onion leaves and bulbs, goutweed leaves, groundnut pods and roots, horse balm seeds and roots, jewelweed stems and seeds, pears, pennyroyal, pine needles and nuts, plums, raspberry leaves, rose hips, sassafras leaves and roots, Solomon's seal root, storksbill, wild beans, wild ginger roots, wild leek roots, wild potato vine root, wintergreen berries and leaves, wood sorrel, yew berries.

For Observation Only

Aniseroot skeletons, black birch cones, false Solomon's seal berries, greenbrier vines and berries, Hercules'-club berries, hog peanut flowers, poison ivy leaves and berries, jewelweed flowers, violet leaves and fertile flowers, wild leek seed head skeletons, wild lettuce seed heads.

PLANTS OF FRESHWATER WETLANDS IN AUTUMN

PICKERELWEED

(Pontederia cordata)

Pickerelweed is a perennial, colonial, herbaceous native. It has large, arrow- or heart-shaped leaves (a southern species, *Pontederia lanceolata*, has lance-shaped leaves) that grow in a rosette and float on the water. The lobes at the base of the leaves are heart-shaped or rounded (*cordata* means "heart-shaped"). Pickerelweed's leafstalks are filled with air chambers for flotation.

The similar-looking arrowhead and arrow arum have pointed lobes at their leaf bases. Also, pickerelweed doesn't have a main midrib in its leaf the way arrowhead does.

Beautiful, blue, tubular, bilaterally symmetrical, two-lipped flowers grow densely clustered on a separate stalk, up to 3 feet tall, arising from the middle of the rosette. They bloom from summer to fall.

The fruits are stiff, oval capsules ⅓ inch long

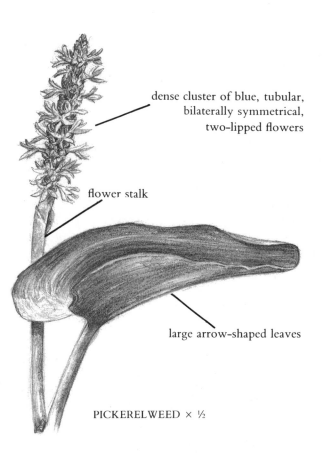

dense cluster of blue, tubular, bilaterally symmetrical, two-lipped flowers

flower stalk

large arrow-shaped leaves

PICKERELWEED × ½

or smaller, enclosed by the tattered remnants of the flowers. Each contains one large seed.

It grows in still, shallow, slow-moving fresh water, where pickerels and other fish might lay their eggs. It grows throughout the East and in the South, but not in the Midwest or on the West Coast. Its genus is named after the early-eighteenth-century botanist Giulio Pontedera.

In early summer, you can add the young, unfurling leaves raw to salads or steam or sauté them lightly. Deer and muskrat also eat them. I strip off the seeds in quantity every autumn. They have a nutty flavor, and the raw seeds are crunchy. I add some to granola, cook others as whole grains, and grind some into flour.

Caution: Although there are no poisonous look-alikes, you must have the water tested for chemical contaminants and pathogens before you eat the leaves.

OTHER PLANTS OF FRESHWATER WETLANDS IN AUTUMN

Edible or Medicinal Plants

Angelica seeds, apples, burdock roots and seeds, cattail roots, coltsfoot leaves, cow parsnip roots, cranberries, curly dock root, elderberry berries, grapes, groundnut pods and tubers, miner's lettuce, Oswego tea, parsnip root, peppermint, rose hips, sheep sorrel, Solomon's seal roots, spearmint, strawberry leaves, sumac berries, sweetgale, watercress, wild mint, wild rice, wintergreen berries, yarrow, yew berries.

For Observation Only

Cattail seed heads, curly dock seeds, false Solomon's seal berries, poison ivy leaves and berries,

thistle, violet leaves and fertile flowers, water hemlock.

PLANTS OF THE SEASHORE IN AUTUMN

Edible or Medicinal Plants

Amaranth, autumn olive berries, bayberries, beach plums, carrot roots, seeds, and flowers, common evening primrose, curly dock roots, wild onion leaves and bulbs, foxtail grass seeds, glasswort, grapes, groundnut pods and roots, lamb's-quarters leaves and seeds, marshmallow root, mugwort, mullein, parsnip roots, prickly pear fruit, pine needles, peppergrasses, rose hips, sea rocket leaves and pods, seaside plantain, strawberry leaves, sumac berries, wild bean pods.

For Observation Only

Russian olive, beach pea, common evening primrose seeds, curly dock seeds, Japanese knotweed seeds, milkweed pods, mullein seeds, pine cones, poison ivy leaves and berries, pokeweed berries, sow thistle flowers and seed heads, yucca seed heads.

PLANTS OF MOUNTAINS IN AUTUMN

Edible or Medicinal Plants

Cranberries, creeping wintergreen leaves and berries, miner's lettuce, pennyroyal, orpine leaves and roots, roseroot leaves and roots, pine needles and nuts, sassafras leaves and roots, sheep sorrel, spicebush leaves and berries, viburnum berries.

PLANTS OF DESERTS IN AUTUMN

PRICKLY PEAR, CACTUS PEAR, INDIAN FIG

(*Opuntia* species)

All cacti do away with normal leaves to retain water. Photosynthesis occurs in very specialized, succulent stems, which are padlike and jointed—clumsy-looking but highly adapted. They're also armed with prickly bristles and large spines which are modified leaves. A plant that dies at the end of the season isn't going to make it in the harsh desert environment, so cacti are perennials, storing food and water in their thick pads—enough to see the plant through the worst drought.

Prickly pears are low-growing cacti with pads separated by joints. Edible species have flattened joints. Dry, woody, inedible species, called chollas, have cylindrical joins. All prickly-pear pads and fruits bear invisible, prickly bristles, so fine they're hard to see. However, you'll have no difficulty feeling them. There are no poisonous look-alikes.

Showy, radially symmetrical, yellow flowers 3½ inches across, with ten to twelve petals, appear on the pads' upper segments in early summer.

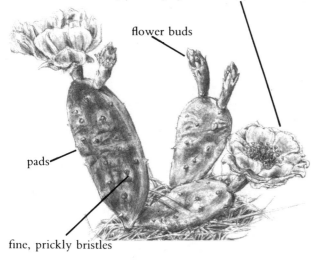

showy, radially symmetrical, yellow flower

flower buds

pads

fine, prickly bristles

EASTERN PRICKLY PEAR IN FLOWER × ⅓

Club-shaped fruits ripen from late summer through autumn. They taper to the bottom and flatten along the top, growing from 1 to 2 inches long and ¾ inch across. They look like dull red or purple cylindrical appendages on the top of the cactus. Their insides are red and pulpy, with many flattened, almost circular seeds.

The eastern prickly pear (*Opuntia humifusa*) is quite small, at most 8 inches long, sprawling along the ground. The small fruit is brick-red.

The western species may be tall or short, and the fruit can be yellow, red, purple, or white. The most common of the western prickly-pear species (*Opuntia phaeacantha*) grows as tall as 3 feet, although it often sprawls along the ground. The pads may be more than a foot across, and the fruits are also larger, as well as juicier, more abundant, and better than those of the similar, smaller eastern species. Engelman's prickly pear (*Opuntia engelmannii*) is also similar but still larger, reaching 4 to 10 feet tall.

Look for prickly pears in sandy or rocky soil. It's a surprise to see cacti, growing at the seashore, but this habitat is very desert-like.

The Eastern species grows along the East Coast down to Texas, as far south as South Carolina, and west to Minnesota. They're rare north of the Great Lakes, since cacti do best in dry, warm climates. You'll encounter other very similar species in the South, and even more on the Gulf Coast. Of course, the best places for these plants are the desert habitats of the Southwest.

Wearing heavy work gloves, rub the fruit vigorously with a damp towel to remove all the bristles, before you even think of handling this fruit: A few years ago, some friends and I ran into two sunbathers at a seashore park they frequented, and they led us to a stand of prickly pears. I made the mistake of mentioning my wish to collect and draw a specimen, even though it would be impossible to collect one bare-handed.

One of the sunbathers insisted on picking the specimen for me—the thorns wouldn't faze him. I

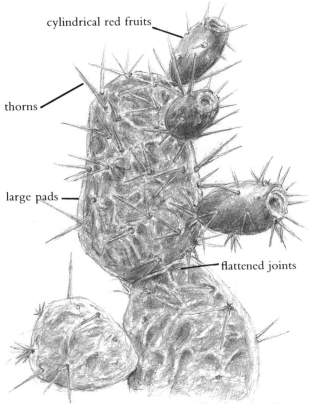

cylindrical red fruits

thorns

large pads

flattened joints

WESTERN PRICKLY PEAR WITH FRUITS × ½

it as is, or remove the seeds and use the pulp in fruit recipes. It's especially good for sauces, puddings, and beverages. Caution: too much prickly pear fruit causes constipation in some people.

You can also dry the seeds and grind them into flour in a grain mill, blender, coffee grinder, or spice grinder. They produce a nutty flavor, especially if pre-roasted. I used them successfully as flour in a bread recipe, and to thicken soup.

In the spring, the tender, young, new pads make a mild, mucilaginous green vegetable, much like okra. Cut away the thorns, remove the bristles with a damp towel, and slice the pads. Or hold the pad over a fire and singe off the needles.

The sliced pads are good in stews and soups, with an okra-like thickening effect. You can also fry them. If you don't like their gummy texture, pre-boil the sliced pads ten minutes and rinse them before further cooking. Caution: Sensitive people may get diarrhea from eating more than a few pads.

The fruit is an excellent source of potassium, and beta-carotene. It also provides plenty of vitamin C, calcium and phosphorus. The pads provide beta carotene and potassium, and the seeds are high in protein and oils.

Eating the pads has a soothing effect on the stomach, and they're also supposed to be good for the lungs and kidneys. Because they retain water, they've also been used as wound dressings. The fruit is a gentle diuretic—good for kidney stones and ulcers. It turns the urine a crimson color, which is frightening if you don't expect it.

couldn't dissuade him, and he grabbed one. When he realized his mistake, he pretended to ignore the unexpected pain. Then, much to my chagrin, he spontaneously shook my hand when we parted.

To prevent most of the pain, I avoided touching anything with that hand until I got home. I laboriously pulled out all the nearly-microscopic bristles with tweezers under a bright light, then placed the specimen on a plate with tongs. In walked my friend, who picked up and examined the strange fruit before I could even open my mouth. The tweezers got another workout.

The flavors range from bland to sweet and delicious. The best Eastern species, *Opuntia humifusa,* tastes like a pulpy combination of watermelon and blackberry. The Western species are best. They have larger, juicier, more abundant, and better fruits than those of the East.

Peel the fruit and remove the sweet pulp. Eat

❧ OTHER PLANTS OF DESERTS IN AUTUMN

Edible or Medicinal Plants

Desert hackberries, mesquite and screw bean pods, pine nuts.

For Observation Only

Pine cones, yucca seed heads.

CHAPTER 5

EDIBLE WILD PLANTS OF LATE FALL
THROUGH EARLY SPRING

Late fall through early spring is not a season you'll find on your calendar, but try telling that to plants that are in season then. Many roots, for example, store food underground in the fall. In mid-spring, they get tough because the food has gone into flower stalks and flowers. In between, you can harvest them. If you live in a warm enough region, root vegetables' basal rosettes will persist, so you can locate the plants, and the ground won't freeze, so you can dig up roots all season. In colder regions, winter may intercede: The basal rosettes die, and the ground freezes.

Many wild greens also grow for this entire period in warm regions. In colder areas, they die in the winter, and return in early spring. Other greens hold out, even in the cold, although their quality isn't as good in bitter cold. You have to trim away yellow, withered parts. So direct your efforts toward foraging during warm spells, when new growth improves plant quality and your own com-

fort. If it's above 40°F, you can take off your gloves, making it much easier to forage.

Some seaweeds grow all year. Weather by the seashore is especially harsh in this season, so wait for mild weather to look for them. A few fruits and berries last throughout this season, although more are in season in the beginning. This is a good time for herb teas. You can use twigs, roots, or the leaves of some deciduous and evergreen trees.

This is also a great time to sharpen your identification skills. When you follow your favorite plants through the year, you'll find they don't all vanish when it gets cold. Many leave "skeletons" that still bear identifying characteristics. They indicate where to look the following year. Trees and shrubs still have their characteristic leaf buds, flower buds, and twigs. Although detailed winter identification is beyond the scope of this book, there are books devoted to winter botany.

PLANTS OF LAWNS AND MEADOWS IN LATE FALL THROUGH EARLY SPRING

COMMON DANDELION

(*Taraxacum officinale*)

The dandelion is a perennial, herbaceous plant with long, lance-shaped leaves. They're so deeply toothed, they gave the plant its name: *Dent-de-lion* means "lion's tooth" in Old French. The leaves are 3 to 12 inches long, and ½ to 2½ inches wide, always growing in a basal rosette. The rosette's immature, tightly wrapped leaf bases just above the top of the root form a tight "crown."

Dandelion's well-known yellow, composite flowers are 1 to 2 inches wide. They grow individually on hollow flowerstalks 2 to 18 inches tall. Each flower head consists of hundreds of tiny ray flowers. Unlike other composites, there are no disk flowers. Reflexed bracts grow under each flower. The flower head can change into the familiar, white, globular seed head overnight. Each seed has a tiny parachute, to spread far and wide in the wind.

The thick, brittle, beige, branching taproot grows up to 10 inches long. All parts of this plant exude a white milky sap when broken.

There are no poisonous look-alikes. Other very similar *Taraxacum* species, as well as chicory and wild lettuce (pages 234 and 246), resemble dandelions only in the early spring. All these edibles also exude a white milky sap when injured, but chicory and wild-lettuce leaves have some hair, at least on the underside of the midrib, while *Taraxacum* leaves are bald. Unlike the other genera, *Taraxacum* stays in a basal rosette. It never grows a tall, central stalk bearing flowers and leaves.

Dandelions are especially well adapted to a modern world of "disturbed habitats," such as lawns and sunny, open places. They were even introduced into the Midwest from Europe to provide food for the imported honeybees in early spring. They now grow virtually worldwide. Dandelions spread farther, are more difficult to exter-

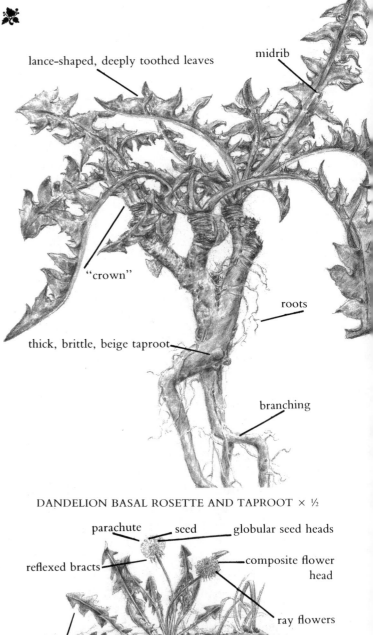

DANDELION BASAL ROSETTE AND TAPROOT × ½

DANDELION ROSETTE FLOWERS AND SEED HEAD × ¼

minate, and grow under more adverse circumstances than most competitors.

Most gardeners detest them, but the more you try to weed them up, the faster they grow. The taproot is deep, twisted, and brittle. Unless you remove it completely, it will regenerate. If you break off more pieces than you unearth, the dandelion wins. "What's a dandelion digger for?" a dandelion asked. "It's a human invention to help us reproduce," another dandelion replied.

Collect dandelion leaves in early spring, when they're the tastiest, before the flowers appear. Harvest again in late fall. After a frost, their protective bitterness disappears. Dandelions growing in rich, moist soil, with the broadest leaves and largest roots, are the best. Select the youngest individuals, and avoid all plants with flowers. Some people eat the greens from spring to fall, when they're very bitter. Others boil out the summer bitterness (and water-soluble vitamins) in two changes of water. It's all a matter of preference.

Dandelion greens are wonderful in salads, sautéed, or steamed. They taste like chicory and endive, with an intense heartiness overlying a bitter tinge. People today shun bitter flavors—they're so conditioned by overly sweet or salty processed food. But in earlier times, we distinguished between good and bad bitterness. Mixed with other flavors, as in a salad, dandelions improve the taste.

I also love sautéing them for about 20 minutes with onions and garlic in olive oil, adding a little homemade wine before they're done. If you're not used to the slight bitterness, cook them with sweet vegetables, especially sliced carrots and parsnips. Boiling dandelions in one or more changes of water makes them milder—a good introduction if you're new to natural foods. Early spring is also the time for the crown—great sautéed, pickled, or in cooked vegetable dishes.

You can also eat dandelion flowers, or use them to make wine. Collect them in a sunny meadow, just before mid-spring, when the most flowers bloom. Some continue to flower right into the fall. Use only the flower's yellow parts. The green sepals at the flower's base are bitter.

The flowers add color, texture, and an unusual bittersweet flavor to salads. You can also sauté them, dip them in batter and fry them into fritters, or steam them with other vegetables. They have a meaty texture that contrasts with other, lighter vegetables in a stir-fry dish or a casserole. A Japanese friend makes exceptionally delicious traditional dandelion-flower pickles, using vinegar and spices.

The taproot is edible all year but is best from late fall to early spring. Use it as a cooked vegetable, especially in soups. Although not as tasty as many other wild root vegetables, it's not bad. I remember finding large dandelions with huge roots growing on the bottom of a grassy hillside. They were only mildly bitter, so I threw them into a potato stock. With the added scallions, tofu, ginger, carrots, and miso, this became an excellent Japanese miso soup.

Preboiling and changing the water, or long, slow simmering, mellows this root. Sweet vegetables best complement dandelion roots. Sautéing the roots in olive oil also improves them, creating a robust flavor. A little tamari soy sauce and onions complete this unusual vegetable side dish.

The leaves are more nutritious than anything you can buy. They're higher in beta carotene than carrots. The iron and calcium content is phenomenal, greater than spinach. You also get vitamins B_1, B_2, B_5, B_6, B_{12}, C, E, P, and D, biotin, inositol, potassium, phosphorus, magnesium, and zinc by using a tasty, free vegetable that grows on virtually every lawn. The root contains the sugar inulin, plus many medicinal substances.

Dandelion root is one of the safest and most popular herbal remedies. The specific name, *officinale*, means that it's used medicinally. The decoction is a traditional tonic. It's supposed to strengthen the entire body, especially the liver and gallbladder, where it promotes the flow of bile, reduces inflammation of the bile duct, and helps get rid of gallstones. This is due to its taraxacin. It's good for chronic hepatitis, it reduces liver swelling and jaundice, and it helps indigestion caused by insufficient bile. Don't use it with irritable stomach or bowel, or if you have an acute inflammation.

The modern French name for this plant is *pissenlit* (*lit* means "bed") because the root and leaf tea acts on the kidneys as a gentle diuretic, improving the way they cleanse the blood and recycle nutrients. Unlike pharmaceutical diuretics, this doesn't leach potassium, a vital mineral, from the body. Improved general health and clear skin result from improved kidney function. One man I spoke to even claims he avoided surgery for urinary stones by using dandelion-root tea alone.

Dandelions are also good for the bladder, spleen, pancreas, stomach, and intestines. It's recommended for stressed-out, internally sluggish, and sedentary people. Anyone who's a victim of excessive fat, white flour, and concentrated sweeteners could benefit from a daily cup of dandelion tea.

Dandelion root's inulin is a sugar that doesn't elicit the rapid production of insulin, as refined sugars do. It helps mature-onset diabetes, and I used it as part of a holistic regime for hypoglycemia (low blood sugar).

Dandelion-leaf infusion is also good at dinnertime. Its bitter elements encourage the production of proper levels of hydrochloric acid and digestive enzymes. All the digestive glands and organs respond to this herb's stimulation. Even after the plant gets bitter, a strong infusion is rich in vitamins and minerals and helps people who are run-down. Even at its most bitter (*Taraxacum* comes from Arabic and Persian, meaning "bitter herb"), it never becomes intolerably so, like goldenseal and gentian.

The leaf's white, milky sap removes warts, moles, pimples, calluses, and sores, and soothes bee stings and blisters.

Unlike most other seeds, dandelions' can germinate without long periods of dormancy. To further increase reproductive efficiency, the plant has given up sex: The seeds can develop without cross-fertilization, so a flower can fertilize itself. This lets it foil the gardener by dispersing seeds as early as the day after the flower opens.

Sexual reproduction leads to greater genetic diversity. This may be important for adapting to predators and parasites, which also change their genetic makeup to increase effectiveness. But the survival criteria vary from niche to niche. Moral: When maximum reproductive speed becomes the key to survival, don't get caught by evolution's lawnmower with your plants down!

A funny thing happened to me when I was collecting dandelions one day in Central Park. Two tour participants were undercover city park rang-

ers. They had used marked bills, surveillance cameras, and walkie-talkies to infiltrate the group. When I ate a dandelion, the entire Parks Enforcement Patrol converged on my group, and I was handcuffed and arrested for removing vegetation from the park. But after I was fingerprinted, they couldn't hold me. I had eaten all the evidence.

I called every newspaper, TV station, radio station, and wire service. The next day, when I went to get the paper, five cops stopped me. They wanted my autograph. I was on page 1 of the *Chicago Sun-Times*, in newspapers and on the radio around the country. *CBS Evening News with Dan Rather* covered the story; national and local talk-show appearances followed.

When I had to appear in court, I served passersby and reporters "Wildman's Five-Boro Salad," complete with dandelions, on the steps of the Manhattan Criminal Court, and the press ate it up once more.

A month later, the city turned over a new leaf, dropped the charges, and hired me as a naturalist, to lead tours teaching people to eat dandelions and other wild foods. I worked for the New York City Parks Department for four years, leaving and resuming free-lance activities after a new anti-environmental administration took office.

❧ OTHER PLANTS OF LAWNS AND MEADOWS IN LATE FALL THROUGH EARLY SPRING

Edible or Medicinal Plants

Chickweed, clover leaves, daylily tubers, common mallow, wild onion leaves and roots, ground ivy, sheep sorrel, sow thistle, yarrow.

❧ PLANTS OF CULTIVATED AREAS IN LATE FALL THROUGH EARLY SPRING

Edible or Medicinal Plants

Chickweed, clover leaves, common mallow leaves and fruits, dandelion leaves and roots, daylily tubers, juniper, Kentucky coffee tree seeds, orpine leaves, sassafras twigs and roots, sheep sorrel, sow thistle leaves.

For Observation Only

Burdock skeletons, orpine skeletons, rose canes, yew, yucca.

PLANTS OF DISTURBED AREAS IN LATE FALL THROUGH EARLY SPRING

COMMON EVENING PRIMROSE

(Oenothera biennis)

This native biennial begins life with a basal rosette, with elliptical to lance-shaped leaves reaching nearly 1 foot long. They're raggedy-edged and hairy, with a prominent whitish midrib tinged with red. Beneath is a fleshy, white taproot from 2 inches to over 1 foot long—often red-tinged in cold weather. The leaves and roots smell and taste like peppers and radishes.

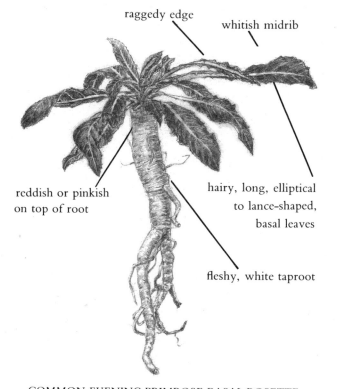

raggedy edge

whitish midrib

reddish or pinkish on top of root

hairy, long, elliptical to lance-shaped, basal leaves

fleshy, white taproot

COMMON EVENING PRIMROSE BASAL ROSETTE AND TAPROOT × ⅓

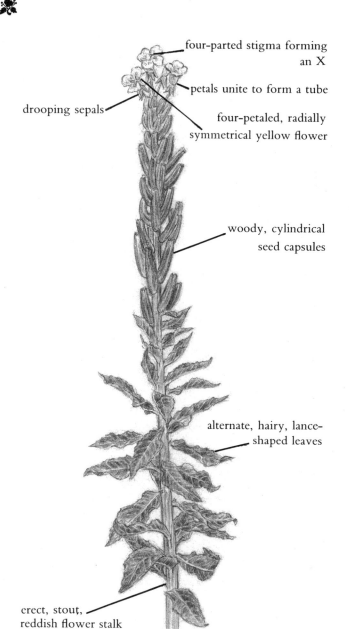

four-parted stigma forming an X

petals unite to form a tube

drooping sepals

four-petaled, radially symmetrical yellow flower

woody, cylindrical seed capsules

alternate, hairy, lance-shaped leaves

erect, stout, reddish flower stalk

COMMON EVENING PRIMROSE FLOWER STALK × ⅕

The plant begins its second year with another basal rosette, except where winters are mild, when the previous year's rosette persists. As the weather warms up, it grows an erect, stout, reddish flower stalk, usually up to nearly 4 feet tall—sometimes as tall as 7 feet. It's either unbranched, or it branches at the base.

Long-stemmed, yellow flowers, 1 inch across, extend from the leaf axils in summer and fall. The bases of the four broad petals unite, forming a long, narrow tube. A distinctive four-parted stigma in the flower's center forms an X shape. Drooping sepals grow beneath the flowers.

After each flower withers, the ovary matures into a woody, cylindrical seed capsule up to 1½ inches long, fully of tiny, hard, reddish seeds. The plant dies in autumn, but the dead flower stalk with the seed capsules persists through the winter.

The flowers are adapted for moth pollination: These insects' long mouth parts can reach down the tube. Moths are active at night, when pale yellow is the easiest color to detect. The flowers reputedly open at dusk and close by midmorning, but unpollinated flowers may remain open longer and, by the time most of the flowers have set seed, the rest tend to remain open all day.

You can use the flowers to rid your garden of Japanese beetles, which congregate inside. Whisk them into a jar early or late in the day, when the insects are sluggish.

Although there are no poisonous look-alikes, common evening primrose is not related to the cultivated garden primrose. There are other very similar edible species and hybrids in this variable genus. *Oenothera hookeri*'s yellow flowers, for example, turn red with age.

This was originally a very mysterious plant. I could easily locate the flower and seeds, when the root was inedible, but I couldn't identify the basal rosette. I mistook it for a nondescript, unidentified dock rosette, which I could never find in flower! One day, I had the mystery rosette in one hand, and a field guide with the primrose flower stalk in the other. The leaves matched up, and I was instantly enlightened.

Since then, I've found this plant in innumerable parks, empty lots, disturbed sites, fields, prairies, and seashore areas across the United States. It does well in poor sandy soil, where its competitors are

at a disadvantage, and digging is easy. Sometimes you need only brush away enough sand to get a good grip on the root, and pull slowly.

Use the taproot of the first-year plant in the fall and early spring, or the second-year plant in early spring before the flower stalk appears, and in the winter if it's warm. By mid-spring it becomes tough and woody.

The root is an excellent cooked vegetable, and some people enjoy the peppery and radishlike flavor raw. The strong, pungent flavor tends to dominate other ingredients, so use it sparingly, parboil to make it milder, or add it to spicy dishes, like chili or curry. You can also create a dish where it's the main ingredient. It's good to pickle, and it will thicken soup, the way okra does.

The tender, young leaves are too hairy to enjoy raw, but they're good simmered 10 minutes in soups. You can make wonderful burgers using chopped primrose leaves. I also dry them, grind them into flour, and combine them with other flours for pasta making or biscuits. The flour is good for savory dishes, but too turnipy for desserts.

When Europeans brought this herb back from America, it became very popular. Primrose-flower tea was supposed to be astringent and mildly sedative—used to calm nervousness, reduce gastrointestinal toxicity, and help with disorders of the lungs, such as spasmodic asthma and spasmodic coughing. It was used for almost any malady before it was forgotten.

Today it has finally regained its former therapeutic glory. The seeds' oil contains gamma-linolenic acid, gamma-linoleic acid (an anticlotting agent), and vitamin F (all essential fatty acids). Gamma-linoleic acid is the precursor of PGEI. This is a prostaglandin, one of many short-lived fatty hormones named because they were first isolated from the prostate gland. It regulates a variety of metabolic functions.

Supported by scientific evidence, common evening primrose oil is sold in health-food stores for premenstrual syndrome, menopause, rheumatoid arthritis, memory loss, alcoholism, weight loss, and to prevent heart disease. Nevertheless, the medical-pharmaceutical establishment's powerful lobbies are trying to remove it from the shelves and require a doctor's prescription to get it. Although it's contraindicated for epileptics, and it gives sensitive individuals headaches, skin rashes, and nausea, it's not dangerous. We need to assert our right to freedom of choice in health care, and oppose such counterproductive legislation.

Unfortunately, nobody has invented machinery to harvest the seeds automatically, so you need a large labor force and a press to extract the oil. You can't obtain this expensive oil from the wild.

However, there are other ways to use this plant to affect hormones: Find a particularly rich stand, walk your date along these plants, and the rest will be easy. You've just taken him or her down the primrose path.

JERUSALEM ARTICHOKE, SUN CHOKE

(Helianthus tuberosus)

This relative of the sunflower doesn't make seeds. It puts its energy into spreading via tuberous roots instead. The whole plant is rough-hairy, with a single, stout, erect stalk 6 to 12 feet tall, branching toward the top. Slender, lance-shaped to oval, three-ribbed, lightly toothed, pointed, long-stalked leaves grow along the stem. They're 6 to 10 inches long and 2 to 4 inches wide, broadest near the base—alternate toward the top of the plant, opposite near the bottom.

Each plant has several flower heads, 2 to 3½ inches broad, that look more like overgrown daisy flower heads than reduced sunflower heads. Each has twelve to twenty yellow rays emerging from a yellow, central disk.

This colonial perennial spreads via rhizomes and tubers instead of seeds. It stores food in tubers, identical to commercial Jerusalem artichokes. These lumpy, beige vegetables are the size of small potatoes, with crisp, white flesh inside.

The plant is easiest to identify in the summer, when it's flowering. It's harder to identify in the winter, when the tubers resemble bulbs, like those of the poisonous iris. However, bulbs are layered, tubers are solid.

Jerusalem artichokes grow in poor, light soil,

but not where there's excessive wetness. Look for them along roadsides, in fields, thickets, waste places, near streams, ponds, railways, and highways. They're native to the Great Plains, but Indians planted them in other regions, so you sometimes find them on the East Coast, and west up to the Rocky Mountains. They're also cultivated for food, and often escape into the wild.

In Italian the plant is called *girasole articiocco*. *Girasole* refers to the way that the plant turns to face the sun, and this word was corrupted to "Jerusalem." *Articiocco* is Italian for "artichoke," but the reason for naming it artichoke is unclear—the plants are unrelated and dissimilar.

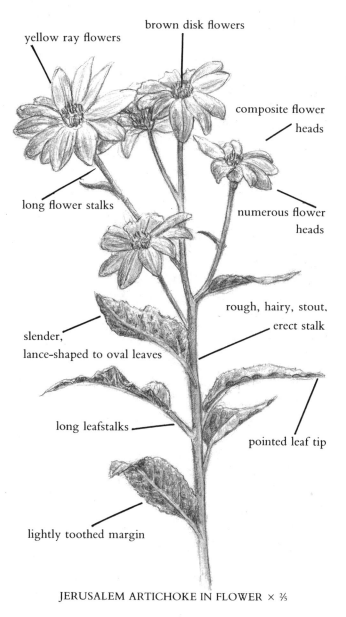

JERUSALEM ARTICHOKE IN FLOWER × ⅔

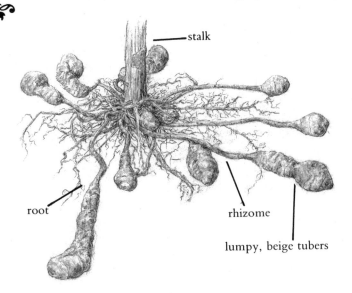

JERUSALEM ARTICHOKE TUBERS × ⅓

The tubers are best from late fall to the early spring. You can dig them up anytime the ground isn't frozen. Raw, they're light and sweet, more like water chestnuts than potatoes. They're also great baked in their skins. In general, you can cook them any way you would cook potatoes. However, they don't fry up crispy like potatoes, and they become creamy when mashed.

They have fewer calories than potatoes, and they're especially high in vitamin A and B-complex, potassium and phosphorus. They contain inulin (explained under Burdock, page 32) instead of starch, which makes them very good for diabetics and hypoglycemics.

COMMON PARSNIP

(*Pastinaca sativa*)

The parsnip is a biennial. In its first year, the stalked basal leaves become huge, often more than 2 feet long. They're feather-compound, with coarse, ragged-edged, stalkless, toothed leaflets on both sides of the midrib.

In mid-spring of the second year, the basal rosette gives rise to a stout, deeply grooved, branching, hairless flower stalk. It grows to a height of 5 feet and produces small, umbrellalike

clusters of tiny, stalked, yellow flowers in the summer. By late summer, they give way to flat, oval seeds about ½ inch long. After that, the plant dies.

The single, thick, whitish, fleshy taproot can be over 1 foot long. Don't confuse it with poison hemlock (page 204), a deadly relative with bundles of fleshy roots.

This fragrant garden vegetable often escapes cultivation, only to reappear on damp disturbed soil, empty lots, roadsides, in wetlands, and in overgrown fields, both inland and near the seashore. It grows across the United States. In the process, a transformation occurs. The root becomes externally rougher and ugly-looking, but the flavor is sweeter, and the cooked root is more tender than its cultivated ancestors.

Eat the parsnip root when the plant is in its basal rosette form, and there's stored food underground—from fall to early spring. Although it's edible raw, large quantities may be laxative, and it has a tough, stringy texture. Cooking tenderizes it and brings out its sweetness. It's a good steamed vegetable, but it's wonderful in soups or stews, or cooked with whole grains.

Pastinaca is Latin for "parsnip," coming from the Latin word for food: *pastus*. *Sativa* is a name for cultivated food plants. If there ever was an edible plant, the parsnip is it.

feather-compound leaves

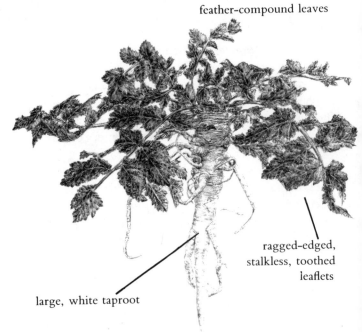

large, white taproot

ragged-edged, stalkless, toothed leaflets

COMMON PARSNIP BASAL ROSETTE AND TAPROOT × ½

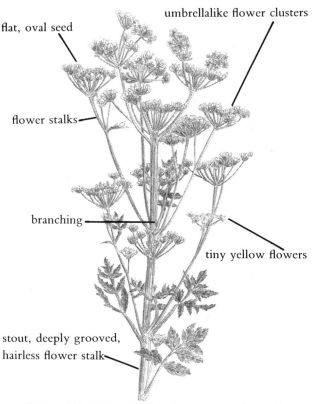

flat, oval seed

umbrellalike flower clusters

flower stalks

branching

tiny yellow flowers

stout, deeply grooved, hairless flower stalk

COMMON PARSNIP WITH FLOWER SEEDS × ⅔

Parsnips provide calcium, potassium, phosphorus, and vitamin C. But they also contain furanocoumarin, which gives sensitive people rashes after touching the leaves, especially with wet or sweaty skin, and exposure to sunlight. This happened to swimmers when parsnips grew along the edges of ponds or lakes. Wear gloves to collect this plant. .

Another warning came from Vincent Titus, a very elderly poet I met on a TV talk show in the 1980s. After he died, I found out he had been Hemingway's drinking buddy, a member of the Abraham Lincoln Brigade, the model for the protagonist in *For Whom the Bell Tolls*, Jean Harlow's lover, and jazz great Charlie Parker's roommate. He told me that medieval monks had poisoned themselves eating parsnip leaves.

Furthermore, the entire second-year plant reportedly becomes poisonous once the flower stalk appears. Some sources recommend a tea of parsnip leaves for diabetes, and for kidney and urinary problems, warning that too much may lower the blood sugar too much. But because of possible toxicity, you should avoid them. Use the root of the plant in the basal rosette form only.

GOATSBEARD, SALSIFY, OYSTER PLANT

(*Tragopogon* species)

This European biennial begins in early spring as a basal rosette, with thick, parallel-veined, grasslike leaves that may reach 1 foot long. Later in the spring, an upright stem with similar alternate leaves appears. The leaf bases partially circle (clasp) the stem, and the long-pointed leaf tips curl over. The long, fleshy taproot and all parts of the plant contain a white, milky sap.

In the spring a giant, solitary, dandelionlike flower tops the stem. Approximately eight distinctive, lance-shaped bracts grow under the flower. After fertilization, the bracts droop like a goat's beard, and the flower becomes a globe-shaped, puffy seed head, like an oversized dandelion.

There are two species, which also hybridize: Yellow goatsbeard (*Tragopogon pratensis*) grows 1 to 3 feet tall, with a yellow flower. Purple goats-

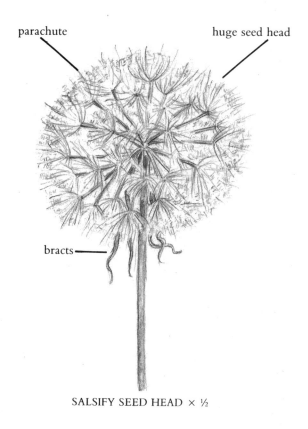

parachute

huge seed head

bracts

SALSIFY SEED HEAD × ½

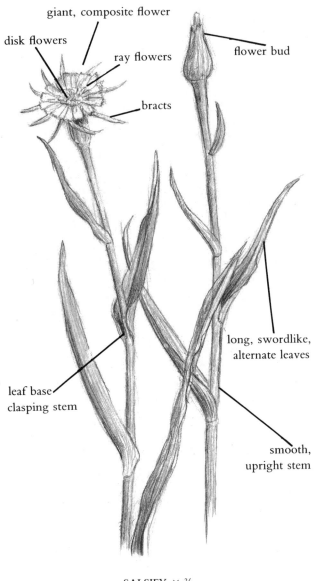

SALSIFY × ⅔

flowers, so locate them in advance. The plant is cultivated in Europe and the roots sold as food.

The leaves and thoroughly cooked roots have an oysterlike flavor, with bitter overtones. Simmer or steam the thinly sliced roots about twenty minutes. They're best in creamy or buttery recipes, with a dash of salt and pepper. A cream of potato soup with celery, onions, and goatsbeard is a delight. Its rich creaminess sets off the slight bitterness. If the roots are tough, add a pinch of baking soda to the cooking water, boil, change the water, and boil again. You can also use the root to make coffee (see Chicory, page 234).

❧

SHEEP SORREL, RED SORREL, FIELD SORREL, SCHAV

(*Rumex acetosella*)

Sheep sorrel starts out in the early spring as a messy-looking basal rosette. The leaves are soft textured and gently curving, up to 4 inches long. The long central lobe is rather thin and arrowlike, over ¾ inch wide. The two narrow, pointed outer lobes, adjacent to the leaf's base, point perpendicular to the central lobe.

Young children always love discovering this plant. Once they see the leaf's similarity to a sheep, with its long "face" and two earlike lobes, they never forget it. When I hand out samples to taste,

beard—salsify or oyster plant—(*Tragopogon porrifolius*) grows up to 4 feet tall, with a tougher taproot and a violet flower.

These relatives of dandelions and chicory grow in fields, on roadsides, and in disturbed areas throughout the United States. Because it is still escaping cultivation, distribution is spotty.

Use the very young leaves and the crowns in early spring, before they become bitter, like dandelion (page 190). Collect the roots in fall and early spring (also in winter where the weather is mild), when the plant is in the basal rosette stage. The grasslike plants are harder to identify without

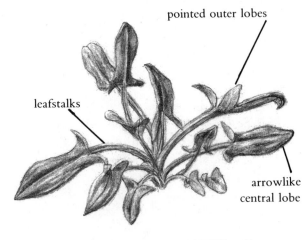

SHEEP SORREL BASAL ROSETTE × ⅔

it becomes the uncontested hit of the day, for young and old alike.

Later in the spring, it develops a somewhat lopsided-looking, smooth, slender, upright flower stalk, that may grow as tall as 20 inches. Many tiny, reddish flowers cluster on long, narrow spikes. Inconspicuous, shiny, yellow-brown fruits, enclosed in papery calyxes, follow. Seen from afar, the flowers look like rusty patches, waving low amid the grass. Drought or frost may also color the leaves rusty red.

A similar, edible close relative, seaside sorrel (*Rumex hastatulus*), dots the sand dunes in beach areas, where the long, perennial rhizomes form thick mats that hold the windblown sand together.

Caution: Sometimes sheep sorrel grows alongside bindweed and the poisonous common nightshade—both with arrow-shaped leaves. But bindweed and common nightshade are vines with different flowers and fruits, while sheep sorrel is an upright herbaceous plant.

Farmers say wherever sheep sorrel is spreading out its low-growing carpet, the soil is worn out. Scrawny patches do subsist on poor, acidic soil, where little else can grow, but I've also seen large, lush patches on rich soil in gardens. Look for it on disturbed sites, in gardens, and on old fields. This European plant grows throughout North America.

I collect sheep sorrel from early spring until the heaviest fall frosts. It doesn't become bitter or tough when the warm weather arrives, although it's easiest to collect in the basal rosette stage. By the time it flowers, the leaves are smaller and mores sparse.

The name "sorrel" comes from *sur*, French for "sour." The familiar, lemony flavor (the species name, *acetosella*, means "little vinegar plant") is more readily enjoyable than more unusual-tasting food, especially on a hot day, when it's cooling and thirst-quenching.

I also enjoy sheep sorrel raw in salads, and as trail nibble. It's tender, tart, and flavorful. Gourmet French cooks make sorrel soup, using a larger, less flavorful, commercial species. Wild sorrel soup is a creamy and tart delicacy. A traditional Jewish version of this soup is called schav (Yiddish for "calf," again referring to the shape of the leaves). Sorrel is also a traditional vegetable to accompany fish and potatoes.

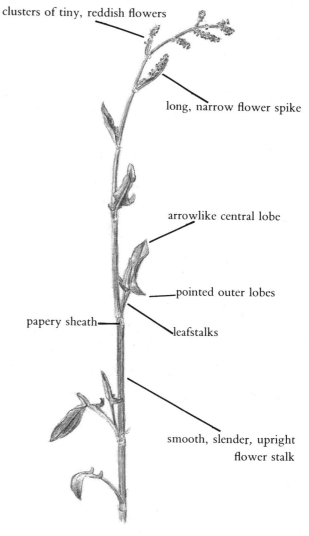

clusters of tiny, reddish flowers

long, narrow flower spike

arrowlike central lobe

pointed outer lobes

papery sheath

leafstalks

smooth, slender, upright flower stalk

SHEEP SORREL IN FLOWER × ½

Sheep sorrel is high in vitamin C. When scurvy was a problem, it was considered a cure. It's high in beta carotene, potassium, and phosphorus.

It also contains oxalic acid, which can interfere with calcium metabolism if you consume immense quantities every day for weeks. So perhaps gourmet cooks are smart to combine sorrel in soups with dairy products, with their extra calcium. In Ireland, people combine sorrel with milk and fish. Still, I've never heard of anyone getting sick from eating normal quantities of sheep sorrel. A European woman I know raves about enjoying sorrel soup with fruit and bread for dinner every day, attributing her good health to this vegetable. However, you should avoid anything with oxalic acid if you have kidney disease or rheumatoid arthritis.

A sorrel infusion is best chilled, like lemonade—sweetened, if desired. Sorrel lemonade will cool you off on a hot day, or if you have a fever. In strong concentrations, it acts as a diaphoretic. A chilled sorrel compress on the forehead is a real fever-cooler. According to the early English herbalist Gerard: "It cooleth the stomacke, tempereth the heat of the liver, openeth the stoppings there of."

The infusion is also astringent, good for diarrhea, and, as a gargle for sore throats. It's also used for internal and external bleeding, such as excessive menstrual bleeding.

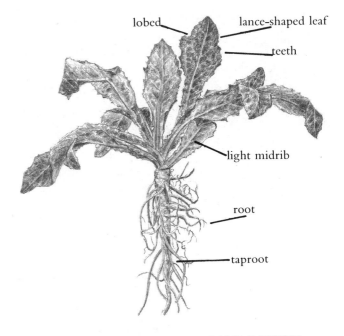

COMMON SOW THISTLE IN BASAL ROSETTE
WITH TAPROOT × ⅓

SOW THISTLE

(Sonchus species)

These European perennials resemble oversized dandelions. Like dandelions, these related plants have toothed leaves, copious white, milky sap, and similar yellow flowers, but dandelions stay close to the ground, each flower stalk has only one flower, and the leaves have larger teeth. Sow thistles have more than one flower per flower stalk, and the various species grow from 1 to 8 feet tall.

Sow thistles' leaf edges are prickly, although some species have such weak spines, you don't have to remove them. The plant owes much of its success to its reproductive proficiency. I've seen flowers as early as April and as late as December. Like dandelions, each flower makes hundreds of seeds, attached to parachutes. If I were to reproduce so efficiently, I'd be spending most of my time defending myself against paternity suits.

Wild lettuces (*Lactuca* species, page 246), are similar-looking edible relatives, but each lettuce leaf's underside's midrib has a thin line of hair or bristles, which sow-thistle leaves lack.

True thistles (*Cirsium* species) are distinct because their leaves and stems are so prickly, they can hurt you, as one of my students learned after I announced, "Let's sit down here and have our lunch break." She took me literally, sat on a thistle rosette, and spent the rest of the lunch period stand-

ing. Ouch! The flowers, which bloom in spring and summer, look like purple shaving brushes, like their nonthorny relative, burdock (page 32). They grow on disturbed soil, and in fields and freshwater wetlands.

The youngest true thistle leaves are supposedly edible, but they taste terrible, and they have so many spines, everyone who ever tried to cut them

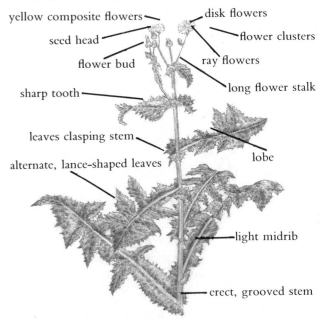

COMMON SOW THISTLE IN FLOWER × ⅔

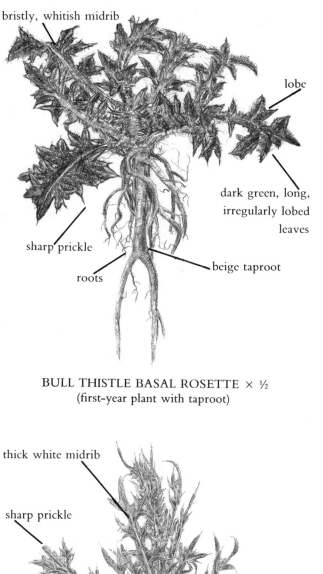

bristly, whitish midrib

lobe

dark green, long, irregularly lobed leaves

sharp prickle

roots

beige taproot

BULL THISTLE BASAL ROSETTE × ½
(first-year plant with taproot)

all off to eat the leaves died of old age first. The taproots of these upright European annual or perennial herbs are supposed to be edible in cold-weather seasons, but I rarely find substantial amounts of food on any thistle roots, and they usually don't taste especially good.

For a short period in mid- to late spring, I trim the immature thistle flower stalks with a knife, wearing heavy-duty work gloves, and eat the celery-flavored vegetable. This greatly impresses any audience. Once the flowers, appear at the top of the plant, the stems become too tough to use.

Sow thistles grow on disturbed soil, along roadsides, in meadows and fields, near the seashore, and occasionally in cultivated areas, throughout North America. Common sow thistle (*Sonchus oleraceus*) grows in most temperate regions worldwide. It's hairless, with long, sharp-pointed, stalked, toothed leaves that clasp the stem. It can grow up to 8 feet tall. *Oleraceus* means "good enough to be a garden vegetable."

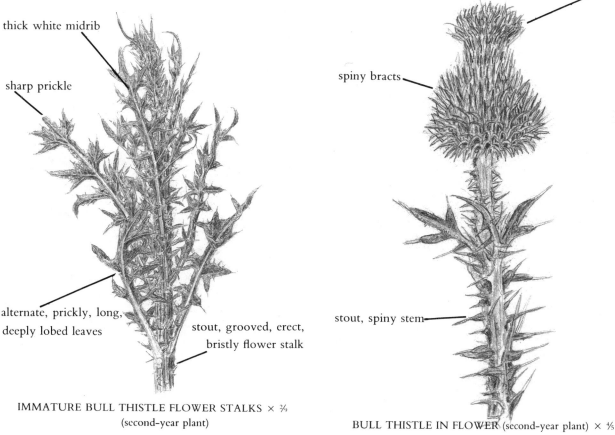

thick white midrib

sharp prickle

alternate, prickly, long, deeply lobed leaves

stout, grooved, erect, bristly flower stalk

IMMATURE BULL THISTLE FLOWER STALKS × ⅔
(second-year plant)

purple shaving-brush-like composite flowers

spiny bracts

stout, spiny stem

BULL THISTLE IN FLOWER (second-year plant) × ⅕

Spiny-leafed sow thistle (*Sonchus asper*) has very spiny, unlobed leaves that curl at their bases and clasp a hairless, ridged stem. You have to remove the spines to eat this species.

Field sow thistle (*Sonchus arvensis*), which grows 2 to 4 feet tall, has weak spines, and a perennial taproot. The leaves resemble dandelion leaves. The bracts, modified leaves under the flowers, are hairy, as are the flower stalks.

All species taste mild if you get them in the cold-weather seasons. I collect it when in the basal rosette stage in early spring and late fall to use in salads, soups, casseroles, and any wild vegetable dish. The mature plant is also good, but only after it's been touched by frost. In warm weather, it's too bitter, although you can reduce the bitterness by boiling it in several changes of water. It's one of the last vegetables to die in late fall, and it takes more than one hard frost to do it in.

Sow thistles get their name because pigs like them and they make nutritious fodder. I can always count on them as an ingredient for my "Wildman's Five-Boro Salad" at my mid-December annual Wild Party. It's an appropriate dish for this event, since some of the guests eat like pigs!

The leaves are a good source of vitamin C. The sap is antinarcotic: Chinese in turn–of–the–century San Francisco used it to help opium addicts break their addiction. The plants were used for treating animals with fever and heart disorders. For people, it was used for "wheezing and shortness of breath" (respiratory disorders), and to "prolong virility in gentlemen."

Today Popeye bolts down spinach for strength. Were he living in ancient Rome, popular belief would have him devouring sow thistle instead. This apparently also works for hares. According to legend, a hare being pursued by hounds will stop, eat some "hare's lettuce" leaves to cool its blood, and escape, reenergized.

Caution: Although there are no poisonous look-alikes, don't collect where people have polluted the soil with excessive nitrates. Because plants need this naturally occurring source of nitrogen, we sometimes add large quantities, as fertilizer. Plants may absorb these contaminants and make you sick.

WILD POTATO VINE, MAN-OF-THE-EARTH, MAN-UNDER-GROUND

(*Ipomoea pandurata*)

This native relative of the sweet potato is an inconspicuous, trailing vine that occasionally climbs. The generic name, *Ipomoea*, means wormlike. It grows 3 to 12 feet long, with strong, branching stems, sometimes tinged purplish. This perennial has alternate, heart-shaped leaves 1½ to 3 inches both across and lengthwise. Some of the leaves are narrowed in the middle, like a violin. *Pandurata* means "shaped like a pandura," a fiddlelike instrument.

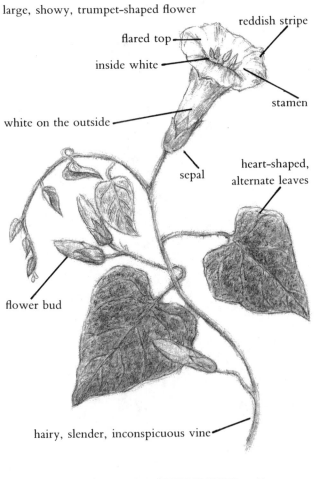

WILD POTATO VINE WITH FLOWER × ½

The showy, trumpet-shaped flower, which blooms from mid-spring to the end of summer, is 2 to 3 inches across. Its top is flared, like the related morning glory. The outside is white, and the inside is white with reddish-purple stripes. The fruit is a two-chambered capsule up to ½ inch long, broadest at the base.

Buried deeply underground is an immense taproot that can be 6 feet long and weigh up to 40 pounds, although the smaller ones are much easier to dig up and supposedly better-tasting. It's called man-under-ground because it reaches the same depth we bury the dead.

You can find the vine in wet or dry, rich or sandy soil in open or partially shaded woods and thickets, along roadsides, and by fences. It grows in the eastern United States, from New York to Kansas and Texas.

Some people say the taproot of the bush morning glory (*Ipomoea leptophylla*), a very similar erect herbaceous plant of the western United States, is excellent; others claim it's inedible. Apparently, this depends on location and other undetermined factors. Other related species' edibility is undocumented.

The bindweeds are closely related plants that resemble the wild potato vine, but without taproots and without all the characteristics listed above. The Japanese eat the flowers, but I find them too bitter.

You can bake, roast, or boil this root like a sweet potato. Never eat it raw, or it will act as a powerful laxative. Also, peel the tough skin. If the vegetable is bitter, boil in several changes of water.

I only recently located this plant, but it was in the summer, and it tasted like cardboard. It's at its best from fall to early spring. Euell Gibbons, the late foraging guru and author, said it tastes a little like a sweet potato, but maintains some crispness when cooked, and that it's one of the best and most nourishing wild foods he ever ate. Contemporary naturalist Doug Elliot, on the other hand, complains of a bitter aftertaste. Apparently, the flavor varies from root to root, possibly due to the age of the root, the soil, the time of year, and your sense of taste.

OTHER PLANTS OF DISTURBED AREAS IN LATE FALL THROUGH EARLY SPRING

Edible or Medicinal Plants

Carrot roots, chickweed, chicory leaves and roots, clover leaves, leaves and roots, curly dock leaves and roots, daylily tubers, wild onion leaves and roots, garlic mustard leaves and roots, ground ivy, horseradish root, mullein leaves, orpine roots, sassafras twigs and roots, winter cress leaves.

For Observation Only

Blackberry canes, burdock skeletons, carrot flower skeletons, common evening primrose seed heads, curly dock skeletons, foxtail grass skeletons, Japanese knotweed skeletons, milkweed skeletons, mullein seed head skeletons, orpine skeletons, parsnip skeletons, poison ivy vines and shrubs, pokeweed skeletons, raspberry canes, rose canes, sumac skeletons, sunflower skeletons.

PLANTS OF FIELDS IN LATE FALL THROUGH EARLY SPRING

WILD CARROT, QUEEN ANNE'S LACE

(*Daucus carota*)

The wild carrot looks like its domestic counterpart. They're different strains of the same species.

If you let garden carrots go to seed for a couple of generations, they'll soon revert to their wild form.

This European biennial's first-year form has lacy, branching, finely divided, feather-compound leaves in a basal rosette. The leafstalks are hairy.

The second year, it begins with a basal rosette, then develops a hairy flower stalk 2 to 3 feet tall,

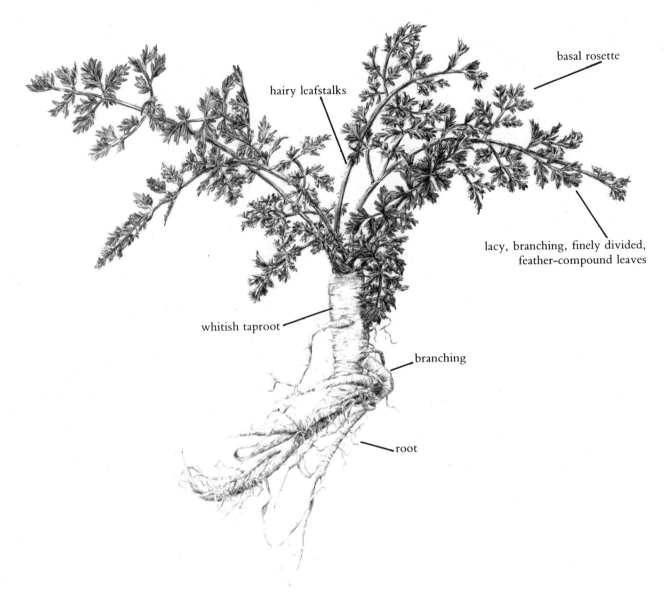

WILD CARROT BASAL ROSETTE TAPROOT × ½

topped by flat-topped clusters of tiny white flowers—Queen Anne's lace—often with a tiny, dark-purple floret in the center. Stiff, three-forked bracts grow under the flower clusters.

The plant emanates from a whitish, carrot-scented taproot 2 to 8 inches long. The root may be as small as a toothpick, or as large as a commercial carrot's. The main root grows downward, and there are often side branches.

Unfortunately, the wild carrot has deadly poisonous relatives that closely resemble it: Poison hemlock (*Conium maculatum*) has similar leaves and flowers, but no hair on the leafstalks, and a bad-

smelling, musky taproot. The hairless flower stalk is 2 to 9 feet tall, and the flowers have no bracts. The stem is spotted with purple.

Poison hemlock is forever associated with Socrates, one of the great philosophers of ancient Athens. The Socratic method involves asking a series of questions to make you think, leading to the truth, which Socrates claimed lay within. When his students began questioning authority, he was tried and convicted of "corrupting the youth." At age seventy, he drank a cup of poison-hemlock tea and, according to Plato, died painlessly.

Fool's parsley (*Aesthusa cynapium*), is also bad-

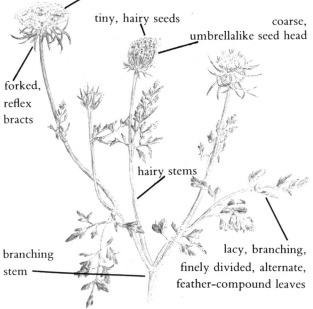

dense umbrellalike flower clusters

tiny, hairy seeds

coarse, umbrellalike seed head

forked, reflex bracts

hairy stems

branching stem

lacy, branching, finely divided, alternate, feather-compound leaves

WILD CARROT (QUEEN ANNE'S LACE) IN FLOWER AND SEED × ¼

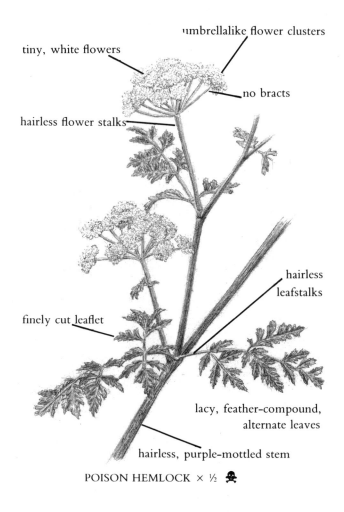

umbrellalike flower clusters

tiny, white flowers

no bracts

hairless flower stalks

hairless leafstalks

finely cut leaflet

lacy, feather-compound, alternate leaves

hairless, purple-mottled stem

POISON HEMLOCK × ½

smelling and poisonous, with finely cut leaves like wild carrot's, but hairless stems. Three-pronged, beardlike bracts dangle from its clusters of white, umbrellalike flowers.

Do not eat wild carrots until you've had extensive foraging experience. Stick to species with no poisonous look-alikes. People have confused carrots with similar-looking poisonous plants and died or became paralyzed.

Wild carrots grow in dry, overgrown fields, sandy soil, open places, and disturbed soil throughout most of the United States. You may have to investigate many stands before you locate one with worthwhile roots. I find the largest carrots in sandy fields, especially near the seashore.

The carrot's taproot is edible before the flower stalk appears. Collect from fall to early spring, when the plant stores food underground. By midspring, with no reserves, it becomes tough and woody. You can still cook it, but it comes out like cooked wood.

Wild carrots taste stronger than commercial carrots. They're chewier—a disadvantage in salads, but a plus in carrot cakes and carrot soups, where they retain a pleasant crunch, and more flavor.

Tender, young carrot leaves are edible in the early spring and fall. I chop them finely and use them sparingly, since their flavor is very strong. You can also eat the flower, which blooms from summer to fall. Add the tender, young leaves or flowers to salads or soups. Use the leaves in stock, and sauté the flowers.

When the flower's petals fall off, the underlying ovaries enlarge into tiny seeds that look like hairy caraway seeds. The seed head, which resembles a bird's nest, persists through the winter and acts as a location marker. When the seeds are still slightly green, in late summer and early fall, you can collect them and use them as a seasoning. They taste like the seeds of caraway, a relative—excellent in breads, soups, and brown rice. The flavor is very strong, so a little goes a long way. You can also dry them for long-term storage.

A strong carrot-seed tea is a carminative and diuretic. It helps the kidneys get rid of excess water, and it's used for kidney stones and bladder disease. It's also been used for chronic coughs, dysentery, liver disease, and jaundice. Much of herbal lore proves correct in laboratory experiment, but not always. Carrot-seed tea is supposed to be an aphrodisiac. A lady friend and I proved that was just a myth: Only six out of ten cups of carrot-seed tea worked!

A mixture of carrot leaves, lemon peel, and cornstarch was used as a surgical dressing by the Germans during World War II, when regular medical supplies were unavailable. The grated roots were also used externally as a poultice for wounds, gangrene, and cancer.

For all its benefits, the wild carrot has one failing. It's one of the few wild plants lacking a nutrient present in the commercial version. Wild carrots have no beta carotene, the precursor of vitamin A that helps prevent cancer. That's why wild carrot roots are white instead of orange.

The flower is named after Queen Anne of England, an expert lace maker. When she pricked her finger with a needle, a single drop of blood fell onto the lace, creating the dark purple floret.

⚜️

WILD ONIONS

(*Allium* species)

The wild onions are herbaceous perennials of the lily family, with basal leaves growing up to a few feet tall. All species have flowers with sepals and petals that add up to six. The flowers arise from a long-stalked, globular, umbrellalike flower head, sitting atop several tissuelike bracts. These native and imported plants grow in a variety of habitats throughout North America, each with its own garlicky odor and flavor. Poisonous look-alikes are odorless.

Field garlic (*Allium vineale*) is easy to recognize. Growing up to 3 feet tall by mid-spring, this European import has long, narrow, hollow, round basal leaves. They emerge from layered, underground bulbs that resemble tiny onions. It grows in bunches because bulb sections split off, propagating the plant underground. Other alliums are similar to field garlic, but most don't have hollow leaves.

Field garlic reproduces in late spring: Small, umbrellalike clusters of tiny, pink to white flowers give way to tiny white bulblets that resemble pearls of barley. These often mature and grow threadlike green leaves while they're still on the seed head. When they fall to the ground, they continue to grow into new plants. Sometimes the plant skips the flower stage and proceeds directly to the bulblets.

Look for field garlic on lawns, in backyards, on disturbed soil, and in open woods throughout

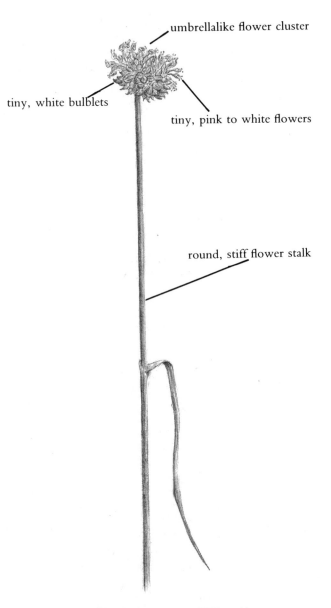

FIELD GARLIC IN FLOWER × ½

eastern North America. *Allium* means "garlic," *vineale* means "pertaining to vineyards," so the plant must also grow in vineyards.

You can find field garlic from the fall, when new shoots begin coming up, until midsummer, when the seeds fall and the plant dies to the ground. The leaves are good until mid-spring, when they become tough. Other alliums generally don't send up autumn shoots.

When I was a beginner, I could never find wild onion species, except in the field guides. Finally, a friend told me where it grew, and I bicycled five

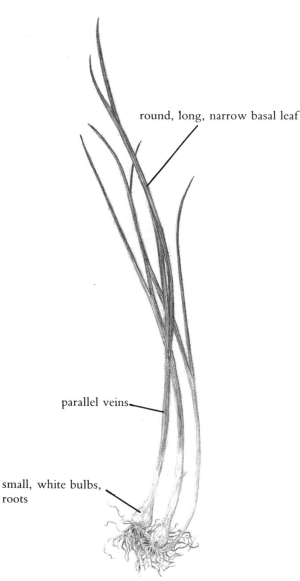

round, long, narrow basal leaf

parallel veins

small, white bulbs, roots

IMMATURE FIELD GARLIC × ½

miles to a park, where I harvested a large bagful of field garlic. Returning home, I was shocked to notice it growing everywhere. It was even thriving in front of my apartment building's entrance. I had probably been mistaking it for grass.

I can't begin to recall how often field garlic has rescued difficult winter tours. More than most other weeds, this European import thrives in cold temperatures. While other cold-weather plants form basal rosettes that hug the ground and spread out in the sun, field garlic grows tall, slumping and arching slightly in the winter—a strangely defiant demeanor in the midst of the otherwise barren terrain.

Wild garlic (*Allium canadense*) is like field garlic, with six-petaled flowers growing out of the bulblets, but with fewer flowers and bulblets. This native species has flat leaves, and doesn't grow quite so high—remaining under 14 inches tall. It grows in open woodlands, fields, and prairies, throughout the eastern half of the United States.

The nodding wild onion (*Allium cernuum*) is also similar, but with a bend in the flower stalk's tip, and no bulblets. It grows in dry woodlands, prairies, and rocky slopes in the Northeast, and south to Tennessee. You can also find it in the Northwest and the Rocky Mountain region.

The wild onion or prairie wild onion (*Allium stellatum*) has a straight flower stalk, a full cluster of lavender flowers, and no bulblets. It grows on rocky banks in the Midwest.

The wild chive (*Allium schoenoprasum*) is a garden escape, descended from commercial chives. The domestic variety is larger—wild chives don't grow much taller than 6 inches. Like field garlic's, the leaves are hollow, but they stand straighter. The flower head is an attractive lavender cluster, much like the wild onion. It grows in moist soil from eastern Canada to Alaska, in the Northeast, and from Wyoming to the Northwest.

Onions and garlic are members of the larger lily family. Many lilies superficially resemble alliums, but some are poisonous. However, **no lilies smell like onions and garlic—most are odorless.** Nothing that smells like onions or garlic is poisonous, unless you're a vampire.

The most dangerous look-alike is fly poison (*Amianthium muscaetoxicum*). It has a layered, on-

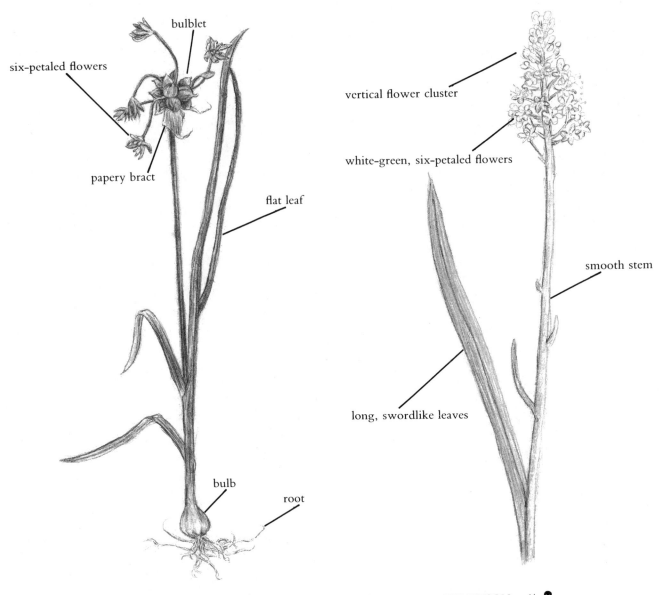

six-petaled flowers

bulblet

papery bract

flat leaf

bulb

root

WILD GARLIC IN FLOWER × ⅔

vertical flower cluster

white-green, six-petaled flowers

smooth stem

long, swordlike leaves

FLY POISON × ½

ionlike bulb which rises up into a smooth stem, topped with white to green, six-petaled flowers. Unlike the alliums, the flowers grow in vertical clusters along the stem. The plant is so toxic, touching your mouth after you handle it will make you sick. Never land on it if you're a fly!

Other poisonous, odorless lilies are death camas (*Zygadenus* species) and star-of-Bethlehem (*Ornithogalum umbellatum*). Only the early, flowerless, spring shoots resemble alliums. Death camas has white-green or pink flowers, and star-of-Bethlehem has white, five-petaled flowers and a whitish-green stripe running down each leaf's midrib.

Use alliums the way you'd use chives or scallions. Chop the leaves finely—they're not as tender as chives and scallions—and add them to salads. Steam or sauté them with other vegetables. I love putting them in sauces and spreads. I make allium-mushroom sauces, spreads with nut butter and miso, or cheese and herbs with alliums.

All parts of these plants offer exciting culinary possibilities. Use the bulbs along with the greens in any recipe that calls for onions. The bulbs are especially good in stews and pickles. They're mildest in the spring, and larger, with full garlic strength by summer.

The French bake commercial garlic in their skins twenty minutes or until soft, then squeeze out the pulp. It has a surprisingly mild flavor. Garlic's hotness occurs after you cut the raw cloves and intercellular components mix and are exposed to air. Cooking it first prevents this process. I've filled baking dishes with large, summer allium bulbs, and mashed the baked pulp with twice as much tofu or ricotta and lemon juice, plus a dash of olive oil. This makes a very unusual and wonderful spread.

Use the young late-spring bulblets like garlic, but collect them soon after they appear, before they get tough. You can also prebake them before you use them, to make them more mild.

If you like garlic oil for cooking or salads, fill a jar one-quarter full with allium bulbs, bulblets, or leaves, and cover with olive oil. Strain out the herb after a week. It lasts as long as the oil stays fresh. Make garlic vinegar the same way.

Alliums provide high levels of protein, beta carotene, vitamin C, vitamins B_1 and B_2, and minerals such as calcium, potassium, magnesium, manganese, copper, zinc, iron, tin, germanium, and selenium.

The biological activity reportedly comes from the sulfur-containing amino acids allicin, allyl sulfide, and allyl disulfides. Alliums don't smell until you cut or crush them. Then they defend themselves by converting the alliin in the oil they contain into allicin. Upon exposure to air, allicin changes into diallydisulphide, the antibiotic component that smells.

The best-known member of the genus, which doesn't grow wild in North America, is commercial garlic (*Allium sativum*). It has so many medicinal uses, entire books have been written about it. The evidence supporting this universally popular folk remedy can no longer be denied, so commercial garlic is one of the few herbs receiving major attention from the medical establishment. The wild species may be even better.

Alliums remove toxic metals, such as lead, copper, cadmium, and mercury, from the body, making it a proven "blood purifier." If they're growing near heavy traffic, they'll also "cleanse" the soil of lead, and become contaminated (although they can be planted to decontaminate soil). Collect alliums even farther from the road than other wild plants.

Garlic is a natural antibiotic. One milligram of allicin is the estimated equivalent of fifteen standard units of penicillin. Garlic species are effective against a wide range of pathogenic bacteria, influenza, meningitis, fungal infections, and vaginitis caused by yeast. They safely kill roundworms, pinworms, tapeworms, and hookworms. Indians even applied bruised wild garlic roots to bee and wasp stings, with good results. Experimental evidence shows that alliums lower cholesterol, reduce blood pressure, and inhibit clotting—useful for heart disease.

Garlic increases circulation to the heart and relieves tight headaches. It's great for some people with poor digestion, although it irritates other peoples' digestive tracts. Russian researchers successfully treated persistent bronchial and lung infections with vaporized garlic compounds.

Unlike pharmaceutical hypertensives, garlic reduces high blood pressure without depleting the body of vital minerals. The only serious side is bad breath—treatable with wild mints or plantain (pages 50 and 227).

The active properties are oil-soluble, so garlic–olive oil infusion has been used for soothing earaches. Before you experiment with home remedies, make sure you don't have a dangerous infection that could permanently damage your hearing.

Deodorized garlic, sold in health-food stores, is also advertised to be effective, even though the sulfur-containing compounds are removed. I prefer the whole plant: Deodorized garlic never gets me a seat in the New York City subway.

OTHER PLANTS OF FIELDS IN LATE FALL THROUGH EARLY SPRING

Edible or Medicinal Plants

Bayberries, chicory leaves and roots, clover leaves, cow parsnip roots, curly dock leaves and roots, goatsbeard roots, groundnut roots, Jerusalem artichoke roots, mullein leaves, orpine roots, parsnip roots, sassafras twigs and roots, sow thistle leaves, yarrow.

For Observation Only

Blackberry cane, burdock skeletons, curly dock skeletons, foxtail grass skeletons, Japanese knotweed skeletons, milkweed skeletons, mullein skeletons, orpine skeletons, parsnip skeletons, pokeweed skeletons, poison ivy shrubs and vines, rose canes, sunflower skeletons, yucca.

PLANTS OF THICKETS IN LATE FALL THROUGH EARLY SPRING

GROUNDNUT, HOPNISS

(*Apios americana*)

This is an inconspicuous, hairless, twining vine with alternate, feather-compound leaves that grows up to 10 feet long. Each leaf consists of five to nine lance-shaped, toothless leaflets from 1 to 2¼ inches long. All the leaflets except the leading one are paired. The vine, which contains a white, milky sap, becomes more noticeable in the winter, when its skeleton turns white.

The showy, pale-maroon or purple-brown, fragrant, pealike flowers bloom from late spring through the summer. They're arranged in clusters that are much smaller, shorter, and tighter—2 to 6 inches long—than other edible legumes with showy flowers, such as wisteria and black locust. The pods are 2 to 4¼ inches long, ¼ inch broad, with edible, lentillike seeds.

Follow the vine underground to locate a necklacelike string (rhizome) of tubers, ranging from the size of a marble to that of a golf ball. Cultivated, this may yield five pounds of tubers.

The perennial vine twines around fences and bushes in moist woodlands, bottomlands, thickets,

fields, near streams, in meadows, and near the seashore. You can find it throughout most of eastern and central North America, from southeastern Canada down to Texas.

This soybean relative's tubers are 25 percent protein—three times that of potatoes. The seeds contain vitamins A, C, and E. Groundnuts were a staple for the Indians, and a lifesaver for early European explorers and the Pilgrims, especially during the winter.

The light, nutty-tasting tubers are edible all year, but they're best in late summer and fall. Boil

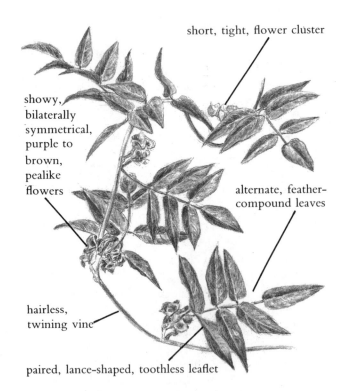

short, tight, flower cluster

showy, bilaterally symmetrical, purple to brown, pealike flowers

alternate, feather-compound leaves

hairless, twining vine

paired, lance-shaped, toothless leaflet

GROUNDNUT IN FLOWER × ⅓

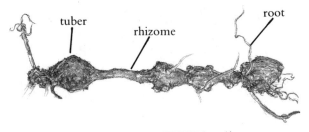

tuber

rhizome

root

GROUNDNUT TUBERS × ⅓

or roast them like potatoes, or slice and sauté them. You can also eat the young pods or ripe seeds, in season from midsummer to fall, as cooked vegetables in the summer. However, they're too small to collect in quantity.

Caution: There are poisonous plants in the legume family. Be certain your plant's characteristics match up exactly with those described above, and don't forage for this plant if you're an unsupervised beginner.

JUNIPER

(*Juniperus* species)

Junipers are woody evergreen plants that range from spreading or erect shrubs to medium-sized trees, with needlelike to scalelike leaves. The fragrant bark is thin and soft, shredding easily. Hard, aromatic, scaly, blue to blue-green berries make identification easy.

Thirty species are native to North America, normally growing in poor, rocky, or mountainous soil. They are especially common in the West, although you can find them throughout the United States. Also, many imported species are planted in cultivated areas throughout the country.

The common juniper (*Juniperus communis*), growing from 2 to 20 feet tall, is one of the very few evergreen bushes native to North America. This alone makes it easy to identify. Its hollow, sharp leaves are ½ inch long, surrounding the stem in whorls of three. There's also a characteristic bright white line on each leaf's upper surface. The reddish-brown bark shreds in papery strips.

The fragrant, light blue-green fruit is dusted with a powdery, waxy bloom. The fruit's three fleshy scales unite at the tip in a fashion reminiscent of a pine cone. This "berry" is actually a type of cone where one of the scales has not become dry and woody.

This low, straggling shrub's branches take root, and send up more bushes, where they touch the ground, creating dense barriers for people and secure havens for small birds and mammals.

The bushes come in separate sexes. Only the females have berries. Wind transports the male flower's pollen to the inconspicuous female flowers. The fruits ripen at the end of the summer, but don't hurry to collect them: They persist through the winter.

Gathering juniper berries creates a satisfying feeling. The tough berries have substance to them, and you can collect a small bag in fifteen minutes. This will last a long time.

Caution: Most species contain a powerful resin (only a few of the western species are sweet and nonresinous). Use them sparingly, as sesaoning or infusions only. Eating more than two fruits of the resinous species is enough to poison a child.

You can dry juniper berries for long-term use, or freeze them. Some Indians ground dried nonresinous species' berries into flour. Mince a single berry, and use it in a main course recipe that serves eight. I've used juniper berries in casseroles with tofu, and with rice, stews, sauerkraut, and vegetables. They add a nice tang. People use them with meat, especially lamb, for flavor, and to stimulate meat digestion, the same way ginger is supposed to complement and help digest beef and chicken.

Juniper berries contain sugars, vitamin C, flavonoids, sulfur, resin, and other substances. They're important medicinally, but you should use them only in very small amounts, for short periods of time.

The herb has also been renowned as a diuretic since the days of Pliny, Dioscorides, and Galen,

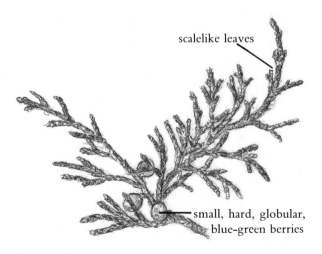

scalelike leaves

small, hard, globular, blue-green berries

JUNIPER × ⅔

who recommended an infusion for water retention. It's also good for cystitis and urethritis. The berries supposedly stimulate and soothe the kidneys, but some sources say they work by irritating the kidneys. Juniper extracts are used in over-the-counter diuretics, which are infamous for their harshness and dangerous side effects to me. Repeated overdoses of juniper infusion may irritate the urinary tract, or even cause kidney failure. Long-term use may interfere with the absorption of iron and other minerals. Never use it during pregnancy.

The herb is also considered a stomach tonic—used to help rid the body of gas and putrefaction, and good for ulcers. Extracts are used in over-the-counter laxatives. Juniper berries are recommended for regulating hypoglycemics' blood-sugar levels. Inhaling the infusion's steam is good for colds, coughs, and mucus.

You can make an oil extract (or flavor vinegar the same way) by soaking two tablespoons of juniper berries in one cup of olive oil, in a warm place, for five to six days. This is taken for urine retention and bladder problems. You can also use it to sauté food: Make a fantastic fried rice with juniper oil, tamari soy sauce, cayenne, and a little thyme.

The juniper oil you get in the herb store is an essential oil, made by distillation in a lab. Never take it internally—it's much too strong—especially dangerous if you're pregnant, or have kidney dis-ease. External contact causes blisters. Diluted, it's great for massaging sore muscles. Rub diluted oil on your temples to soothe a nervous, jumpy mind. You can also rub in diluted juniper oil to ease the pain of arthritis, sciatica, and neuralgia. **Caution:** Be careful not to get any in your eyes. Juniper is an antiseptic and astringent. For coughs and colds, the Indians would chew on a few berries, swallow the juice, and spit out the pulp. This could lead to overdose, so stick to a dilute infusion.

OTHER PLANTS OF THICKETS IN LATE FALL THROUGH EARLY SPRING

Edible or Medicinal Plants

Hog peanut seeds, Jerusalem artichoke roots, sassafras roots and twigs, spicebush twigs, viburnum bark, highbush cranberries, wild potato vine root.

For Observation Only

Blackberry canes, grapevines, Japanese knotweed skeletons, mugwort skeletons, poison ivy vines and shrubs, pokeweed skeletons, raspberry canes, rose canes, sumac skeletons.

PLANTS OF WOODLANDS IN LATE FALL THROUGH EARLY SPRING

ANISEROOT, SWEET CICELY, SWEETROOT

(Osmorhiza species)

Aniseroot is a soft plant with multiple compound leaves that look like broad-leaf parsley and smell like licorice. The lower leaves can reach 1 foot in length, and the entire plant grows from 1 to 3 feet tall. Different species may be hairy or hairless, with variations of aroma and flavor.

Aniseroot is an umbelliferous plant, related to carrots, parsnips, and parsley. It displays small, sparse, flat-topped, umbrellalike clusters of five-petaled white flowers in the spring. The seedpods, which appear in the summer and persist into autumn, are about 1 inch long, thin, curved, and tapered—black when mature. The gnarled, sometimes clustered, perennial, fleshy rhizomes are light beige, up to 6 inches long.

Coming upon aniseroot as I walk along a forest path is always a delight. The plant's fragrance and

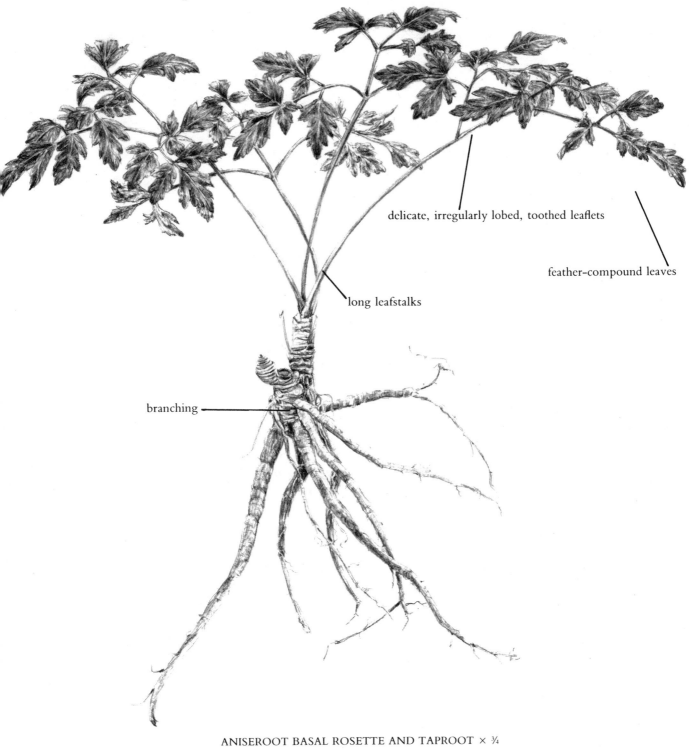

delicate, irregularly lobed, toothed leaflets

feather-compound leaves

long leafstalks

branching

ANISEROOT BASAL ROSETTE AND TAPROOT × ¾

feathery air always seem in perfect harmony with the enchantment of the woods. It grows in partially open areas of moist woods throughout the United States and Canada. Aniseroot is a common ground cover under one of my favorite black walnut trees,

and I've become very cautious about walking through the aniseroot when I gather nuts in autumn. Otherwise, the hard, sharp seedpods catch on to and pierce my clothing like botanical needles.

Use the "roots" in early spring and fall. They're

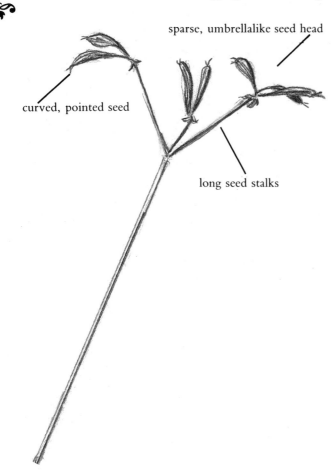

curved, pointed seed

sparse, umbrellalike seed head

long seed stalks

ANISEROOT SEED HEAD × ½

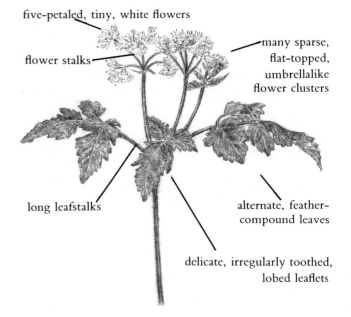

five-petaled, tiny, white flowers

flower stalks

many sparse, flat-topped, umbrellalike flower clusters

long leafstalks

alternate, feather-compound leaves

delicate, irregularly toothed, lobed leaflets

ANISEROOT IN FLOWER × 1

hard to clean, but a little of this strongly flavored treat goes a long way. They tend to be stringy and tough, so they're best finely grated and cooked twenty minutes. Try them mixed with other vegetables or grains as well as in desserts. Add some to fruit sauces, or cook them in fruit juice and puree them in the blender, using the liquid in baking. I've made some of my best anise-flavored cookies and cakes this way.

The leaves, flowers, flower stalks, and young green seedpods are great raw and in salads, and delicious lightly steamed or simmered in soups. They impart a licorice flavor, and are surprisingly tender. They're especially good with Japanese knotweed, moderating its extreme tartness. Freeze the cooked plant to store it, but don't dry aniseroot. When I tried, the whole apartment smelled like anise, as the volatile oils that impart all the flavor vanished into thin air.

After aniseroot flowers and goes to seed, in the early summer, it dies to the ground. You can still locate the plant if you've observed its life cycle and learned to recognize the dead flower stalks and lingering seeds. Otherwise, look for the basal leaves when they reappear in autumn.

You can steep the leaves, flowers, roots, or immature roots to make a delicious stomach-strengthening tea, which is also gently carminative—good for indigestion and flatulence. Various Indian tribes once applied fresh or dried powdered roots to sores, cuts, boils, and wounds. Perhaps it has antimicrobial properties. Some diabetics and hypoglycemics also use the root as a sugar substitute.

Caution: Don't confuse this plant with water hemlock (page 128) and poison hemlock and related poisonous plants (covered on page 204).

BLACK BIRCH, SWEET BIRCH

(*Betula lenta*)

Black birch is one of my favorite edible trees. This tall woodland species grows from 50 to 80 feet high. It has dark, smooth, shiny, largely non-peeling bark, although some older trees' bark shows large, irregular cracks. The bark has horizontal light-gray lenticels that look like dashes.

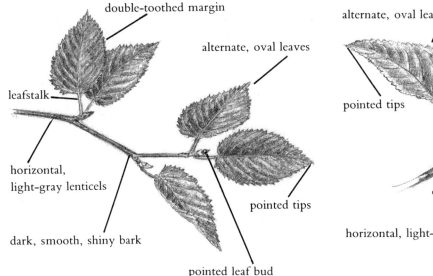

double-toothed margin

alternate, oval leaves

leafstalk

horizontal,
light-gray lenticels

pointed tips

dark, smooth, shiny bark

pointed leaf bud

BLACK BIRCH × ¼

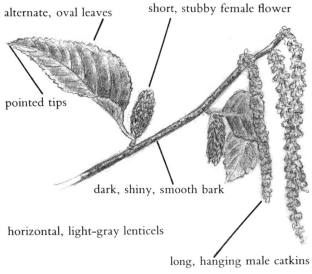

alternate, oval leaves

short, stubby female flower

pointed tips

dark, shiny, smooth bark

horizontal, light-gray lenticels

long, hanging male catkins

BLACK BIRCH IN FLOWER × 1

Black cherry's similar-looking bark has lots of cracks.

The simple, alternate, oval, pointed, stalked leaves, 4 to 6 inches long, are bright green and shiny above. They're especially distinctive because they're double-toothed. Male and female flowers grow in small, stalkless catkins, separately on the same tree, in spring. The male catkins are the longer ones. The fruits are small, upright cones a little over 1 inch long with many tiny, winged seeds inside. Look for them in late summer to fall.

Black birch is easy to identify any time of year. Scratch the twigs, and you'll smell wintergreen. The tree contains the same oil as wintergreen (page 222). **Caution:** Never take pure oil of wintergreen, sold in herb stores, internally. Such concentrations are toxic.

There are no poisonous look-alikes. Other birch species range farther south, and you can use them the same way, although their flavor is weaker. Yellow birch (*Betula lutea*), for example, has yellow to silver-gray bark that peels in narrow curls.

Black birch grows on protected slopes, in open woodlands, moist areas, and deep, rich soil in the northeastern United States and Canada. This easy-to-propagate native tree grows quickly, so environmentally oriented landscapers can use it to restore woodlands and to encourage other native life-forms to take residence.

You can chew the twigs for a delicious trail nibble all year—a sure cure for wild-garlic breath. Black-birch-twig infusion makes a very popular beverage. Don't boil the twigs, or you'll drive off the volatile oils.

I start many classroom presentations by asking whether anyone has ever eaten a tree. When the teacher survives a "forced" nibble, everyone wants a taste. Kids of all ages always love it, although some remark that it tastes like toothpaste or chewing gum or wintergreen Life Savers.

You can also cut away the outer bark and remove the delicious inner bark with a knife. Collect after a tree's been knocked down by a storm, so you don't injure living trees. This emergency survival food, in season all year, tastes like wintergreen candy. Eat it as is, grind it into flour, or add it to soups.

Birch bark is surprisingly nutritious. It provides beta carotene, calcium, vitamins B_1 and B_2, calcium, copper, iron, manganese, phosphorus, potassium, sodium, and silicon. It also contains xylithol, a harmless natural sugar substitute. Other constituents include betulinal and a glucoside.

A strong inner-bark or twig infusion is used to treat edema, bladder infections, kidney stones, gout, rheumatoid arthritis, and eczema, and a compress is good for bathing skin eruptions (page 9).

You can also tap the tree in late winter and use

the sap as emergency water or to make a maplelike syrup. Although the sap is natural, boiling it down to less than one fortieth of its original volume transforms it into one of the most highly processed forms of sugar, as unhealthful as any other form of concentrated sugar.

HOG PEANUT

(*Amphicarpa bracteata*)

This slender, sparsely branched, hairy, climbing, native herbaceous plant grows up to 4 feet tall. It has palmate-compound leaves consisting of three oval, short-pointed, toothless leaflets. It produces two kinds of flowers and two kinds of seeds.

Aboveground, racemes originate from the leaf axils, bearing small, pale-purple or white pealike flowers, ½ inch across. They bloom from summer to early fall. They're followed by small, flattened bean pods, each containing one seed.

Additional tiny petalless blossoms arise from the roots on threadlike, creeping branches. These hidden, self-fertilizing flowers turn into delicious seeds or "peanuts." (The true peanut's flowers and seeds are also subterranean.) The pealike seeds are light brown, up to ½ inch in diameter with a tough outer covering. When they're ripe, you'll find them among masses of white stringy vines entangled along the ground, just below the leaf litter. Sometimes I can cut away a whole layer of them and take home a cup or two of seeds.

They grow on rich soil in woods and thickets in the eastern half of North America. Collect the seeds in fall and early spring (in the winter, in warm regions). The pods ripen in the summer.

The outer shell comes off easily, especially after soaking. Use them like beans. They cook in about ten minutes, and they're much tastier than com-

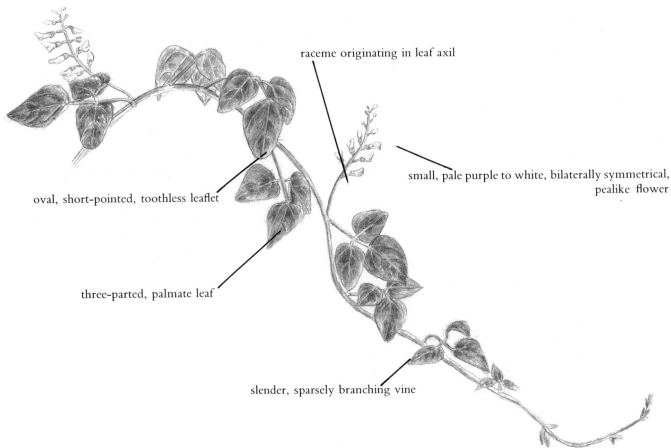

raceme originating in leaf axil

oval, short-pointed, toothless leaflet

small, pale purple to white, bilaterally symmetrical, pealike flower

three-parted, palmate leaf

slender, sparsely branching vine

HOG PEANUT × ½

mercial beans. Like beans, they're high-protein, around 25 percent. You can also eat the young pods, while they're still tender enough to pinch through. However, they're so small, they're not good for more than a trail nibble.

PINE

(*Pinus* species)

Pine trees are easy to identify, common, and edible. You recognize these evergreen trees by their long, thin, round needles. If you mistakenly collect spruce or fir needles, you can use them the same way. Just don't collect yew needles (page 143), which are flat instead of round, and poisonous.

Pines are very easy to distinguish from other evergreen, cone-bearing trees. They're the only trees with long, slender needles in bundles of one to five. Unfortunately, it's not always easy to tell one pine from another, since many species have two to three needles per bundle.

The distinctive white pine (*Pinus strobus*) is 150 feet tall, with five needles per bundle, and slender, cylindrical cones 5 to 6 inches long. The cone's scales open wide. This eastern species grows straight and tall, with dark, deeply furrowed bark. It was invaluable for making masts for sailing ships.

The pitch pine (*Pinus rigida*) is a small, gnarled, misshapen-looking species. It's especially hardy, growing on dry, rocky soil other trees can't tolerate. It's one of the first trees to reappear after fires, requiring intense heat to melt the resins that seal the seeds.

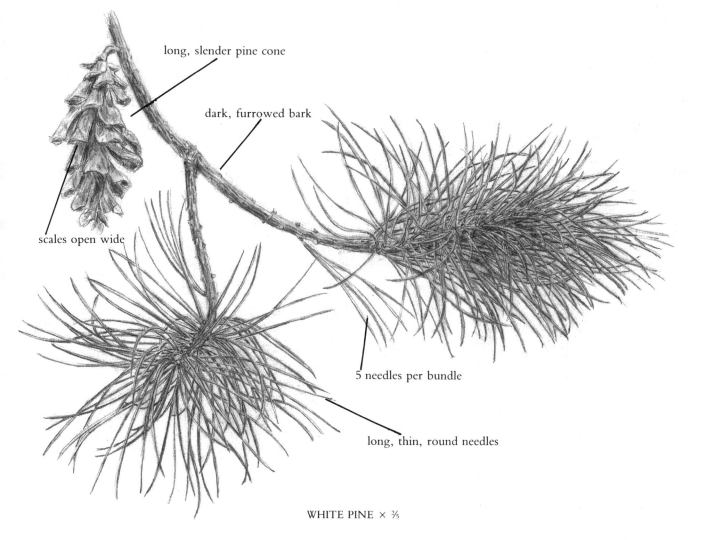

long, slender pine cone

dark, furrowed bark

scales open wide

5 needles per bundle

long, thin, round needles

WHITE PINE × ⅖

The tall red pine (*Pinus resinosa*) is named after its reddish-brown bark, which has broad, scaly plates. There are two needles per bundle. It grows scattered in mixed forests in the Northeast.

The shortleaf pine or yellow pine (*Pinus echinata*) is a large tree with horizontally spread branches, reddish-brown, irregularly scaled bark, and medium-sized needles in twos or threes. It's very common in eastern North America, especially in the South.

Pines are the needle-bearing members of the conifers, which in turn are a major part of a group of trees and shrubs called the gymnosperms. Gymnosperm means "naked seed." They have catkins and cones instead of flowers and fruits, and the seeds usually aren't surrounded by a fruit or shell. In the flowering plants (angiosperms), the seed is enclosed in an ovule—the seed cover or fruit.

The seeds of most species are too tiny to eat, but you can still show them to children, who are usually fascinated by pine cones and love taking things apart. If you live in the West, there are species of piñon pines, with edible pine nuts, growing on slopes and canyons just above the sagebrush deserts, that ripen in late summer or fall.

The one-leafed piñon or single-leaf piñon (*Pinus monophylla*) is a spreading species with coarsely furrowed, dark-brown bark. It's the only native species with one needle per bundle. The egg-shaped cones are 2 to 3 inches long, consisting of thick, stiff, blunt, light-brown scales. Each scale contains two brown, oval, wingless seeds ¾ inch long. It usually produces a bountiful crop every few years. It grows in Utah and adjacent portions of nearby states.

The piñon pine (*Pinus edulis*) is a small, spreading, bushy southwestern species with thin, irregularly furrowed, scaly, grayish to reddish-brown bark. It has two needles per bundle. The cones are egg-shaped, 2 to 2½ inches long. Their thick, blunt, yellow-brown scales each have two egg-shaped seeds, ½ inch long. They produce large crops every three to four years.

The sugar pine (*Pinus lambertiana*) grows over 200 feet tall, on mountain slopes in the West Coast states. The trunk is straight, and the reddish-brown bark is thick with deep, irregular furrows. The needles are five per bundle. The egg-shaped cone, 1½ feet long, contains oval, blackish seeds ½ inch

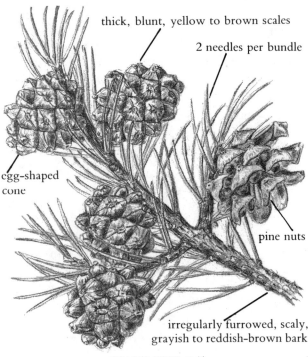

thick, blunt, yellow to brown scales

2 needles per bundle

egg-shaped cone

pine nuts

irregularly furrowed, scaly, grayish to reddish-brown bark

PINON PINE × ½

long. They're often difficult to collect because they grow so high.

The digger pine (*Pinus sabiniana*) is a spreading western tree that grows to 50 feet tall. The blackish bark is irregularly scaly, and there are three needles per bundle. The spiny, egg-shaped cones are as long as 10 inches. The dark-brown to blackish, oval seeds are ¾ inch long.

If you've ever eaten commercial or wild pine nuts, you'll know how good they are. Quality in the wild varies—some trees have sweet and buttery nuts, others are more bitter, but there are no poisonous look-alikes. All are rich in protein (14.5 percent) and fats, with lots of calories.

Store the cones in a dry place, so they release their seeds. Roasting the dry cones facilitates seed removal. Shell the nuts and use them raw or roasted, the same as commercial pignolia nuts, or any nuts. The high oil content makes shelled pine quite perishable, so refrigerate or freeze them, or use them soon.

Pine inner bark is full of starches, sugars, vitamins, and minerals, but it tastes terrible—a survival food for dire emergencies. Multiple boiling will make it more digestible and tasteless—a vast improvement!

You can also eat the pollen cones (aments). I've scraped the youngest, most tender growth, and it's as bad as the inner bark. Although some individuals like it, the next time I consume it will be when I'm lost and starving, with a case of selective amnesia that makes me forget every other wild food.

The inconspicuous male cones or catkins produce tasty, nutritious pollen, which you can collect in the spring. (See directions under Cattail, page 67.) Look for a fine, yellow powder underneath pine trees that have cones, during the first year of the two-year cone-growth cycle.

People also eat the young shoots that appear at the tips of the twigs in the spring—straight from the tree, peeled, or cooked. I think they're awful, but some people like the tender texture. Early settlers candied them by boiling them in heavy sugar syrup until transparent.

From the Indians, we also learned to chop and infuse the eastern species' needles, in season all year, using one part needles to two parts water. They used the delicious, vitamin C–rich tea (which also contains beta carotene) to save the lives of scurvy-ridden European explorers. It's also an expectorant and diuretic. You can even pour some into the bathtub for a pine-scented bath. **Caution:** Livestock have reportedly been poisoned by eating needles of ponderosa pine (*Pinus ponderosa*) and loblolly pine (*Pinus taeda*)—western species—so use only eastern pines for tea.

Because pine resin is an effective natural antibiotic, the inner bark of pine trees is good for wounds, sores, and insect bites. You can either use it as a bandage, tied in place with string or cloth and kept moist, or you can make a poultice. In past times, people ground the bark, soaked it in liquor, and put it on wounds. It's supposed to stimulate new, healthy skin to grow.

You can make pine-bark syrup by filling a jar with white-pine bark and hot water. Steep for a few hours. For every cup of water, add three quarters of a cup of vodka or brandy. Let it infuse for a day, then strain out the bark. You can add one-half to one cup of honey, to improve the taste and preserve it. Some traditional cough syrups also contain wild-cherry bark, sassafras root, and other herbs, but pine alone is simple and effective.

Indians applied a salve made of pine pitch mixed with lard and tallow to boils, external ulcers, and wounds caused by splinters, with good results.

Turpentine and pine tar are strong resinous constituents of the pine tree, extracted by distillation. People in the eighteenth and nineteenth centuries used them for everything from worms to chronic rheumatism and skin diseases. However, they're dangerous and highly carcinogenic. Large quantities of turpentine taken internally can cause vomiting, convulsions, shock, and death. Avoid even inhaling turpentine fumes.

All pine species are very useful in other ways: Native people used the wood for building shelters, tools, and boats, and they used the thick, resinous sap for glue. Pines were among the first species the European settlers overexploited. They used them for building ships and houses and making industrial products such as turpentine and pitch.

SASSAFRAS

(*Sassafras albidum*)

If there's one wild edible many people already know before attending my tours, it's this one. Sassafras is one of the easiest trees to identify. It's a medium-sized tree, 10 to 125 feet tall, with irregularly furrowed, red-brown bark. The long, flattened, irregular vertical ridges are randomly split by horizontal cracks.

The three different kinds of stalked, toothless, fragrant leaves, all 3 to 5 inches long, are most distinctive. One leaf is oval, one has two lobes like a mitten, and one has three lobes. The mulberry tree also has various kinds of leaves, but they're toothed and odorless.

Sassafras's tiny, yellow, five-petaled, radially symmetrical flowers bloom in the spring, just as the leaves are beginning to form—clusters of males and females on separate trees. The blue-black, egg-shaped fruits, about ½ inch long, appear in red, long-stalked cups in the summer.

Sassafras is even easy to identify in the winter. This native "weed tree" can't compete with taller trees in a maturing forest, so it reproduces prodigiously before it gets shaded out. Wherever you find mature trees, you'll find many immature saplings. They're all green, and the twigs on mature and immature trees arch upward like a candelabra.

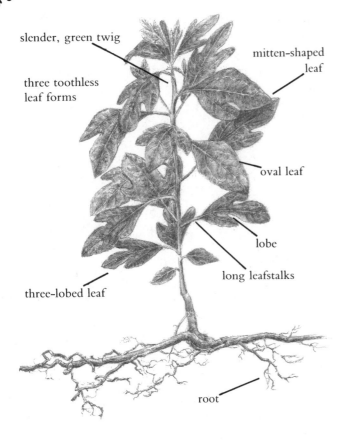

slender, green twig

mitten-shaped leaf

three toothless leaf forms

oval leaf

lobe

long leafstalks

three-lobed leaf

root

SASSAFRAS SHOOT × ¼

The green twigs are hairy under magnification, and all parts of the tree have a pleasant fragrance when you scratch or crush them. If you scratch the bark of a mature tree, it's reddish underneath.

Sassafras grows in old fields, on the borders of woods and in open woodlands, on bottomlands, in thickets, on lower slopes, and in many urban parks. It's also an occasional street tree. It grows throughout the eastern half of North America.

You can gather sassafras all year. The leaves are in season from spring to fall, but the root is good all year. So are the twigs, although I usually collect them only in the winter, when the ground is too frozen to remove any roots.

Any part of this tree—leaves, twigs, or roots—makes a wonderful reddish-brown herb tea, although the inner bark of the root is best. Two or three people pulling steadily can easily uproot some of the smaller saplings near the larger parent trees. There's usually plenty to go around. Choose saplings growing in the shade, which can't survive to maturity anyway.

Sassafras is one of the few aromatic herbs you simmer rather than steep (although you can make an infusion with the leaves). Wash the soil off the sapling's root, simmer 20 minutes in a covered pot, strain, and serve. The longer you simmer, the stronger the tea. The tea is also good chilled, and kids love making root beer by adding seltzer water and sweetener to it. You can also make jelly from the tea by adding agar and sweetener—great on pancakes.

You can also dry whole roots, or the root's scraped-off inner bark. Use this for making tea, or powder the dried inner bark in the spice grinder or blender, and use it in place of cinnamon.

You can dry sassafras leaves and store them, or grind the dried leaves into a powder, called gumbo filé—an expensive gourmet item. When you've finished cooking enough soup to serve six, turn off the heat and stir in half a cup of filé. It flavors and slightly thickens the soup. This is how Southerners make gumbo, although they sometimes use okra instead.

Sassafras contains oils, fats, resins, wax, camphor, albumin, starch, gum, lignin, tannin, salts, and sassafrid.

Indians used sassafras-root decoction as a blood purifier or detoxifier, and as a traditional spring tonic. It was one of our first export crops, outselling even tobacco in seventeenth-century Europe. People used it to treat virtually everything.

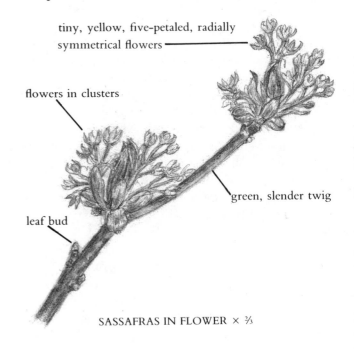

tiny, yellow, five-petaled, radially symmetrical flowers

flowers in clusters

green, slender twig

leaf bud

SASSAFRAS IN FLOWER × ⅔

small, shiny, black berries

SASSAFRAS × ½

Sassafras compresses are good for external infections, burns, and poison ivy rash. (It's also good internally for poison ivy.) The tea is also a diaphoretic and alterative. People use it for colds, fever, chronic arthritis, gout, high blood pressure, kidney problems, stomachaches, and skin diseases such as eczema. The aromatic oil distilled from the root, and safrole, a phenolic ester, are supposed to relieve the pain of menstrual obstruction, and parturition (childbirth).

Before antibiotics were discovered, sassafras was unsuccessfully used, along with sarsaparilla, to treat syphilis. Suddenly, Europeans stopped drinking sassafras tea, so they wouldn't be suspected of having venereal disease.

The U.S. Food and Drug Administration removed sassafras from the health-food stores because astronomical quantities of concentrated synthetic safrole caused cancer in rats. Rats convert safrole, which is not carcinogenic, into a carcinogen. Humans do not. Nevertheless, beer, with a carcinogenicity index fourteen times that of sassafras for rats, is still on the market, and gourmet stores sell gumbo filé powder. People have enjoyed sassafras in moderation for thousands of years with no ill effects and consumed it in root beer before it was replaced by artificial chemicals. I think sassafras is safe.

WILD GINGER, INDIAN GINGER

(Asarum species)

Wild ginger is a stemless, native herbaceous plant that usually grows under 1 foot tall. This native's beautiful leaves are its most conspicuous characteristic—heart-shaped, elegantly veined, and softly fuzzy. They're dark-green on top and lighter beneath, from 4 to 7 inches broad, and about the size of the palm of your hand. The slender, long, hairy leafstalks display them in pairs.

The slender-stemmed, deep brown-purple three-petaled flowers, which bloom from early to mid-spring, aren't readily visible: They're usually borne against the ground, in the crotch of the two leafstalks, or even buried by forest debris—an ideal location, since they're pollinated by ants. There are no petals, just the three reddish sepals that unite to form a protective layer around the cream-colored ovary.

The brittle, branched rhizomes, about as thick as a pipe cleaner, run very close to the ground's surface, connecting the dense colony. They're crisp and tender, with a clear ginger odor and taste—more subtle than the unrelated commercial ginger.

Three very similar wild ginger species (*Asarum canadense*, *Asarum acuminatum*, and *Asarum caudatum*) grow from British Columbia to California.

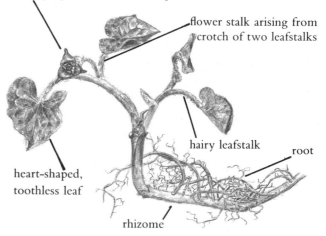

deep-purple to brown, three-parted flower

flower stalk arising from crotch of two leafstalks

hairy leafstalk

root

heart-shaped, toothless leaf

rhizome

WILD GINGER WITH FLOWER AND RHIZOME × ¼

Asarum canadense is the eastern species. It grows in all the eastern states except the Deep South, and west to Minnesota.

Look for wild ginger in undisturbed, rich woods. Collect the rhizomes from spring to fall. If you know where to dig, you can also collect in the winter.

The rhizomes make a refreshing nibble, and they're one of my favorite seasonings. I use them in about twice the quantity of the more pungent commercial ginger. You can also dry them. Collect a limited quantity from large, dense stands, so you don't damage the plant.

American Indians used wild ginger rhizomes the way Asians use commercial ginger. It was an important seasoning for many tribes, also making wild game and fish more digestible.

They used an infusion for gastric upset, indigestion, nausea, coughs, cramps, and fevers. They inhaled the dried, powdered rhizome like snuff for sinus congestion, and they applied the crushed rhizome to inflamed and infected cuts and wounds. The Meskawakis even poured a cooled decoction directly into the ear, for infections.

Indian tribes consistenly used wild ginger for protection. It kept evil away, especially if a sorcerer was trying to put harmful medicine into food. Warriors mixed it into dried fruits and meats before long journeys, to keep them fresh and ward off evil.

Ongoing research has uncovered antibacterial and antifungal properties, plus aristolochic acid—both antibacterial and antitumor. Because wild ginger is a good circulatory stimulant, it's included in many herbal formulas, to increase the other ingredients' effectiveness.

WINTERGREEN, CHECKERBERRY, TEA LEAF, DEERBERRY

(Gaultheria procumbens)

It's always exciting to chance upon this beautiful little plant. Its small, glossy, dark-green leaves make it stand out from a forest of trees. The red berries are a pleasure to behold, and even better to eat. When you're looking for something calming and woodsy, have some wintergreen on hand.

There are quite a few low-growing evergreen plants with bright berries that grow in the same areas as wintergreen, but none has the distinctive spicy fragrance that permeates all parts of this plant. Wintergreen has shiny, thick, slightly toothed, leathery, oval, evergreen leaves, 1 to 2 inches long and lighter in color underneath. They're clustered at the top of a heavy, almost woody stem. Each of the branches, 3 to 6 inches tall, looks like an individual plant, but they're all attached to a perennial stem that creeps along the surface of the ground, or just below it. The young, softer leaves are yellowish-green, tinged with rusty red. Some leaves are all red. The older leaves are dark-green and leathery.

Tiny, solitary or paired, white, bell-shaped flowers appear in midsummer. They hang from the upper leaf axils on drooping stalks. The tips of the flowers are divided into five lobes, and there are five sepals at the base of the flower, indicative of wintergreen's relationship to blueberries and cranberries—all in the heath family. Later, the bright-red, solitary berries, ⅓ inch across, hang in place of the flowers, ripening in late fall and

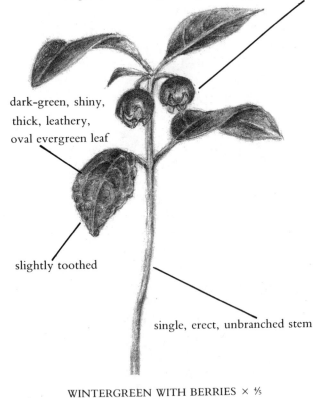

bright-red berries growing from upper leaf axil

dark-green, shiny, thick, leathery, oval evergreen leaf

slightly toothed

single, erect, unbranched stem

WINTERGREEN WITH BERRIES × ⅕

persisting through the winter to the following spring.

Wintergreen grows in poor, acidic soil in woods, bogs, and clearings in the Northeast, and around the Great Lakes region.

There are also two western varieties of wintergreen: salal or western wintergreen (*Gaultheria shallon*), and creeping wintergreen (*Gaultheria humifusa*). Salal grows along the Pacific coast from northern California to southeastern Alaska. The shrub grows over 8 feet tall, its many branches often forming dense thickets. The oblong to elliptical alternate, lightly toothed, evergreen leaves are 1 to 4 inches long, somewhat larger than eastern wintergreen's. The white or pinkish flowers grow in narrow clusters of five to fifteen near the tops of the shrub. The purple-black fruits that ripen in summer are not quite as large as the eastern variety, but more plentiful.

Creeping wintergreen grows in subalpine and alpine areas in the Northwest, and on Pacific Northwest mountains. It's a creeping shrub reaching 2 to 6 inches in height, with short, spreading branches. The thick, alternate, evergreen leaves are broad near the base, and small—either toothless or with tiny teeth. When they're young, you can add the leaves raw to salads, or cook them. The solitary, white, urn-shaped flowers appear in the junctions of leaves. The berries are red and dry—not as good as other species.

Inedible, odorless plants from a totally different family are also sometimes called wintergreens, probably because that's their group name in the heath family. Here, using scientific names avoids confusion. The imposters are all *Pyrolaceae* species—low, evergreen, herbaceous plants, without wintergreen's slightly woody stem or wintergreen odor. Their leaves grow near the base of the plants, not near the top like wintergreen's. They also have tall, leafless flower heads. Some of their common names include pipsissewa, spotted cone flower, "wintergreen," and round-leaf "wintergreen."

Wintergreen berries taste great, and they go a long way in salads and fruit dishes. They're best combined with other fruits, since they need juiciness and sweetness for balance. Don't cook them, or you'll destroy their flavor. My favorite wintergreen recipe is a blended fruit shake including a handful of the berries, pears, bananas, and water.

It's creamy, sweet and wintergreen-fresh. You can also make wintergreen jam, but cook the agar in fruit juice first, let it cool partially, then add raw pureed wintergreen berries. It's incredible.

Wintergreen berries are so high in vitamin C, they were used to treat scurvy. But don't dry the berries or leaves: The oil volatizes, making the house smell like a candy factory, but leaving you with bland, leathery dried fruit and leaves.

Steep lots of fresh or dried wintergreen leaves or berries in water to cover, for a mild, pleasant, refreshing tea. Women drink it for delayed, irregular, or painful menstruation, and to ease the pain of childbirth. It's used externally for hives, swelling, arthritis, and sore nipples.

Oil of wintergreen, the active ingredient, doesn't dissolve well in water, so some people greatly increase the flavor by steeping it for a day or two, until the tea begins to ferment, and bubbles appear. The resultant alcohol helps dissolve the oil. This is how people used to make root beer with herbs, only they added yeast.

Wintergreen tea is warming—great for a plant that ranges into areas that can be frigidly cold in winter. The tea is perfect if you feel run-down or chilled in winter. The fermented tea is wonderful for soothing and strengthening unsettled stomachs—the Indians considered it a carminative.

Chewing wintergreen leaves or eating the berries not only refreshes your breath but soothes irritated gum and canker sores. It's great to use after a visit to the dentist, or when you're suffering from a toothache. You can also use the tea as a mouth rinse or gargle.

Wintergreen oil is great for aches and pains. It's been an ingredient in massage oils for decades. I've even heard of people applying it externally to successfully relieve intense headaches and arthritis

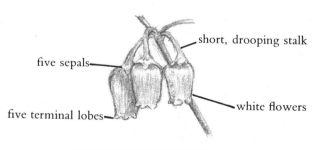

WINTERGREEN FLOWERS × 3

flare-ups. In the old days some people believed you could relieve any pain simply by inhaling the vapor of steaming wintergreen.

Oil of wintergreen contains methyl salicylate (also present in black birch, page 214), a compound similar to aspirin (acetylsalicylic acid). Like aspirin, it's an analgesic or pain reliever, and a diaphoretic. (The Indians drank wintergreen infusion for arthritis.) The two substances are so similar that people who are allergic to aspirin should avoid wintergreen. These salicylates were first chemically analyzed in the nineteenth century, and they've been used for pain, fevers, and arthritis ever since.

Very small amounts of wintergreen oil are also used to flavor drugs, toothpaste, and lotions. The oil is very powerful, and is not to be used frivolously: Chewing on some leaves is fine, but never use the oil internally. It's much more toxic than aspirin—six milliliters can kill an adult, and only tiny amounts are put in massage oil. Even so, the massage oil will burn like crazy if even a trace gets in your eyes.

According to Lenape Indian legend, wintergreen's origin is connected to the extinction of the mastodons. These pachyderms were put on the earth for the people to use. But they became too destructive and unruly. A great, bloody battle ensued in a bog, with all the other animals fighting the mastodons. The mastodons lost, and to this day, you can sometimes still find their bones in the bogs. The Great Spirit compensated for the loss of meat by transforming the spots of blood into wintergreen berries, which still dot the bogs with red to this day.

OTHER PLANTS OF WOODLANDS IN LATE FALL THROUGH EARLY SPRING

Edible or Medicinal Plants

Cow parsnip roots, daylily tubers, wild onion leaves and roots, garlic mustard leaves and roots, groundnut roots, highbush cranberries, horse balm roots, Kentucky coffee tree pods and seeds, Solomon's seal roots, spicebush twigs, viburnum bark, wild leek roots, wild potato vine root, wintergreen leaves and berries.

For Observation Only

Blackberry canes, grapevines, greenbrier vines, poison ivy vines and shrubs, pokeweed skeletons, raspberry canes, rose canes, wild leek seed head skeletons, yew.

PLANTS OF FRESHWATER WETLANDS IN LATE FALL THROUGH EARLY SPRING

CRANBERRIES

(*Vaccinium* species)

Cranberries are creeping native perennials forming mats that may cover a whole field. The plants, which may be as long as 2 or 3 feet, are connected by underground rhizomes. The small, oval, alternate leaves are leathery, with smooth-edged leaves that curl under ever so slightly. Their undersides are paler than the tops.

The small, pink flowers appear in the late spring and summer. They have four petals, turned straight back, and a long yellow stamen that looks like a bird's bill. The early colonists used this to give the plant its original name, crane-berry. Cranberries like the cold. They thrive in the Arctic and Canada, extending only through the northern states and into mountainous regions farther south. There are three major species:

The small cranberry (*Vaccinium oxycoccus*), rarely over 1 foot long, has thinner leaves and stems than other species. The fruits grow only on the tips of the plant, not along the leaf axils. It grows in

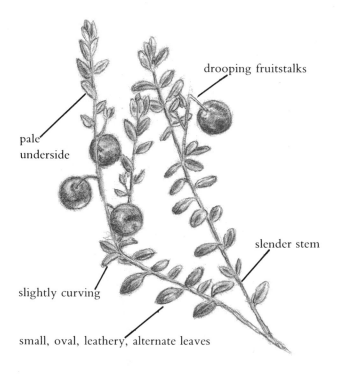

pale underside

drooping fruitstalks

slender stem

slightly curving

small, oval, leathery, alternate leaves

LARGE CRANBERRY × 1

eastern North America as far south as North Carolina.

The large cranberry (*Vaccinium macrocarpon*) may be 3 feet long. Its fruit grows along the leaf axils, not at the tip of the plant. It grows in southeastern Canada, around the Great Lakes and east through New England, and farther south only on mountains.

The mountain cranberry (*Vaccinium vitis-idaea*) stands erect rather than creeping. It grows as far south as northern Minnesota and New England.

Cranberries are adapted to poor, acidic soil. Finding places to collect cranberries isn't always easy. You need pine barrens, glaciated regions, or bogs. I was once surprised to find them near the seashore, in a bog that had formed behind the dunes. The small and large cranberries grow in bogs, peaty soil, and swamps. The mountain cranberry grows in cool, moist, acidic woodland areas and on rocky mountains.

Cranberries ripen in the early fall and become tastier after late fall frosts. Because they persist through the winter and need no storage, they were very popular with the colonists.

When you're fortunate enough to find wild cranberries, use them like commercial cranberries.

Since cranberry cultivation is relatively new, there's been little genetic change, and they're quite similar. They're not very good raw—they're rather acidic and mealy. Cook them with apples, honey, and walnuts. This is so good, you'll regain the wonderful lively feeling you get when you first glimpse the tiny glimmers of red, nestled in the springy, green, moss-covered hills or bogs.

Most people enjoy them cooked and sweetened, in traditional cranberry sauce. It's delicious, and it aids fat and protein digestion. The Inuit mixed an Arctic variety with rose-hip pulp and a sweetener. For a quick lemonadelike drink, mash or blend cranberries with water and a sweetener. Cranberries work best combined with other fruits, in filling for pies and cobblers. They contain lots of pectin, so they help thicken any dish. Because they're cold-adapted, they freeze well raw.

Cranberries are high in vitamin C and provide beta carotene, magnesium, potassium, and calcium.

Cranberries help cool down fevers and soothe skin inflammations. The juice is good for the urinary tract. Research shows it prevents infection by stopping pathogenic bacteria from attaching themselves to the urinary tract's mucous membranes. It's also a gentle, effective diuretic. To cleanse your kidneys, bladder, and urinary tract, and help prevent urinary-tract infection, drink a few glasses of cranberry juice every day.

Unfortunately, herb books, usually written by men, recommend cranberry juice for acute cystitis, a painful inflammation of the lower female urinary tract. The idea is to acidify the urine and kill the germs. Not only doesn't this work, but applying acid to a very sensitive, sore region aggravates the pain and inflammation. A teaspoon of baking soda dissolved in water, to deacidify the urine and soothe inflamed tissues, makes more sense.

OTHER PLANTS OF FRESHWATER WETLANDS IN LATE FALL THROUGH EARLY SPRING

Edible or Medicinal Plants

Cattail roots, cow parsnip roots, curly dock leaves and roots, groundnut roots, miner's lettuce, parsnip roots, Solomon's seal roots, sweet gale, watercress, winter cress leaves, wintergreen leaves and berries.

For Observation Only

Blackberry canes, burdock skeletons, cattail seed heads, curly dock skeletons, grapevines, parsnip skeletons, poison ivy vines and shrubs, rose canes, yew.

PLANTS OF THE SEASHORE IN LATE FALL THROUGH EARLY SPRING

Edible or Medicinal Plants

Bayberries, carrot roots, common evening primrose leaves and roots, curly dock leaves and roots, wild onion leaves and roots, groundnut roots, marshmallow roots, mullein leaves, parsnip roots, pine needles, sheep sorrel, sow thistle leaves, yarrow.

For Observation Only

Blackberry canes, carrot seed head skeletons, common evening primrose skeletons, curly dock skeletons, foxtail skeletons, grapevines, Japanese knotweed skeletons, milkweed skeletons, mullein skeletons, parsnip skeletons, pine cones, poison ivy vines and shrubs, pokeweed skeletons, prickly pear pads, rose skeletons, sumac skeletons, yucca.

PLANTS OF MOUNTAINS IN LATE FALL THROUGH EARLY SPRING

Edible or Medicinal Plants

Black birch twigs, juniper, miner's lettuce, mountain cranberries, pine needles, roseroot roots, sassafras roots, sheep sorrel, spicebush twigs, viburnum bark, creeping wintergreen leaves and berries.

For Observation Only

Roseroot skeletons, orpine skeletons, pine cones.

PLANTS OF DESERTS IN LATE FALL THROUGH EARLY SPRING

For Observation Only

Pine cones, prickly pear pads, yucca.

CHAPTER 6

EDIBLE WILD PLANTS OF EARLY SPRING

We arbitrarily differentiate early spring from late fall through early spring, and from mid-spring, although they overlap. We're going to focus on those shoots and greens that first emerge, or provide the best foods, when the worst of the cold is just over. These plants differ from the fall through early-spring shoots and greens because they don't also provide edible parts in autumn. Many will be out of season by mid-spring, because they'll be tough or bitter. This is the end of the season for most roots, so they're covered in the late-fall-through-early-spring section.

Identifying plants is not easy now, since important characteristics are missing. This is why we cover this season last. Now that you're more experienced, revisit areas where you identified plants too late to use last year.

PLANTS OF LAWNS AND MEADOWS IN EARLY SPRING

PLANTAIN

(*Plantago* species)

Romeo: Your plantain-leaf is excellent for that.
Benvolio: For what, I pray thee?
Romeo: For your broken [cut] shin.

—WILLIAM SHAKESPEARE, *Romeo and Juliet*

Plantains grow in basal rosettes. The leaves, hairless or slightly hairy, all have distinctly parallel veins. A green, central flower stalk extends above the leaves in late spring, summer, or fall. Early on, it's covered with many tiny, greenish-white flowers, each with four transparent, paper-thin petals. At some point in the summer or fall, the flower heads turn into seed heads. Minuscule individual seeds are packed into tiny green or brown capsules.

Plantains (*Plantago* species) are not related to the dissimilar banana-family plantain, or the inedible water plantains (*Alisma* species), which have similar-looking leaves.

Plantains are terrestrial, growing in open, sunny meadows, lawns, waste places, and nooks and crannies in concrete. They're so common, practically everyone takes them for granted, even though they're important sources of food and medicine.

Several species thrive throughout North America, without poisonous look-alikes. Most were inadvertently brought from Europe by the first settlers, and by the early 1700s, people already thought they were native plants. Part of their success comes from the fibrous root system that can find the smallest traces of moisture and nutrients in poor soil. Also, their short stature thwarts lawn mowers.

Common plantain (*Plantago major*) is my favorite species, with the most tender leaves. It has broad, irregularly rounded to oval leaves, 1 to 6 inches long, with smooth, wavy, or toothed edges; three to eleven ribs run its length. The generic name comes from the word *planta*, Latin for "sole." Indians, first seeing it near European settlements, called it "white man's footprint." Each leaf has a long, fibrous leafstalk. As a child, I tried to see how many fibers I could pull out. Dense clusters

of inconspicuous flowers, then tiny seed capsules, cover the central flower stalk, which appears in mid-spring. It grows from 3 inches to 1 foot long.

English plantain—long-leaf plantain, or buckhorn—(*Plantago lanceolata*) has long, narrow (or slightly rounded), lancelike leaves that tend to stand at attention instead of lazing around like common plantain's. The leaves originate from a short leafstalk. They grow from 2 inches to 1 foot tall, and up to 1½ inches wide, tapering to a point, with three to five ribs. The wiry flower stalk grows to almost 1 foot tall, and supports a short, stubby, hairy cylindrical flower head. The tiny flowers are bronze, with white, projecting stamens.

Another tasty species, seaside plantain (*Plantago juncoides*), looks like English plantain but with fleshier leaves and fewer veins. It grows near the seashore along the East coast.

English plantain tends to come up in very early spring, almost the only plant to brave the elements so boldly. The smallest, youngest leaves are best. They taste fresh, green, and cleansing, like a spring tonic.

Common plantain comes up a little later, but by mid-spring, the plantains become tough and stringy—you'll love eating them at this stage, but only if you're a rabbit or gosling. The fibrousness comes from the heavy veins. You can try eating around them or cutting them out of common plantain, but most foragers look for different plants once spring is in full swing. The best food use for mature plantain is in vegetable stock, to get the minerals and avoid the tough fibers. Here, clover leaves, potatoes, onions, and herbs are good companions.

Plantain provides beta carotene and calcium. All species contain mucilage—a carbohydrate fiber. This fiber reduces both the L.D.L. (low-density lipoprotein) cholesterol and triglycerides, helping to prevent heart disease. Plantain also contains monoterpene alkaloids, glycosides, sugars, triterpenes, fixed oil, linoleic acid, and tannins.

Plantain is astringent and soothing, also creating a cooling effect. It works internally and externally, soothing, rebuilding, and helping to waylay and prevent infections. An infusion of the leaves or mashed seeds treats sore throats, gastritis, diarrhea, bronchitis, fevers, inflammations, kidney and bladder disorders, and cystitis. Chinese herbalists use

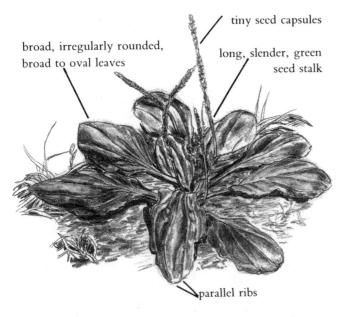

broad, irregularly rounded, broad to oval leaves

tiny seed capsules

long, slender, green seed stalk

parallel ribs

COMMON PLANTAIN BASAL ROSETTE × ⅓

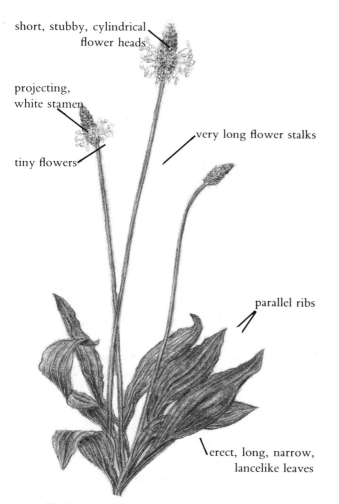

short, stubby, cylindrical flower heads

projecting, white stamen

tiny flowers

very long flower stalks

parallel ribs

erect, long, narrow, lancelike leaves

ENGLISH PLANTAIN BASAL ROSETTE × ½

probably because of its chlorophyl content. Next time your date reaches for a Certs, grab a plantain leaf from the lawn.

According to Pliny the Elder, you should beware of unscrupulous herbalists who sometimes remove a portion of plantain seeds from prescriptions and plant them instead. With recurring symptoms induced by this abuse of magic, patients end up returning to buy more of the diminished remedy.

COMMON BLUE VIOLET

(*Viola papilionacea*)

A basal rosette of stalked, shallow-toothed, heart-shaped leaves, each up to 5 inches across, first appears in early spring. The violet-colored flowers of mid-spring have five petals. They're bilaterally symmetrical and roughly butterfly-shaped (*papilionis* means "butterfly"). The two side-petals' beards are spectacular under magnification. The three lower petals are all deeply veined—the lowest being the longest. The whole plant grows 3 to 8 inches tall. Underground grow gnarled rhizomes, the thickness of pencils.

The common blue violet grows as a cultivated ornamental in gardens. It's also common in meadows, damp woods, rich soil, and along the edges of trails and paths. The richer the soil, the more plants will bloom. It's the most common of close to eighty cultivated and wild violet species, plus innumerable hybrids, growing throughout most of North America.

The common blue violet is a well-loved spring wildflower, but few people know it's one of the best-tasting, most abundant wild foods of springtime. The leaves are mild, sweet, and slightly peppery at the same time, with a mucilaginous (slightly gummy) texture. They're wonderful in salads, and they have a slight thickening effect in soups. They're great sautéed, and they're fine as ingredients in casseroles and vegetable dishes.

I hold bunches of violet leaves, along with some edible flowers, in one hand, and break them off with the other. That way I get lots of leaves, some flowers, and not too many of the slightly tougher stems. **Caution:** Violet rhizomes are delicious if

it for hepatitis. The infusion is also a good douche for yeast infections, and the fresh juice is an effective diuretic.

Plantain's astringent qualities make it a scientifically verified remedy for insect bites and skin irritations. Shred the leaf with your fingernails until it releases juice, and repeatedly rub it on the trouble spot. Relief comes within fifteen minutes. If you've touched poison ivy, this will probably prevent a rash. You can also make an ointment or salve from this herb.

Plantain seeds' mucilaginous fiber, like related psyllium seeds', is wonderful for cleansing and soothing the intestine. It acts as a gentle laxative. A European species is the main ingredient of the laxative Metamucil. This bulk fiber also helps control appetite, and reduces intestinal absorption of fat and bile.

Plantain also makes a great breath freshener,

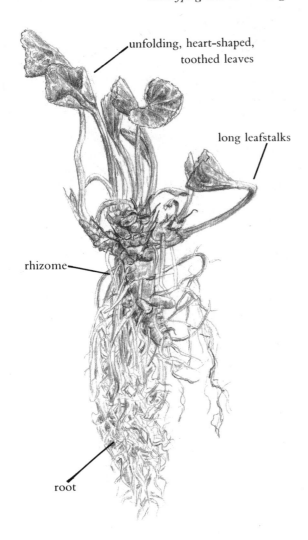

YOUNG VIOLET AND RHIZOME × ⅗

unfolding, heart-shaped, toothed leaves

long leafstalks

rhizome

root

drink with a violet-embedded ice cube, because it looks so appealing. People also used to boil violet flowers in sugar syrup to candy them.

Violet flowers and leaves are considered blood purifiers or detoxifiers. They're a traditional treatment for cancer, especially breast cancer, taken internally and applied externally. They're also supposed to be good for fibrocystic (lumpy) breasts. I don't know whether they really help.

Violets contain rutin, which strengthens the capillaries, as well as vitamin C. Thus, violets should be good for hemorrhoids and varicose veins. Violet-leaf tea is also supposed to be good for lung congestion and infections, and a violet-leaf poultice is soothing for all kinds of skin irritations, small wounds, and rashes.

Violet flowers look so different from other flowers because they're sterile. The showy flowers you see in the spring don't set seed. An inconspicuous brown autumn flower growing at ground level makes tiny, round, black, fertile seeds. This flower is self-pollinating: It doesn't open until the seeds are mature. This botanical curiosity is not edible, but it's well worth seeking out and observing. Violets spread mainly via underground rhizomes, which give rise to dense colonies of plants. The seeds are mainly for establishing new colonies.

Violets are important in literature and mythology. Zeus was carrying on with a water nymph called Io. His wife, Hera, would not have appreciated the situation had she found out, so Zeus transformed Io into a white heifer, for her own safety.

you're a wild turkey, and a turkey you will be if you eat them, because they're poisonous to humans. They'll make you throw up.

I begin collecting violets after they've been up a few weeks, when they're large enough to make collecting them worth it. By the end of spring, the leaves get tough and unpalatable, but sometimes tender new leaves come up in the fall. Fortunately, the leaves and flowers are easy to collect in quantity and dry, so I can enjoy this wild food all year.

Common blue violet flowers have less flavor than the leaves, and they take a long time to collect in quantity, but they beautify any dish. You can fill an ice-cube tray halfway with water, freeze it, place one violet flower in each compartment. fill with water, and refreeze Individuals who are completely phobic about wild foods will often accept a

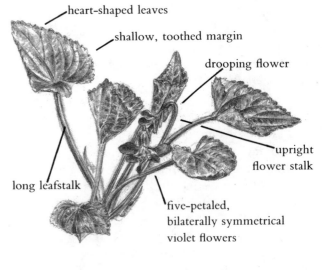

heart-shaped leaves

shallow, toothed margin

drooping flower

upright flower stalk

long leafstalk

five-petaled, bilaterally symmetrical violet flowers

VIOLET IN FLOWER × ⅗

Io didn't like the rough grass she had to eat, and began to cry. Zeus felt sorry for her, and transformed her tears into sweet-smelling flowers later called violets. The Greek word for "violet" is *ion*.

More recently, Napoleon's followers adopted the violet as their symbol while their leader was in exile. Napoleon had promised to return in the spring with the violets, and he was true to his word, arriving in Paris on March 20, 1815.

In poetry and literature, I often hear of violets' wonderful fragrance ("Sweet thief [referring to the violet], whence didst thou steal the sweet, if not from my love's breath?"—SHAKESPEARE), yet I couldn't smell a thing, even when I practically stuck the flower up my nose. This is because European species are the most fragrant. They smell so good, the scent is used in perfume. Most American species are odorless.

Fragrant or not, blue violets are a symbol of faithfulness. If you bring your lover a bouquet of violets with a single red rose, it means never-ending love.

Here are another two violet species: The marsh blue violet (*Viola cucullata*) has flowers that are taller than their heart-shaped leaves. The lower petal of the flower is the shortest, not the longest.

The birdfoot violet (*Viola pedata*) has deeply cut leaves that resemble a bird's foot. The upper petals curve backward, and no petals have beards. There are conspicuous orange stamens in the flower's center.

There are also violet species with white and yellow flowers. Avoid the yellow ones, which are usually too rare to pick, and may be cathartic.

Caution: There are a couple of poisonous plants that bear some resemblance to violets. Dwarf larkspur or spring larkspur (*Delphinium tricorne*) has an erect spur sticking out of the back of its violet-colored, five-petaled flower. Monkshood (*Aconitum uncinatum*) has a large, helmetlike upper sepal that covers two petals. Both plants are very poisonous. Also, don't eat the African violets. These inedible greenhouse plants are not violets.

When violets' flowers are out of season, violets resemble many other plants, some of which are poisonous. Collect violets only when their distinctive flowers have been in season for a couple of years and you've become completely familiar with them.

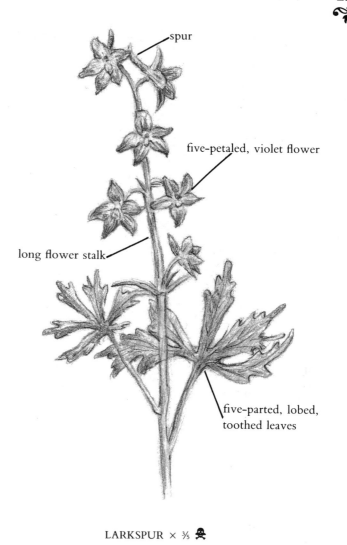

spur

five-petaled, violet flower

long flower stalk

five-parted, lobed, toothed leaves

LARKSPUR × ⅗ ☠

OTHER PLANTS OF LAWNS AND MEADOWS IN EARLY SPRING

Edible or Medicinal Plants

Chickweed, clover leaves, dandelion leaves, flowers, and roots, daylily shoots, common mallow, wild onion leaves and roots, ground ivy, peppergrasses, sheep sorrel, shepherd's purse leaves, sow thistle, storksbill, yarrow.

For Observation Only

Thistle.

PLANTS OF CULTIVATED AREAS IN EARLY SPRING

DAYLILY

(*Hemerocallis fulva*)

Daylily shoots are among the first wild vegetables to come up in early spring. The light green leaves are swordlike, with parallel veins. **Caution:** They resemble poisonous lilies, irises, and daffodils. The best verification method is unearthing the

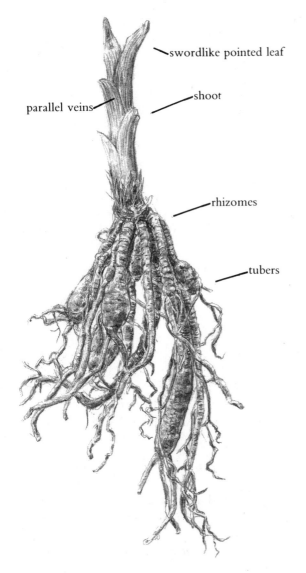

DAYLILY SHOOT AND TUBERS × ¾

roots. Underground, you'll find potatolike tubers with tiny, hairy roots, attached to the shoot's base by tubelike rhizomes, a little over ⅛ inch thick. If you unearth one long, thick rhizome without tubers, or a single bulb, you've got the wrong plant. Daylilies spread underground, forming dense colonies, making collecting fast and easy.

Wild daylily flowers bear no seeds (although some cultivated varieties do). They spread mainly by rhizomes and tubers. Sometime in their history, this process became so successful that varieties that relied more and more on vegetative propagation won out over those that wasted energy on sexual reproduction, and the latter vanished. Still, new stands of daylilies appear in the wild. Perhaps squirrels or other rodents dig up the tubers and rebury them elsewhere.

To confirm your ID, revisit the plants just before the summer solstice. The swordlike leaves are all basal, but now they're several feet long, bowing to the ground under their own weight.

Each plant will bear six to fifteen large, dazzlingly beautiful, erect, orange, short-stemmed, funnel-shaped flowers rising above the leaves on a 3-foot-tall, leafless flower stalk. The three reflexed petals, and three nearly identical reflexed sepals, are about 4 inches long. Six long, prominent, pollen-covered stamens surround a single pistil in each flower.

The daylily gets its scientific name from the Greek. The words *hemera* and *kallos* mean "day beauty." It gets its common name because each flower lives only one day, opening in the morning and wilting at night. Because each plant has many flowers, they're in season for a few weeks every summer. You'll usually find the papery, withered remnants of dead flowers along with open and unopened flowers. The unopened flower buds are thumb-sized cylinders that begin to turn orange just before they open.

No poisonous plant resembles the daylily in flower. The tiger lily's (*Lilium tigrinum*) edible flowers are spotted, and its flower stalks bear leaves.

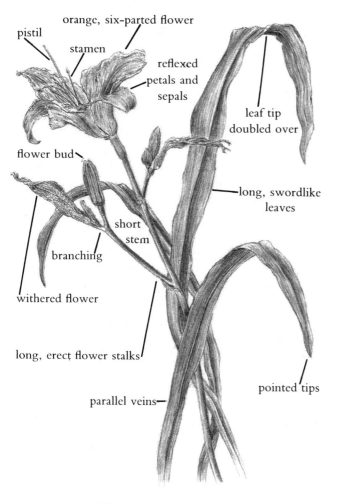

orange, six-parted flower

pistil

stamen

reflexed
petals and
sepals

leaf tip
doubled over

flower bud

long, swordlike
leaves

short
stem

branching

withered flower

long, erect flower stalks

parallel veins

pointed tips

DAYLILY IN FLOWER × ⅓

showy, bilaterally symmetrical
violet flower

parallel veins

deeply veined sepal

long, swordlike leaves

IRIS, BLUE FLAG × ½ ☠

Daylilies grow in meadows, along roadsides and streams, in gardens and disturbed habitats, and in open woodlands. It's officially naturalized throughout eastern North America, but people plant it in their gardens throughout the country, and I'm sure it's escaping cultivation elsewhere.

Pick shoots no taller than 6 inches, or they'll be tough and acrid. **Caution:** Daylilies, especially raw, don't agree with everyone. About one out of fifty react with nausea, vomiting, or diarrhea, especially after they pig out. If you eat a very small amount and can't tolerate this food, you'll experience minimal symptoms.

The young shoots are superb raw, in salads or dips. They have a mellow, pleasant, oniony flavor (the onion is a relative), and a crunchy texture. If you eat them raw by themselves, they may leave a slight burning sensation in the back of your throat.

I often sauté them, especially with olive oil, garlic, and a dash of salt. Or sauté them in oil with thinly sliced carrots and garlic, and add this to rice, with a drop of lemon juice or vinegar. They're also excellent in soups, and they're ideal in stir-fry dishes. They're incredible pickled.

Daylily flowers are sweet, tender, and a bit zingy. They're a traditional gourmet delicacy in Asia and Italy, where people dip them in batter and deep-fry them. You can also eat the petals raw. They're beautiful in salads against a mass of leafy greens. Be sure to cut off and discard the green part toward the flower's base—it's acrid.

Daylily flowers are wonderful in soups. Add at the end of the cooking time, so they keep their tender orange freshness. They're sold dried in New York's Chinatown for that purpose, as "golden needles." Even wilted flowers from the previous day reconstitute wonderfully in soups.

The unopened flower buds are delicious too.

They taste a little like string beans. They're too strong-tasting to use raw, but you can simmer, sauté, or steam them 10 to 15 minutes. They make great pickles: A friend sometimes serves them during my nature walks' lunch breaks.

Daylily's tubers are edible, but they're small and labor-intensive. The best time to collect them is in the fall and early spring, when the most food is stored underground, although they're edible all year.

You can thinly slice young, firm tubers and add them raw to salads. They have a light, crunchy nuttiness. Or cook them 10 to 15 minutes in soups. They're like tangy potatoes, but lighter and less starchy.

If the skin is loose and thick, the tuber is too tough to eat, and they're too small to make peeling worthwhile, although you could cook them, snip off an end, and squeeze out the pulp like toothpaste.

When a student took a course on cultivated daylilies at a prestigious botanical garden, she was amazed to learn that these hundreds of varieties of "weeds" were all worth a lot of money. Each plant was treated like a rare and precious china plate. At the end of the course, she asked if all the different varieties were edible. The teacher looked aghast at this horrible suggestion, but finally admitted that they were, in a tone suggesting that you could also use museum wood carvings for firewood. Perhaps she didn't know that daylilies had been eaten in Asia for millennia, and Westerners only recently began breeding them as ornamentals.

Daylily flower buds are good sources of beta carotene and vitamin C. The blossoms and tubers are high in protein and oils, and you can make a pleasant-tasting tea from the buds, shoots, and flowers. This infusion is supposed to be relaxing to the stomach and detoxifying.

In China, daylily tuber tea is used as a diuretic for urinary problems, and for jaundice, nosebleeds, vaginal yeast infections, uterine bleeding, and as a poultice for mastitis. Chinese laboratory research confirms the diuretic effect, and indicates that root extracts are antibacterial, and effective against blood flukes. However, they also report finding potentially toxic substances that may be cumulative. People have been eating these plants for centuries without being poisoned, so enjoy them in moderation.

OTHER PLANTS OF CULTIVATED AREAS IN EARLY SPRING

Edible or Medicinal Plants

Apple blossoms, chickweed, clover leaves, common mallow, dandelion leaves, flowers, and roots, juniper, Kentucky coffee tree pods and seeds, orpine leaves and tubers, sassafras roots and twigs, sheep sorrel, sow thistle, violet leaves.

For Observation Only

Cherry blossoms, cornelian cherry flowers, Juneberry flowers, peach flowers, yew flowers, yucca.

PLANTS OF DISTURBED AREAS IN EARLY SPRING

CHICORY, BLUE-SAILORS, WILD SUCCORY

(Cichorium intybus)

You can recognize this European perennial in the early spring by its rosette of deeply toothed basal leaves, 3 to 6 inches long, with midribs or leafstalks that are often reddish. Chicory has a long, deep taproot, in season all year. There are no poisonous look-alikes.

It differs from its relative, the common dandelion (page 190) because most of the leaves have irregular hairs on them, while the dandelion is hairless. Wild lettuce (page 246) is similar, but with fibrous roots.

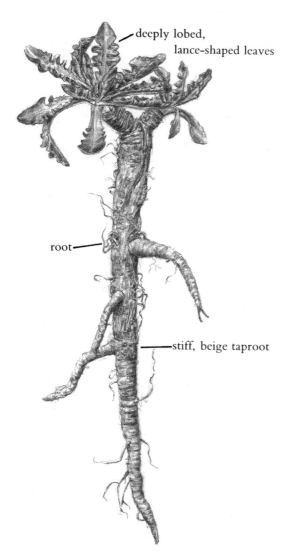

deeply lobed,
lance-shaped leaves

root

stiff, beige taproot

CHICORY TAPROOT AND BASAL ROSETTE × ⅖

Young chicory leaves taste similar to dandelion and wild lettuce in early spring. In England, chicory is considered superior to its cultivated relative, endive. The French prefer the tender inner crown of the leaves, cultivating chicory indoors, so the leaves don't become green. This makes them less bitter and less nutritious.

Use chicory the same way you'd use dandelion greens, before the mid-spring flower stalk appears, when it turns very bitter. Some people reduce this bitterness by boiling it in two changes of water, but I turn to other greens.

The edible flowers wilt soon after you pick them, and they're not especially good. But they're so plentiful, you might try dipping them in batter and deep-frying them—a process that makes virtually anything taste good.

The taproot is too tough and bitter to eat, but you can roast it in a 250°F oven 2 to 4 hours until it's dark-brown, brittle, and fragrant. Then grind it in the blender to make a homemade, caffeine-free coffee substitute. Put it in a coffeemaker like freshly roasted coffee. Not only does it taste like

Later in the season, all similarities cease. Chicory grows a rather rigid, somewhat hairy flower stalk nearly 4 feet tall, with alternate, lance-shaped, toothed leaves. The leaves get much smaller toward the top.

From late spring to early fall, chicory bears many conspicuous, stalkless, composite flower heads, over 1½ inches across, originating in the upper leaf axils. They consist of fifteen to twenty sky-blue ray flowers with distinctive fringed edges. Only a few bloom at once. So vivid and beautiful are they, that you hardly notice the weedy-looking stems and sparse leaves.

Chicory grows in disturbed habitats, along roadsides, and in fields, throughout the United States.

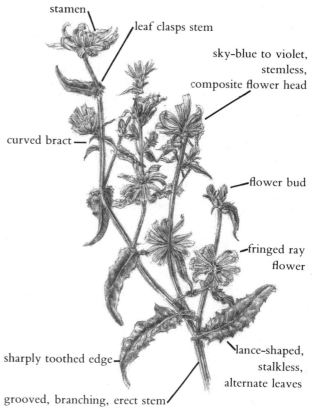

stamen

leaf clasps stem

sky-blue to violet,
stemless,
composite flower head

curved bract

flower bud

fringed ray
flower

sharply toothed edge

lance-shaped,
stalkless,
alternate leaves

grooved, branching, erect stem

CHICORY IN FLOWER × ⅓

coffee, but, according to the folklore, it counteracts the nervous overexcitement that coffee creates.

Medicinally and nutritionally, chicory has properties similar to the common dandelion. If you're a farmer, you can turn a field of mature chicory into hay, cutting it four times in one season, with a greater yield than alfalfa.

CURLY DOCK, YELLOW DOCK

(*Rumex crispus*)

In early spring, you can recognize curly dock by its basal rosette of long, narrow leaves, nearly 2 feet long, and no more than 3½ inches wide, with conspicuously curled, wavy margins. (*Crispus* means "curly" in Latin.) Each leaf has a long, central vein with smaller veins heading toward the leaf's edge, then curving in again. They meet smaller veins and make a rough pattern on the leaf's edge.

In the summer, this European perennial grows a flower stalk up to 5 feet tall, with smaller alternate leaves similar to the basal leaves. Papery sheaths surround the flower stalk and connect it to the leafstalks—typical of the buckwheat family. The flower stalk soon becomes covered with dense clusters of tiny flowers emanating from the leaf axils. Each flower consists of six green, white, or pinkish sepals. The flowers give way to seed clusters. Each seed is enclosed in rusty-brown, three-angled papery wings, like buckwheat groats. One plant may have as many as 40,000.

Herbalists call it yellow dock because of the long, stout, yellow taproot. (*Dock* is the word for the solid part of an animal's tail, and *to dock* means to remove the tail, as was done to sheep and dogs. All "weeds" were eventually called "dock" in English, perhaps because people tried to remove, or "dock," them.) Any but the most concerted attempts to eradicate curly dock fail: Little bits of roots will regenerate entire plants.

Other related docks, all with similar seeds and yellow roots, are more or less edible, depending on their bitterness; Broad-leaf dock or bitter dock (*Rumex obtusifolius*), for example, has wider leaves without curly edges.

People sometimes compare curly dock with the dandelion and its relatives because of their long leaves and basal rosettes. Dandelions are smaller, with toothed leaves and white milky sap.

When burdock is in its basal rosette form, its leaf is broader, rougher-looking, and woolly underneath. It also tends to be enormous. Even though both plants sport "dock" in their names, they're unrelated. Dock is a common name for coarse, weedy plants. Burdock belongs to the arti-

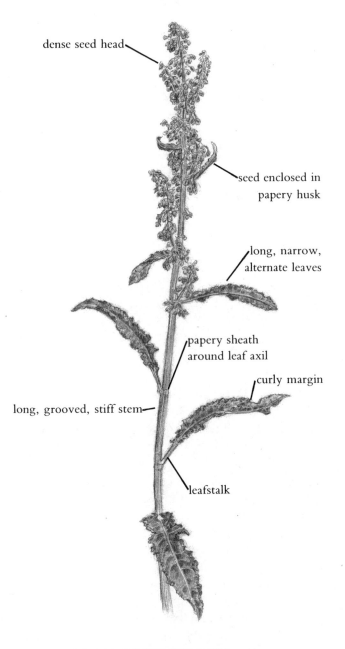

dense seed head

seed enclosed in papery husk

long, narrow, alternate leaves

papery sheath around leaf axil

curly margin

long, grooved, stiff stem

leafstalk

CURLY DOCK WITH SEEDS × ½

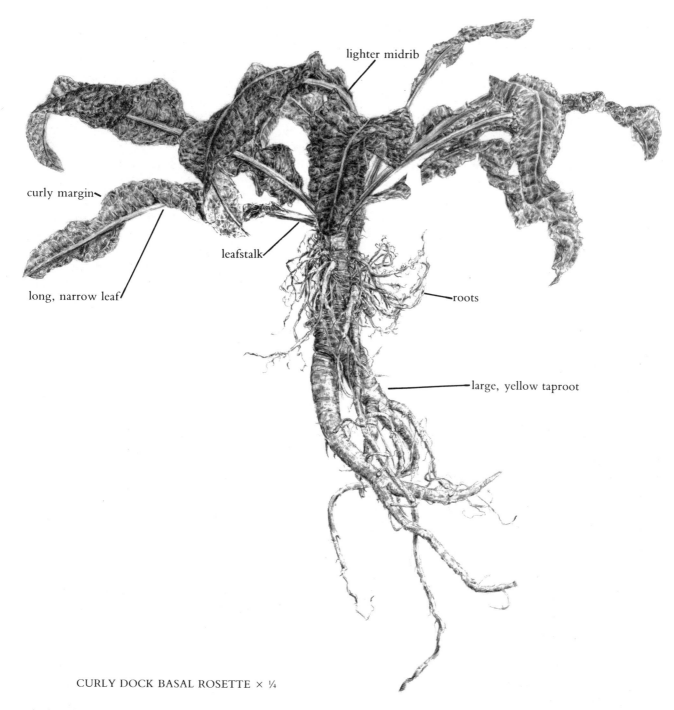

lighter midrib

curly margin

leafstalk

long, narrow leaf

roots

large, yellow taproot

CURLY DOCK BASAL ROSETTE × ¼

choke family, while curly dock belongs to the buckwheat family.

Curly dock and its relatives thrive in fields near the beach, on disturbed soil, along roadsides, in waste places, and along riverbanks. They grow from coast to coast.

I eat curly dock's young basal leaves in the early spring. By mid-spring, they're too bitter. They lose the bitterness in late fall, after a frost. Eventu-

ally, with enough bitter cold, curly dock retires for the winter. In the South, it never calls it quits.

The leaves have a distinctive, sour flavor, which also makes it first choice for rabbits. Collecting a shopping bag of these large leaves sometimes takes me only a few minutes. Nothing is better than a wild salad fully of curly dock at its best, with some chickweed, wild onions, and dandelions.

Curly dock makes a great sour, spinachlike side

dish, steamed 5 to 10 minutes. I love curly dock's slight thickening effect on soups and stews. It adds flavor and creamy body and never shrinks with cooking. The green leaves turn a yellow-green color, and they're more gooey.

I learned just how gooey this vegetable becomes when an experimental recipe I tried surprised me. Because dock leaves were so big, I thought they would be great stuffed and baked. I made a tomato and rice filling with Italian seasonings and nuts and rolled this into curly dock leaves. The baked dock unexpectedly turned mucilaginous. My vege-rolls were delicious, but unsightly and messy. The dock and stuffing should have been baked in bell peppers.

Caution: Very young curly-dock leaves, when they've just started to come up in early spring, contain enough chrysophanic acid to irritate your mouth if you eat them without washing them first. After one tour, where everyone sampled this delicious green in the field, our tongues were a little numb for nearly twelve hours.

Just as the leaves start getting bitter in midspring, the lemony-flavored flower stalks are ready to eat. Cut them off at the base, and peel away the tough outer layer. Add them raw to salads, or cook them 1 to 5 minutes. They have a celerylike texture raw, and 1 minute of parboiling makes them soft.

Unfortunately, the seeds aren't a practical food. They're minuscule, surrounded by the inedible, three-sided husks, and huge quantities of chaff.

Curly dock outdoes spinach nutritionally, with a third more protein, iron, calcium, potassium, beta carotene and phosphorus, and more than double the vitamin C. It's not surprising that this became an important vegetable during the Great Depression, when people were hungry. A root decoction provides the minerals potassium and manganese, plus lots of iron—people used it for iron-deficiency anemia for centuries. It's better than poorly absorbed iron pills, which can irritate the digestive tract.

Curly-dock root is too bitter and tough to eat, but it's been an important herbal medicine since at least 500 B.C. It's a blood purifier or detoxifier, especially useful for skin conditions, glandular inflammation, and swelling. It contains rumicin, a liver decongestant (liver disorders are often related to skin problems) that also stimulates the liver to produce bile. The decoction works as a bitter tonic for the stomach, and another active ingredient, emodin, stimulates peristalsis, making this herb a good remedy for constipation. Curly dock is a very safe laxative, getting things rolling while simultaneously strengthening the colon. Large overdoses may cause gastrointestinal disturbance, leading to nausea or diarrhea.

Its tannin content makes curly dock astringent, good for wounds and bleeding. In a *Star Trek— The Next Generation* episode, Dr. Crusher had Captain Picard crush a yellow, bitter-tasting taproot to her injuries, to stop the bleeding. Apparently, curly dock also grows on other planets!

A compress works on eruptions and skin irritations. A decoction stops diarrhea. Curly dock contains anthraquinones, good for ringworm and other fungal skin infections. It's also been used externally and internally for eczema and psoriasis, which are hard to treat medically, as well as skin cancer and leprosy, which are best treated conventionally. You can also cure nettle stings by rubbing curly dock leaves on the rash. The British say: "Nettles in, dock out."

JAPANESE KNOTWEED

(*Polygonum cuspidatum*)

Japanese knotweed has a distinctive hollow, jointed, woody stem, like bamboo, even though it's related to buckwheat, curly dock, sheep sorrel, and lady's thumb. It grows up to 10 feet tall in one season, then dies to the ground. The yellow, spreading rhizomes are perennial.

The stout, jointed shoots look like overly thick asparagus stalks—colored magenta and yellow-green. Beige-reddish, papery membranes wrap around enlarged nodes. Locate the shoots among the dense tangle of tall, hollow, woody, jointed stalks of last year's growth.

The mature plant's alternate, simple, stalked leaves are about 4 inches long, triangular, with a long, pointed tip, wavy margins, and broad, straight base. The tiny, greenish-white flowers, which appear in late summer, grow on long, lacy

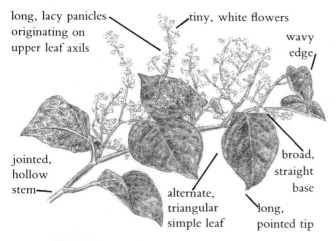

long, lacy panicles
originating on
upper leaf axils

tiny, white flowers

wavy
edge

jointed,
hollow
stem

alternate,
triangular
simple leaf

broad,
straight
base

long,
pointed tip

JAPANESE KNOTWEED IN FLOWER × ¾

panicles originating on the upper leaf axils. They're so beautiful that the Asian plant was cultivated as an ornamental, only to escape and spread. The fruits, which appear in autumn and sometimes persist into winter, are tiny, dark-brown nutlets, surrounded by whitish, papery wings.

Look for dense stands of Japanese knotweed on disturbed soil, in overgrown fields, vacant lots, roadsides, thickets, and near the seashore. It grows anywhere it can, and displaces everything else. There are other similar edible species in the genus, and no poisonous look-alikes.

Collect the shoots in early spring, and discard the leaves, which are tough. For maximum efficiency, collect shoots 6 to 8 inches tall. You can still use them 1 foot tall, but you have to peel away the tough rind (which you can include in marmalade recipes). By late spring, the entire plant becomes tough and woody, and the knotweed season ends.

Your first taste of this plant might surprise you. It's tart, like rhubarb, but with a flavor all its own. If the flavor is too strong for you, use it to enhance other dishes, like rhubarb. Add sliced knotweed to soup for extra zip, or steam it 10 minutes with carrots and parsley and serve on rice.

Like rhubarb, you can use Japanese knotweed's tart flavor to offset the sweetness of fruit in jams or desserts. Use about 8:1 fruit to knotweed. I've made terrific apple-knotweed pies and pear-knotweed puddings with these proportions. Foraging guru and author Euell Gibbons made a 100 percent knotweed pie, but he had to add tons of unhealthful sugar.

Japanese knotweed, like many sour foods, is a good source of vitamin C. Large quantities may act as a gentle laxative, like rhubarb. This mucilaginous vegetable is very soothing to the stomach and intestines, and its mild acidity helps break up fat and stimulates digestion.

Eating Japanese knotweed lowers your risk of heart attack, and Japanese knotweed has been used to treat cardiovascular disease in Japan for centuries. It contains resveratrol, the same substance beneficial to the heart that has recently been isolated from red wine, but without the unhealthful alcohol. (See the discussion under Grapes, page 165, which contain the same substance.)

After years of fruitless searching, I first learned to recognize the plant during a mushroom tour.

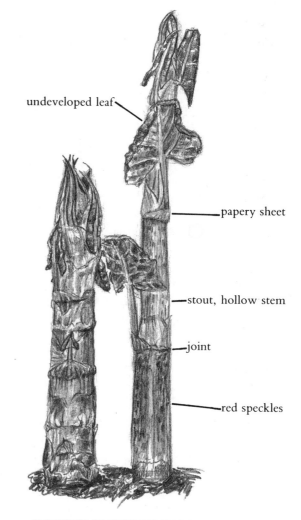

undeveloped leaf

papery sheet

stout, hollow stem

joint

red speckles

JAPANESE KNOTWEED SHOOTS × ¼

As we passed through a cemetery, another partici-pant, who thought I knew all the edible plants, commented on the abundance of Japanese knot-weed. I nonchalantly agreed, then discreetly pock-eted a sample to study. Soon after, I found some by a playground two blocks from home, and be-gan seeing it in parks and neighborhoods every-where.

MUGWORT, MOXA

(Artemisia vulgaris)

This downy-stemmed perennial herbaceous Asian import grows from 2 to 12 feet tall. The stalked, pointed, lower leaves are deeply divided into many finely cut lobes. The stalkless upper leaves are less divided. The leaves, which grow from 1 to 4½ inches long, are dark green above, white and woolly beneath. Many tiny, green, in-conspicuous flowers appear on long, leafy, upright spikes in summer and fall. It's a composite related to the wormwoods.

Mugwort thrives in sunny, disturbed areas, overgrown fields, and thickets throughout eastern North America, and it's probably spreading to other regions. If you're an urban dweller in eastern North America, chances are you've passed stands of mugwort thousands of times without recogniz-ing them.

Mugwort spreads quickly, aggressively dis-placing other species, even secreting an herbicide from its roots to create a monoculture. Many peo-ple hate it because it's hard to get rid of, looks raggedy, and produces allergenic pollen, like its relative, ragweed. On the other hand, I appreciate any vigorous plant that thrives in the city. I'm sure small animals value a tall dense, mugwort refuge in the noisy metropolis.

Mugwort is one of the first plants to appear in early spring, and you can find it in autumn until a series of hard frosts kills it to the ground.

In mid-spring, when it's only 6 inches tall, Asian people collect it to deep-fry. Other people use it like parsley. I've never cared for its flavor, and I'm allergic to its pollen, so I tend to think of this widespread plant more as a medicinal herb. The name mugwort comes from England, where

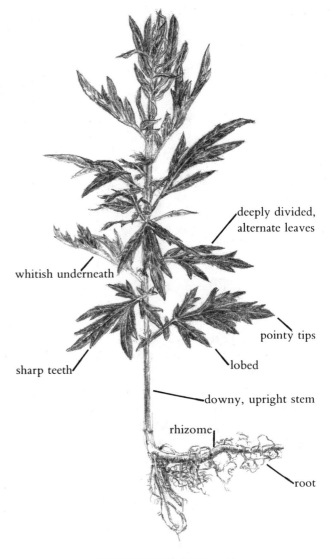

deeply divided, alternate leaves

whitish underneath

pointy tips

sharp teeth

lobed

downy, upright stem

rhizome

root

YOUNG MUGWORT × ⅕

wort is an old word for plant or herb, and the mug refers to the old practice of flavoring a mug of beer with this herb.

An infusion contains a volatile oil, the bitter principle absinthin, flavonoids, and tannin. It has a strong, warming effect on most people. One cup before dinner tones the stomach, relaxes the ner-vous system, and promotes digestion. Drinking a strong infusion, and taking a bath containing the infusion, helps alleviate suppressed menstruation. Hippocrates noticed mugwort's ability to stimulate uterine contractions. It's a uterine tonic, so you can't go wrong using mugwort tea for menstrual cramps and other disorders of the female reproduc-tive system. Although Chinese herbalists use it to

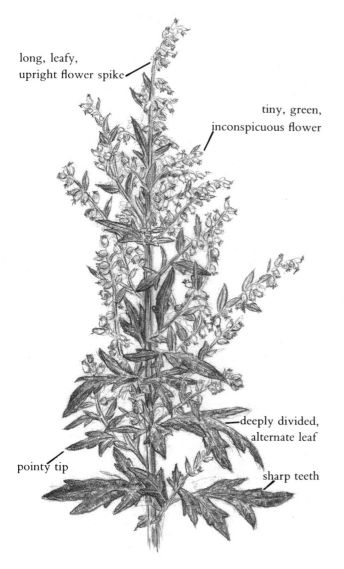

long, leafy, upright flower spike

tiny, green, inconspicuous flower

deeply divided, alternate leaf

pointy tip

sharp teeth

MUGWORT IN FLOWER × 1

mugwort with her dinner, reported the most colorful, dramatic dreams ever. Even after I corrected the mistake, she continued eating mugwort and experiencing exceptionally vivid dreams. Apparently there is more to this common plant than we know. Fresh or dried mugwort in the closet is supposed to drive away moths—maybe it gives them nightmares.

The late political activist Abby Hoffman, researching drug use across America for a book he was writing, once asked me about psychoactive herbs. I told him about mugwort, and when the book was published, I was erroneously described as the man who could find plants in Central Park to get people high!

NETTLES

(*Urtica* species; *Laportea* species)

These annual or perennial native and European herbaceous plants are distinctive for many reasons, as you'd quickly discover if you ever encountered them wearing shorts. Nettles are covered with tiny, nearly invisible stinging hairs that produce an intense, stinging pain, followed by redness and skin irritation. The generic name comes from the Latin word *uro*, "I burn." Nevertheless, they're superb, nonstinging cooked vegetables.

Nettles usually appear in the same places year after year. Look for them in rich soil, disturbed habitats, moist woodlands, thickets, along rivers, and along partially shaded trails. They grow throughout most of the United States. Here are a few of the most common species:

Stinging nettle's (*Urtica dioica*) rather stout, ribbed, hollow stem grows 2 to 4 feet tall. The somewhat oval, long-stalked, dark-green, opposite leaves are a few inches long, with a rough, papery texture, and very coarse teeth. The leaf tip is pointed, and its base is heart-shaped.

This is a dioecious plant, with male and female flowers growing on separate plants. The species name, *dioica*, means "two households" in Greek. By late spring, some plants have clusters of tiny, green female flowers hanging from the leaf axils

prevent miscarriage, it can cause problems in pregnancy without proper supervision.

Mugwort is also a nerve tonic and diuretic. People use the infusion for colds, bronchitis, fever, sciatica, kidney disorders, hysteria, epilepsy, and acute nervous tension. Chinese acupuncturists also burn incense over mugwort that's on the skin, a process called moxabustion, for arthritis caused by cold, damp situations.

Mugwort under your pillow is supposed to produce vivid dreams. It works for some people, not for others. Some students even get vivid dreams if they steep mugwort in vinegar and use mugwort-flavored vinegar in salad dressings. One woman who misunderstood me and ate some

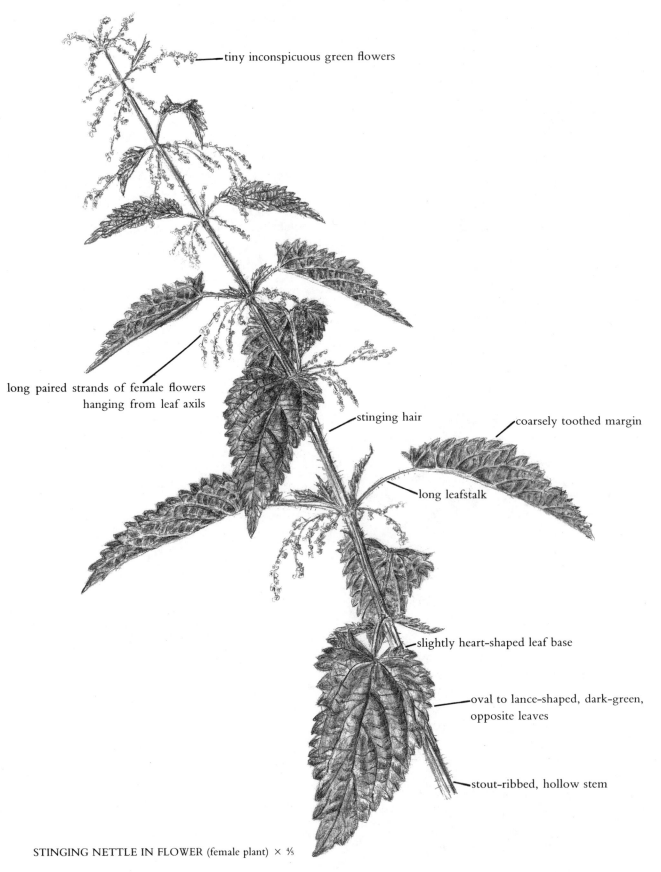

tiny inconspicuous green flowers

long paired strands of female flowers
hanging from leaf axils

stinging hair

coarsely toothed margin

long leafstalk

slightly heart-shaped leaf base

oval to lance-shaped, dark-green,
opposite leaves

stout-ribbed, hollow stem

STINGING NETTLE IN FLOWER (female plant) × ⅕

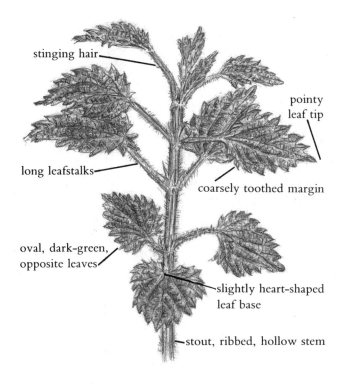

stinging hair

pointy
leaf tip

long leafstalks

coarsely toothed margin

oval, dark-green,
opposite leaves

slightly heart-shaped
leaf base

stout, ribbed, hollow stem

STINGING NETTLE SHOOT × ⅗

in paired strands. Other plants possess diagonally upright male flower strands, poised at the tops of the plants. Slender nettle (*Urtica gracilis*) is similar, with sparse stinging hairs and slender, opposite leaves.

Wood nettle (*Laportea canadensis*) has fewer stinging hairs. The leaves are alternate rather than opposite—larger and wider, with more rounded bases than the ones stinging nettles have. Wood nettle has flower clusters on top as well as in the leaf axils. Other true nettle species are also edible.

You'd think the stinging hairs would make nettle identification easy. Nevertheless, I once ran into some people in the woods who insisted that clearweed (*Pilea pumila*), a similar-looking, non-poisonous relative, with a translucent stem and no stinging hairs, was stinging nettles. They had been eating this nontoxic plant, which I had always rejected as unpalatable, all summer.

Sometimes nettles grow near catnip, another similar-looking plant. Mints, of course, have no stinging hairs, and catnip is fragrant. Catnip and nettles are an excellent combination for herb tea.

Collect nettle leaves before they flower in spring. They may be bad for the kidneys after they

flower. New nettles come up in the fall, and you can pick them before they're killed by frost.

People have been using nettles for food, medicine, fiber, and dyes since the Bronze Age. Collect them using work gloves, and wear a long-sleeved shirt. If you happen upon nettles when you have no gloves, put your hand inside a bag. The young leaves are the best part of the plant. They come off most easily if you strip them counter-intuitively, from the top down.

Whenever any of my groups find nettles, I announce that someone will volunteer to get stung, to demonstrate how jewelweed (page 73) cures the rash. Sure enough, someone accidentally gets stung, and we cure it. Once I was the careless one who got stung, but I kept my mouth shut and treated myself surreptitiously. Plantain and dock (pages 227 and 236) also work. Surprisingly, some people (masochists?) actually find nettle stings invigorating, and use them to wake up the body. Some Pacific Northwest Indians stung themselves with nettles to stay awake during long whaling voyages. The watchman was allowed to use healing herbs on his many stings only after a whale was sighted.

I have to travel quite a distance to find a place

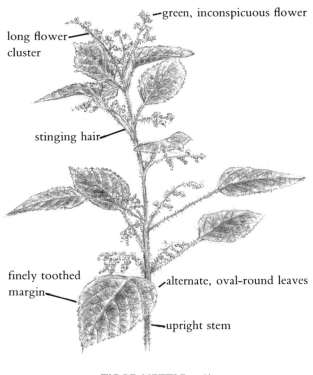

green, inconspicuous flower

long flower
cluster

stinging hair

finely toothed
margin

alternate, oval-round leaves

upright stem

WOOD NETTLE × ¼

where they grow like "weeds." As you can imagine, I pick in quantity, steam them, freeze them, put them in soups, stews, and other dishes. I dry them, tincture them in alcohol, and sometimes get stung by them. They get used up quickly—everyone loves them—and I'm back at the nettle patch.

Clean and chop nettles wearing rubber gloves. Once you've cooked them a little, the stingers are deactivated, and the plant becomes edible.

Nettles have a bad reputation as an unpleasant-tasting survival food in some circles. That's because people don't know how to prepare them. They often boil them, which is awful. Nettle leaves are good simmered in soups 5 to 10 minutes, but my favorite method is the waterless steaming method, recommended for spinach in a 1699 cookbook by John Evelyn, and described in the cooking section, page 281.

I enjoy nettles as a vegetable side dish with rice and beans. Sometimes I make creamed nettles—much more satisfying than creamed spinach. Because nettles have the richest, hardiest taste of any green, I often combine them with lighter ingredients, such as celery, zucchini, lemon juice, or tomato sauce.

I also dry nettles for winter use and tea. Sitting here writing this book, I frequently sip on warm nettle tea. It's one of my favorites. It doesn't taste like a normal tea—not bitter, spicy, minty, or lemony. It's more like a strong stock of a rich, deep, green plant essence, and it's one of the most nourishing drinks of all. Whenever I feel run-down, tired, or even irritable, I make myself some.

As food, this tonic is good for rebuilding the system of chronically ill people. Nineteenth-century literature is full of "constitutionally weak" people, who usually die on the last page. In Russia, they were given freshly squeezed nettle juice—a tonic loaded with iron and other nutrients—for iron-deficiency anemia. This often worked.

Many of the benefits are due to the plant's very high levels of minerals, especially calcium, magnesium, iron, potassium, phosphorus, manganese, silica, iodine, silicon, sodium, and sulfur. They also provide chlorophyll and tannin, and they're a good source of vitamin C, beta carotene, and B-complex vitamins. Nettles also have high levels of easily absorbable amino acids. They're 10 percent protein, more than any other vegetable.

The substances in the stingers have medicinal uses. In the late 1980s, scientists studying the differences between dried and freeze-dried herbs accidentally discovered that freeze-dried nettles cured one of the researcher's hay fever. Subsequently, a randomized double-blind study at the National College of Naturopathic Medicine in Portland, Oregon, showed that 58 percent of hay-fever sufferers given freeze-dried nettles rated it moderately to highly effective. Nettles are a traditional food for people with allergies.

Nettles sting you because the hairs are filled with formic acid, histamine, acetylcholine, serotonin, 5-hydroxytryptamine, and unknown compounds. Some of these substances are destroyed by cooking, steeping, or drying, but not by freeze-drying or juicing. Unfortunately, you need a vacuum chamber to freeze-dry herbs. However, you can purchase freeze-dried nettles in capsules for hay fever.

As an expectorant, it's recommended for asthma, mucus conditions of the lungs, and chronic coughs. Nettle tincture is also used for flu, colds, bronchitis, and pneumonia.

An infusion is a safe, gentle diuretic—considered a restorative for the kidneys and bladder, and used for cystitis and nephritis. It's also recommended for weight loss, but you may shed more pounds of water than fat.

Nettle-tea compresses or finely powdered dried nettles are also good for wounds, cuts, stings, and burns. The infusion was also used internally to stop excessive menstruation, bleeding from hemorrhages, bloody coughs, nosebleeds, and bloody urine. It helps blood clot, but major bleeding is dangerous—indicative of a serious underlying condition. Consult a competent practitioner in such cases. Use home remedies for minor cuts.

Other uses include treating gout, glandular diseases, poor circulation, enlarged spleen, diarrhea and dysentery, worms, intestinal and colon disorders, and hemorrhoids. Nettles are usually used along with other herbs that target the affected organs.

German researchers are using nettle-root extracts for prostate cancer, and Russian scientists are experimenting with nettle-leaf tincture for hepatitis and gallbladder inflammation.

Eating nettles or drinking the tea makes your

hair brighter, thicker, and shinier, and makes your skin clearer and healthier—good for eczema and other skin conditions. Commercial hair- and skin-care products in health-food stores often list stinging nettle as an ingredient. Nettles have cleansing and antiseptic properties, so the tea is also good in facial steams and rinses.

Nettles' long, fibrous stems were important in Europe for weaving, cloth making, cordage, and even paper. Native Americans used them for embroidery, fishnets, and other crafts. You can even extract a yellow die from the roots.

Nettle tea is given to house plants to help them grow, but the strangest use I've ever heard is for severe arthritis. You must whip the victim over most of the body until an extensive rash develops. This flagellation or "urtication" may stimulate the weak organs, muscles, nerves, and lymphatic system, and increase circulation. Or maybe it causes so much pain, the victim forgets about the arthritis.

STORKSBILL, FILAREE, ALFILARIA

(Erodium cicutarium)

Growing over 3 to 12 inches tall, this busy-looking European annual has one to several hairy, reddish, spreading stems with swollen nodes. The fernlike feather-compound leaves are twice divided; two to four pink-purple, long-stemmed, radially symmetrical five-petaled flowers grow in several, small branched clusters. The fruits are narrow, elongate cranelike capsules, up to 1½ inches long. They open from the base to the tip. When the five seeds separate and curl upward, they retain a part of a long style that looks like a crane's or stork's bill. Even the generic name, *Erodium*, means "crane."

Storksbill grows throughout the country, in woodlands, clearings, meadows, and disturbed soil. It's most plentiful in arid, grassy areas in the Southwest, where it's so abundant, it makes a useful addition to the diet even in the winter.

Although the young leaves of early spring are best, you can eat the whole plant, including the

flower buds and mature stems, in the spring. Use the young leaves and flower-bud clusters raw in salads, or cook them 2 to 3 minutes. The plant keeps flowering until the fall, and it sometimes produces new growth in autumn. You can also peel the stems when they get tougher and use them in soups, or steamed. Storksbill provides vitamin K, necessary for proper clotting, as well as beta carotene, thiamin, riboflavin, niacin, and calcium.

The mature leaves also make a nice tea. Folk cultures use it to induce sweating, as a diuretic, and to stop uterine hemorrhages.

Don't use filaree until you're thoroughly familiar with the plant and you've observed it throughout at least one growing season. Also, be sure it's hairy. Plants with similar fernlike leaves in the carrot family, such as poison hemlock and fool's parsley (page 204), are deadly poisonous, but hairless. Members of the carrot family all have umbrellalike flower clusters, and are usually more upright than filaree. The confusing species name, *cicutarium*, used to mean "resembling poison hemlock" (see page 203 under Wild Carrot). Due to changes in nomenclature, it now means "resembling water hemlock," although storksbill still resembles poison hemlock.

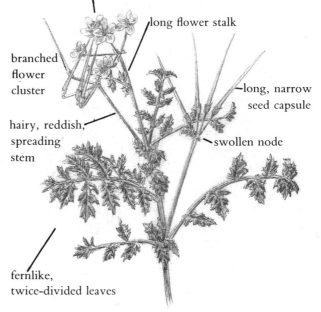

pink to purple, five-petaled, radially symmetrical flower

long flower stalk

branched flower cluster

long, narrow seed capsule

hairy, reddish, spreading stem

swollen node

fernlike, twice-divided leaves

STORKSBILL × ¾

WILD LETTUCE

(*Lactuca* species)

Common commercial lettuce is a hybrid bred from wild lettuce, but its appearance, flavor, and texture are so different, it's hard to believe that they're related. The only visible link is a bit of the characteristic white milky sap in commercial lettuce's core.

When you find wild lettuce in the early spring, chances are you'll mistake it for other lettuce species, dandelion, or chicory (pages 190 and 234). They all have similar basal rosettes of toothed leaves with white milky sap in early spring, but they're all edible, without poisonous look-alikes. In Latin, *lac* means milk, and *lactuca*, which means "milk-giving," is an ancient name for lettuce.

Dandelion leaves are never hairy, while hairy lettuce (*Lactuca hirsuta*) has hairs on the undersides of the leaves, and prickly lettuce (*Lactuca scariola*) has prickles. Wild lettuce or tall lettuce (*Lactuca canadensis*) is hairless, but the leaves are light green, and the stem is powdered with a blue-green, waxy bloom. Chicory has a thin line of hair along the

underside of the midrib of the leaf. Wild lettuces are usually a lighter green than their relatives, and the leaves are more widely variable—some deeply lobed, others barely toothed. The leaves of some (edible) mustard species resemble lettuce's, but never have white sap.

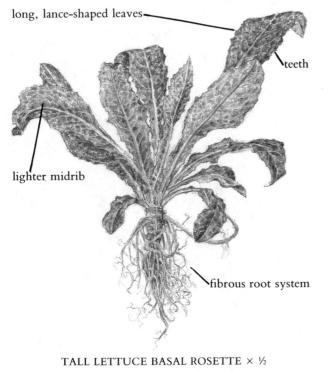

TALL LETTUCE BASAL ROSETTE × ½

TALL LETTUCE IN FLOWER × ½

The wild-lettuce rosette soon grows a tall, scraggly flower stalk. The leaves toward the top tend to be less toothed. Some species don't grow over 1 foot tall, especially where mowing occurs; others reach 10 feet. They look like a dandelion gone crazy, with smaller yellow composite flowers arising from branching stalks at the top of the stem. Blue lettuce (*Lactuca pulchella*) is similar, but with blue-violet flowers.

Look for wild-lettuce species in disturbed habitats or open fields, along walls, at the edges of meadows, in thickets, and in openings in the woods, throughout North America. Use the leaves and flowers the same way you use dandelions. The nutritional content is also similar.

Most species are annuals, with no underground investment. The leaves are most useful, because the roots are usually fibrous—not taproots. (Dandelions and chicory have usable taproots.)

Collect wild lettuce soon after the rosette appears, in early spring, when it's tender and mild. Use the leaves like dandelions or chicory, raw or cooked. They soon become bitter, and you'll have to boil them in several changes of water to make them palatable.

Wild lettuce has been used medicinally for centuries: After it supposedly cured Emperor Augustus of Rome of a dangerous illness, he built an altar and erected a statue in the plant's honor.

The very bitter, white, milky sap in the leaves and roots contains the mild sedative lactucarium—chemically similar to opium—that has been used in cough medicine. People cut into the flower stalk, collect the sap that drips out, and use it for pain. It isn't narcotic, but it's supposed to be relaxing—especially calming for overwrought nerves. I gave a strong wild-lettuce-leaf infusion to a friend suffering from endometriosis (a very painful gynecological disorder). It had no effect whatsoever, even after she ate large quantities of the bitter dried leaves. I think the association with opium is 99 percent wishful thinking, 1 percent chemistry.

Wild-lettuce tea is also supposed to be cooling and diaphoretic: People use the infusion for gastric spasms, fevers, and insomnia, and apply the sap to rashes, warts and acne. On the other hand, old herbal books say it cools off "excessive" sexual excitement.

OTHER PLANTS OF DISTURBED AREAS IN EARLY SPRING

Edible or Medicinal Plants

Carrot leaves and roots, chickweed, clover leaves, coltsfoot flowers, common evening primrose leaves and roots, wild onion leaves and roots, dandelion leaves, flowers, and roots, daylily shoots and roots, garlic mustard leaves and roots, goatsbeard root, ground ivy, horseradish roots, mullein leaves, nettles, orpine roots, parsnip roots, peppergrasses, sheep sorrel, shepherd's purse leaves, sow thistle leaves, spearmint, storksbill, wild potato vine, winter cress leaves, flower buds, and flowers.

For Observation Only

Carrot skeletons, mullein skeletons, peach blossoms, poison ivy leaves, vines, and shrubs, thistle, sumac skeletons.

PLANTS OF FIELDS IN EARLY SPRING

MULLEIN, JACOB'S-STAFF, FLANNELLEAF

(*Verbascum thapsus*)

After exposure to countless associations between this common, widespread medicinal herb and Native American culture, I mistakenly thought this herb was native. I was astonished when I finally learned that mullein is Eurasian. After its early arrival on these shores, the Indians adapted it. They had discovered the same healing properties that made it a mainstay in European folk medicine for thousands of years.

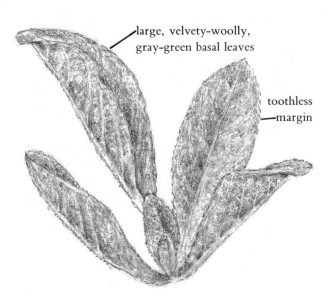

MULLEIN BASAL ROSETTE × ⅓

large, velvety-woolly, gray-green basal leaves

toothless margin

alkaline soil, so it's especially common near the seashore. Archaeologists sometimes look for Indian sites where there's lots of mullein, because the lime from the Indian shell piles increases soil alkalinity, encouraging this plant to proliferate.

Mullein tea provides vitamins B_2, B_5, B_{12}, and D, choline, hesperidin, PABA, sulfur, magnesium, mucilage, saponins, and other active substances.

People use the tea as a beverage, but it's best known as one of the safest, most effective herbal cough remedies. Mullein is an expectorant, and a tonic for the lungs, mucous membranes, and

The name *mullein* has two possible derivations. It comes either from *mollis*, which means "soft" in Latin, or the Latin word *mulandrum*, which comes from *melanders* and means "leprosy"—an illness this plant was used to treat. *Verbascum* means "mullein" in Latin. It derives from the word *barbascum*, which means "with beard." Roman men shaved, barbarians didn't, and mullein is certainly as woolly as any barbarian you'll ever encounter. The species name is *thapsus* because mullein resembles the European genus *Thapsia*, named after an ancient town in present-day Tunisia.

Mullein is a biennial: The first year the leaves form a basal rosette, with strikingly large, flannellike, velvety-woolly, long-oval, gray-green leaves up to 2 feet long. When I bring this plant to school classes, the children first think it's artificial.

The second year, the basal leaves precede a stout, erect flower stalk that may reach 2 to 8 feet in height. The stalkless flowers bloom sequentially from mid-spring to early fall, growing in long, tight spikes. They're yellow, with five radially symmetrical petals, about ¾ to 1 inch across. The pointed, elongated, globular fruits are five-parted woody capsules, ⅜ inch long, opening toward the tips. The dead, brown fruit stalks stand out in the winter.

Mullein grows in old fields, roadsides, and disturbed habitats throughout the United States. It does well in dry, sandy conditions, especially in

five-petaled, yellow, radially symmetrical flower stalk

five-parted, woody seed capsules

tall, stout, erect flower stalk

toothless margin

alternate, large, velvety-woolly, gray-green, basal leaves

MULLEIN FLOWER STALK × ⅙

glands. An infusion is good for colds, emphysema, asthma, hay fever, and whooping cough. Strain the infusion through a cloth, or the hairs may get stuck in your throat and make you cough even more. Laboratory tests have shown that it's anti-inflammatory, with antibiotic activity, and that it inhibits the tuberculosis bacillus. The Indians smoked dried mullein and coltsfoot (page 71) cigarettes for asthma and bronchitis, and indications are that it's effective: I've observed it working for bronchitis.

The tea is also an astringent and demulcent. It's good for diarrhea, and it's been used in compresses for hemorrhoids since it was recommended by Dioscorides centuries ago. It's also supposed to help other herbs get absorbed through the skin. Pliny of ancient Rome, Gerard in sixteenth-century England, the Delaware Indians, and country folk in the South used the heated leaves in poultices for arthritis.

A tincture of the flowers is used for migraine headaches. An oil extract of the flowers, which contains a bactericide, is used for ear infections, although you should consult with a competent practitioner first, to avoid the possibility of permanent hearing loss if the herb doesn't work.

Roman ladies used the flowers to dye their hair blond. Roman soldiers dipped the flower stalks in tallow to make torches. Mormon women—forbidden to use makeup—rubbed the rough leaves of this rubrifacient on their cheeks to create a beautiful red flush. People who spend time in the woods are attracted to mullein's large, velvety leaves when they run out of toilet paper, again creating a beautiful red flush on their cheeks.

MUSTARDS

(Brassicaceae)

This is a very large, widely distributed family of native and exotic herbaceous plants full of edible species. It's sometimes difficult to identify a mustard as to species, and we don't have the room to describe them all, but the family is easy to recognize, especially when it's in flower, and no mustards are poisonous.

Mustards usually begin as basal rosettes. The leaves may be a variety of shapes, but they're usually partially divided into lobes, and most are also toothed. Unlike dandelions and its relatives, which have similar leaves and different flowers, there's never a colored sap. Also, mustard leaves' teeth point toward the leaf tip, whereas dandelions and their relatives' teeth point toward the leaf base.

Different species flower from spring to fall. These herbs have small to tiny white, yellow, or lavender flowers in terminal clusters. Each flower has four petals arranged like a cross, which is why the family is called cruciferous. There are usually four long and two short stamens in the flowers. The flowers give way to long and narrow seedpods, or flat and circular seedpods—all filled with many tiny seeds.

The leaves are edible when they first appear. Some species become bitter; some disappear later on. Others are good all season. Some have edible seeds and pods, while others' pods are too small or tough for food. Some species have edible roots. All mustards are pungent-tasting, some more than others.

These highly nutritious vegetables provide lots of calcium, potassium, and vitamins B_1 and B_2. Research has shown that all mustards, even commercial ones, contain concentrated substances which help prevent cancer, including isothiocyanates, beta carotene, vitamin C, and fiber. The flower buds—edible in most species—are good sources of protein.

The mustards are a very successful family. As such, they've diversified to occupy many habitats and niches. They often grow in open, sunny places, disturbed soil, and fields, although some species specialize in different habitats. Here are a few of the best varieties:

Black mustard (*Brassica nigra*) is a European annual growing 2 to 4 feet tall. It has bristly, irregularly lobed lower leaves and narrow, hairless, wavy-toothed upper leaves. It has showy terminal clusters of yellow flowers that bloom from summer to fall. Its seedpods, which are lobed, narrow, linear capsules, point upward and end in a beak. Collect in summer and fall.

Field mustard, rape, or wild mustard (*Brassica rapa*) is similar to black mustard, but it's smooth, succulent gray-green—covered with a whitish

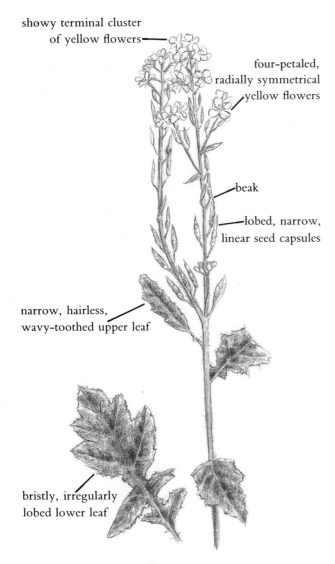

showy terminal cluster
of yellow flowers

four-petaled,
radially symmetrical
yellow flowers

beak

lobed, narrow,
linear seed capsules

narrow, hairless,
wavy-toothed upper leaf

bristly, irregularly
lobed lower leaf

BLACK MUSTARD × ½

bloom (a protective waxy powder). The upper leaves, which have earlike lobes, clasp the stem.

Both brassicas' leaves are best in the spring, when young, before they flower and become bitter. The seeds are good ground with vinegar, like commercially prepared mustard. They're widespread in open, sunny areas, disturbed habitats, and overgrown fields throughout the United States.

Brassicas and horseradish contain a fixed (nonvolatile) oil, proteins, and mucilage. The enzyme myrosin works on the glycoside singrin in the presence of water to produce mustard oil (isothiocyane), the active ingredient that creates the pungent taste.

Mustard oil is strongly antibacterial and antifungal. People ground mustard seeds and applied them externally as plasters. This stimulation was good for stubborn coughs and arthritic joints, although it blistered the skin if left on too long. Soothing herbs in the poultice, such as slippery elm, lessened this effect. Mustard-seed footbaths increase circulation and benefit chilblains (cold water–associated frostbite), and upper respiratory mucus.

Hedge mustard (*Sisymbrium officinale*) is a hairy species I first find in mid-spring, with a basal rosette of lobed leaves. Next comes the flower stalk, which grows up to 4 feet tall—stiff and upright, with a few branches. This European immigrant flowers by late spring and dies in the summer. The small, yellow flowers grow on long, narrow racemes, to be followed by small, dark-brown, oval, inedible seeds. It grows in disturbed habitats. Eat the deliciously pungent basal leaves and the young leaves and flowers from the flower stalk, raw or lightly cooked. They taste like Chinese mustard.

Winter cress (*Barbarea vulgaris*) is a smooth-stemmed, European import with glossy, dark-green, toothless, hairless, deeply lobed basal leaves. It looks a little like watercress, but it grows on moist soil and in muddy places, rather than in the water.

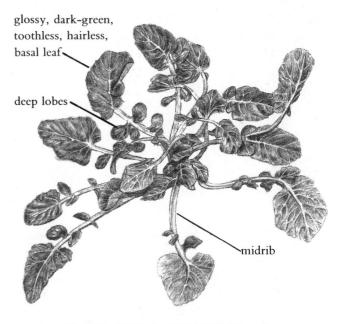

glossy, dark-green,
toothless, hairless,
basal leaf

deep lobes

midrib

WINTER CRESS BASAL ROSETTE × ⅓

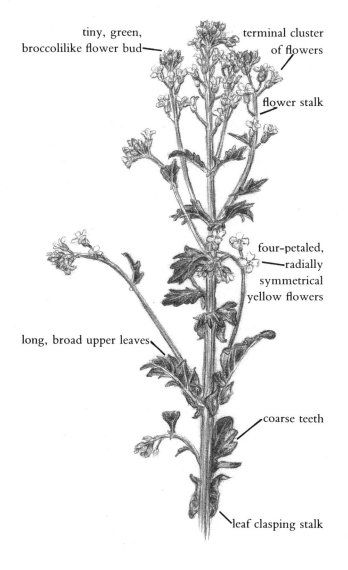

tiny, green, broccolilike flower bud

terminal cluster of flowers

flower stalk

four-petaled, radially symmetrical yellow flowers

long, broad upper leaves

coarse teeth

leaf clasping stalk

WINTER CRESS IN FLOWER × ½

The upper leaves on the flower stalk are long to broad, toothed, and clasping. The lower leaves have four to eight earlike lobes. By the time they appear, in early spring, the plant is too bitter to eat, although the small flower buds, which taste and look like broccoli, are edible, if labor-intensive to collect. The four-petaled, yellow, stalked flowers, which open soon after, grow in terminal clusters, like other mustards. The narrow, cylindrical, unlobed seedpods are too tough to eat. The plant grows 1 to 2½ feet tall.

Look for winter cress in disturbed sites, vacant lots, roadsides, and ditches. it thrives in wet places, and grows from coast to coast.

This highly nutritious vegetable is good only in the cold weather, from late fall to early spring, when it tastes like a pungent cabbage. At other times, it's horribly bitter. Many people, especially children, dislike it even when it's at its best.

There are no poisonous look-alikes. A similar edible species with deeper, more even lobes—scurvy grass or creasy greens (*Barbarea verna*)—grows in the southeastern United States, where it's sold in markets.

The peppergrasses (*Lepidium* species) are mustards of lawns and disturbed soil with tiny, white or yellow flowers that usually have four, sometimes two stamens. The seedpods are flat and circular. These annual or biennial herbs begin as basal rosettes in early spring and flower soon thereafter. They generally grow on lawns, in fields, near the seashore, and in disturbed habitats—in season from early spring to fall. They're high in vitamin C, and good sources of iron. Various species, all tasty and easily recognizable as peppergrasses, grow across the United States. Many are good from early spring to fall.

Cow cress or field peppergrass (*Lepidium campestre*) has coarsely toothed basal leaves over 3 inches long that embrace a densely hairy stem. The nearly circular stem leaves, which are lobed near the base, are much smaller.

The plant grows up to 2 feet tall, bearing white flowers from mid-spring to early fall. The round, flattened seedpods are about ⅒ inch in diameter, slightly notched at their tips. The leaves and seedpods are very good but quite pungent. I use them sparingly, even though I love spicy food.

We find poor man's pepper (*Lepidium virginicum*) so often on my field walks, some of my students have begun calling it "Wildman's" pepper. It's similar to cow cress, but the leaves are more deeply toothed, and they're stalked. It sometimes reaches 3 feet in height, and it flowers from spring to late fall.

Both species have edible leaves and seedpods. They taste pungent, and although there are fewer leaves after the plants flower, they never become bitter. New rosettes occur in autumn.

Shepherd's purse (*Capsella bursa-pastoris*) is similar to the peppergrasses, but the flat seedpods are heart-shaped, reminiscent of the purses of old-time shepherds. There are other very similar species in

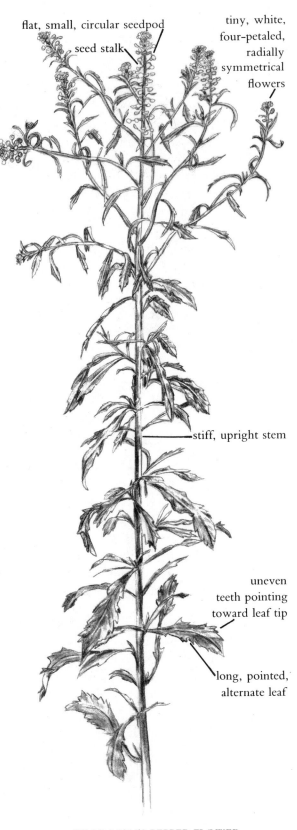

flat, small, circular seedpod

seed stalk

tiny, white, four-petaled, radially symmetrical flowers

stiff, upright stem

uneven teeth pointing toward leaf tip

long, pointed, alternate leaf

POOR MAN'S PEPPER FLOWER
AND SEED STALK × ¾

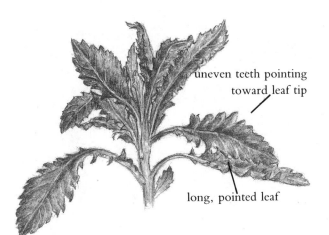

uneven teeth pointing toward leaf tip

long, pointed leaf

POOR MAN'S PEPPER IMMATURE FLOWER STEM × ½

the genus, all edible, with no poisonous look-alikes.

The basal leaves of this European annual are toothed like a dandelion's, but without the milky sap, and the stem leaves clasp the stem. The basal leaves have a mild pungency early in the spring, but the plant soon flowers and the leaves become tough and tasteless.

The tiny, stalked, white flowers grow in terminal clusters. The seeds inside the purselike pods are reputed to be edible and peppery, but I've always found them flavorless. Furthermore, they're dangerous: Don't go near them if you're an insect smaller than the period at the end of this sentence.

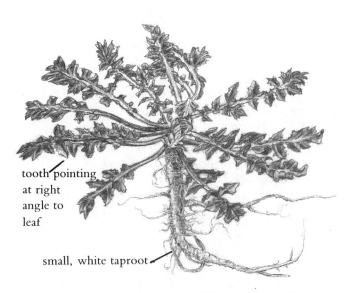

tooth pointing at right angle to leaf

small, white taproot

SHEPHERD'S PURSE BASAL ROSETTE AND ROOT × ½

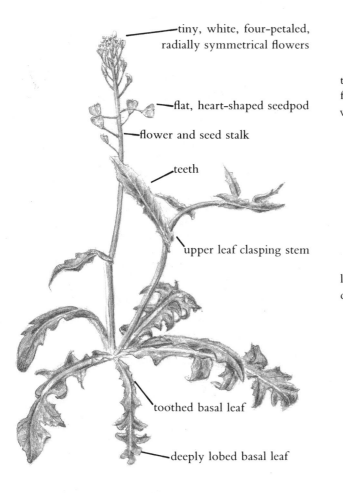

tiny, white, four-petaled, radially symmetrical flowers

flat, heart-shaped seedpod

flower and seed stalk

teeth

upper leaf clasping stem

toothed basal leaf

deeply lobed basal leaf

SHEPHERD'S PURSE × ¾

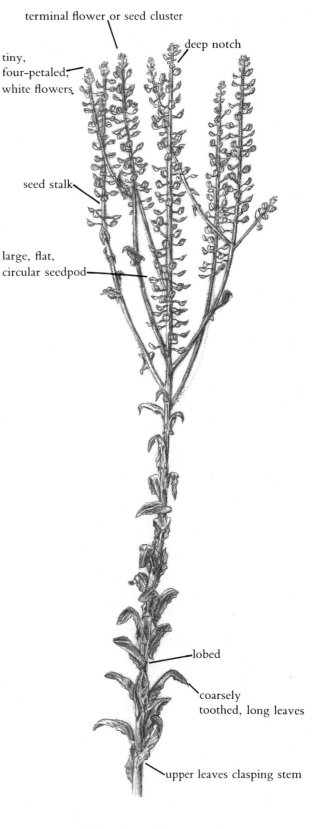

terminal flower or seed cluster

deep notch

tiny, four-petaled, white flowers

seed stalk

large, flat, circular seedpod

lobed

coarsely toothed, long leaves

upper leaves clasping stem

FIELD PENNYCRESS × ⅓

They're so sticky, you'll get stuck, and the carnivorous baby plant will begin life by absorbing and assimilating you. Because of this supplemental nitrogen, shepherd's purse can grow in poor, sunny soil and disturbed areas, although it also grows on lawns.

Shepherd's purse is mildly astringent. It contains saponins, choline, acetylcholine, and tyramine. It's supposed to be good for long and painful menstrual periods. It's also supposed to be good for stopping bleeding in general. Before today's pharmaceuticals were discovered, people used it for hemorrhages of the stomach, lungs, uterus, and kidneys. I tried dried seedpods on a paper cut, but it didn't seem to work. Maybe a tincture would have been better.

Field pennycress (*Thlaspi arvense*) is similar to shepherd's purse, but the seedpods are large, flat, circular, and more deeply notched than cow cress.

The basal leaves are toothed, but they don't resemble dandelion leaves. The plant grows up to 2½ feet tall. The branched stems are smooth.

The seedpods are very pungent, and I use them instead of cayenne hot pepper whenever possible. The leaves are also very hot-tasting. I've found this plant in overgrown fields, especially near the seashore. It flowers and bears seed from early spring through summer.

Garlic mustard (*Alliaria officinalis*), also called Jack-by-the-hedge and sauce-alone, defends itself from insects by smelling like garlic, which insects don't like. Of course, if a swarm of Italian insects finds it, the plant soon becomes extinct.

This erect European herb of open woodlands and disturbed soil has dark-green, heart-shaped,

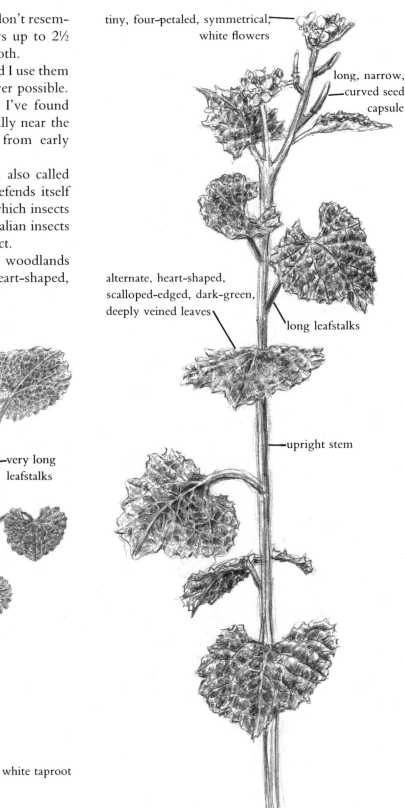

tiny, four-petaled, symmetrical, white flowers

long, narrow, curved seed capsule

alternate, heart-shaped, scalloped-edged, dark-green, deeply veined leaves

long leafstalks

upright stem

heart-shaped, scalloped-edged, dark-green, deeply veined leaves

very long leafstalks

white taproot

root

GARLIC MUSTARD BASAL ROSETTE × ¼

GARLIC MUSTARD IN FLOWER × ¾

scallop-edged, deeply veined, long-stalked basal leaves that grow up to 5 inches across. The stalked stem leaves are smaller and more triangular. The garlic odor is apparent when you crush a leaf. Look for the basal rosettes from fall to early spring. The leaves survive the winter, and you can even find them under the snow. They contain natural antifreezes that lower the freezing point of water. **Caution:** Never put garlic mustard leaves into a car radiator. It's not that kind of antifreeze.

This biennial grows up to 3 feet tall in midspring of its second year, flowers, produces inedibly tough, long, narrow seedpods, and dies. A schoolchild recently showed me that the seeds in the disintegrating seedpod skeletons are good to eat in autumn (and possibly earlier). They taste like mustard. The summer basal leaves of the surrounding first-year plants remain horribly bitter until it gets very cold again. Violets and gill-over-the-ground (pages 229 and 55) look like garlic mustard until you hold the leaves side by side, or crush and smell the plants. There are no poisonous look-alikes. Other similar *Alliaria* species—all edible members of the garlic mustard group—may also have sneaked in from Europe.

The pungent, mildly bitter, garlic-flavored basal leaves are good from late fall to early spring. They taste great to some people, while others find them too bitter unless cooked, or mixed with milder vegetables.

Many plants become more bitter as they mature. But garlic mustard's young, new, arrowhead-shaped stem leaves are more pungent and less bitter in the spring than the basal leaves were in the cold. They carry overtones of sweetness and are easy to strip off, so you can collect bagfuls, along with the terminal clusters of tiny, tasty flower buds and white flowers.

Garlic mustard is great raw in salads, mixed with more mild greens. It's also good steamed, simmered, sautéed or in sauces. Cook only 5 minutes, or the leaves will become mushy.

Sometimes you'll find garlic mustard with exceptionally large leaves. These may have large, whitish, fleshy taproots, which taste like horseradish. They're good from late fall to early spring, before the flower stalks appear. Use them like horseradish, grated into vinegar, as a condiment. I chop these roots into slices and hand them out to children in the classroom. Chaos reigns as the kids rush to the water fountain. Then they want seconds.

Garlic mustard is antiseptic: Juice from the leaves is used for cleansing skin ulcers and eruptions. People also crush the leaves until moist and rub them on aching limbs, for a feeling of warmth.

Another special wild mustard is watercress (*Nasturtium officinale*), adapted to fresh running water. You can find it forming dense mats along streams, in springs, and in shallow pools of running water throughout the United States. It looks like the watercress you get in the store, but it's much better-tasting—more pungent (*nasturtium* means "nose-twister"), and less bitter.

The delicate, dark-green, alternate leaves are from 1½ to 6 inches long. They're divided into lobes, with a large leading lobe at the tip, and three to eleven pairs of progressively smaller lobes continuing along the midrib.

Many white, stalked flowers grow in long, narrow clusters. Each flower is only ⅕ inch across. The fruits are narrow, slender capsules, about ¾ inch long. The base of the plant is covered with fine, white roots.

Caution: Never eat wild watercress raw unless you've had the water tested. Otherwise, you risk serious infection by pathogenic microorganisms. I've had the water tested in a freshwater spring I visit, and I've been collecting watercress (and

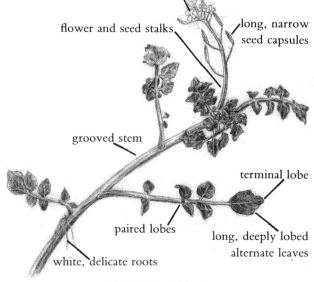

tiny, four-petaled, radially symmetrical, white flowers

flower and seed stalks

long, narrow seed capsules

grooved stem

terminal lobe

paired lobes

long, deeply lobed alternate leaves

white, delicate roots

WATERCRESS × ⅖

drinking the water) for a decade. Watercress used to be in season in the Northeast from mid-spring to late fall, but now that the winters are milder, I find it all year. However, it's more labor-intensive to collect and clean in winter, when frost kills it to the waterline and debris from other dead plants gets mixed in.

Watercress flowers from late spring to early summer, when there are fewer basal leaves, making it harder to collect. Also, it then becomes even more sharp-tasting and somewhat bitter, so I prefer gathering it in mid-spring and autumn.

Note: Watercress is not native to the United States. Anytime you find it, you know someone probably ate a watercress sandwich somewhere along the waterway, and got sloppy. Whenever you collect it, let a few pieces with roots fall back into the water. The plant propagates vegetatively, so it will float downstream and start new stands.

Along with watercress, you'll often find a very tiny plant with paired leaves a fraction of the size of your fingernail, with no lobes. This is duckweed (*Lemna* species), distinguished by having smaller flowers than any other North American flowering plant. It looks a little like chickweed (page 137), but chickweed doesn't grow in the water. Although some cultures around the world to eat duckweed species, I've turned up little information about its edibility here, and it's too small to be worthwhile anyway.

Use wild watercress the way you would commercial watercress. It's great in salads and sandwiches, and it makes wonderful soups, especially combined with potatoes.

Along with dandelions and lambs'-quarters, watercress is one of the world's most nutritious vegetables, excellent for convalescence. It's an outstanding source of beta carotene, iron, and calcium. It also contains vitamin C, bioflavonoids, vitamins E, B_1, and B_2, and the minerals phosphorus, sodium, iodine, manganese, sulfur, zinc, copper, cobalt, and vanadium.

Hippocrates used watercress as a stimulant and expectorant. It's been used for coughs and bronchitis for hundreds of years. Cooked and pureed, watercress also acts as a gentle diuretic: It's good for the kidneys, bladder, and urinary tract, and by relieving fluid retention, it helps the heart. You can also puree it with sea salt and apply it externally,

for gout and arthritis. Watercress even relieves indigestion and dispels gas. It's a favorite of fishermen because it provides a hiding-place for trout, and a home for the invertebrates that trout eat.

Sea rocket's (*Cakile edentula*) fleshy, wedge-shaped, hairless, wavy-toothed, alternate leaves are about 3 inches long and 1 inch wide. Their succulence helps it conserve water—a great advantage by the sea, where fresh water is at a premium.

The small, pink-purple flowers, which grow in terminal clusters, are typical of the mustards, but the cone-shaped seedpods are unique. They look like the nose cones of rocket ships. They're round at the base and pointed at the tip. The resemblance is coincidental. Rocket is an old name for mustard.

You find this salt-tolerant plant closer to the sea than almost anything that's not a seaweed. Its roots actually help hold the dunes together, so harvest with care: Don't take too much of any one plant, and leave the inedible roots alone. Sea rocket grows along beaches throughout eastern and western North America. A similar edible Floridian sea-rocket species (*Cakile lanceolata*) has narrower leaves.

Collect sea rocket from spring to early fall. The leaves are milder at first. Gather the immature pods in midsummer.

This hot mustard rockets your mouth into orbit. I love pungent foods, so this is one of my favorites. The leaves, flowers, and tender young

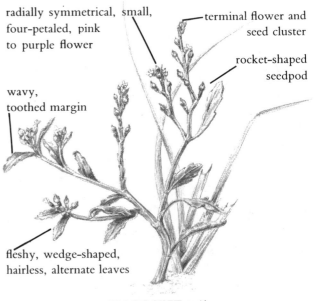

radially symmetrical, small, four-petaled, pink to purple flower

terminal flower and seed cluster

rocket-shaped seedpod

wavy, toothed margin

fleshy, wedge-shaped, hairless, alternate leaves

SEA ROCKET × ¼

seedpods (the seedpods get tough as they mature) are wonderfully spicy, becoming more pungent as the season progresses. Mix with milder greens (almost anything is mild compared to this one) in salads, or include it in cooked vegetable recipes. Use it sparingly, or it may overpower the other vegetables and your mouth. It cooks in about 10 minutes. The young seedpods are great pickled, and they make a wonderful prepared mustard (page 290).

Cuckooflower or lady's-smock (*Cardamine pratensis*) is a perennial, hairless, erect mustard, growing from 8 to 20 inches tall. Its distinctive, fernlike, feather-compound leaves consist of three to seven paired leaflets, and one terminal leaflet. Basal

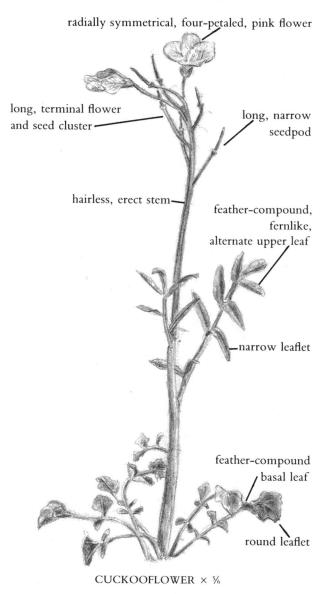

radially symmetrical, four-petaled, pink flower

long, terminal flower and seed cluster

long, narrow seedpod

hairless, erect stem

feather-compound, fernlike, alternate upper leaf

narrow leaflet

feather-compound basal leaf

round leaflet

CUCKOOFLOWER × ⅝

leaflets are rounder, while upper leaflets are more narrow. The pink flowers are typical of mustards. They bloom in spring; then the plant disappears, so you have to collect early. Look for it in moist places, such as swamps, springs, wet meadows, and wet woods, throughout the northern half of North America. Its peppery taste makes it a favorite of everyone who tries it. Use it raw or cooked.

Toothworts (*Dentaria* species) are named for their toothed leaves. They have long clusters of small, pink, purple, or white flowers typical of the mustard family. The unbranched, erect stems, usually hairless, are normally leafless below, with a few leaves above. Basal leaves arise from the horizontal rhizomes—shallowly buried, thick, fleshy, and white—sometimes attached to tubers. These colonial plants grow in moist woods in the eastern and western United States.

The rhizomes and leaves are peppery-tasting and quite good. Enjoy the leaves raw or cooked, like the other mustard greens. Use the rhizomes in salads, ground finely as a condiment, or grated and preserved in vinegar, like horseradish.

There are a number of species: Cut-leaved toothwort (*Dentaria laciniata*), which grows in the Northeast, has narrow, deeply toothed leaves in groups of threes, arranged in whorls around the stem. It grows from 8 to 14 inches tall, and flowers from early to late spring.

Toothwort or pepperwort (*dentaria diphylla*) has long-stalked basal leaves and opposite stem leaves, each divided into three broad, coarsely toothed leaflets. Otherwise, it's similar to the preceding species. Large toothwort (*Dentaria maxima*) is similar to toothwort, but its three leaves grow at different levels on the stem.

Horseradish (*Armoracia lapathifolia*) is a coarse-looking European garden escape that can grow up to 4 feet tall. The huge, ovate to lance-shaped leaves are close to 2 feet long and over 4 inches wide, superficially like curly dock's, but with slight teeth and without curly margins. Also, curly-dock leaves are more nearly parallel to the ground, while horseradish leaves point steeply upward.

The white flowers, which bloom from midspring to midsummer, are larger than most wild mustards'. The egg-shaped seedpods are tiny. Virtually everyone has tasted the fiery, whitish taproot, which can reach several feet in length.

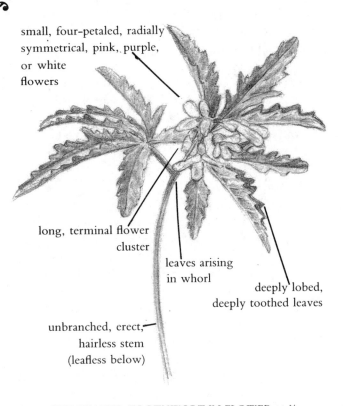

small, four-petaled, radially symmetrical, pink, purple, or white flowers

long, terminal flower cluster

leaves arising in whorl

deeply lobed, deeply toothed leaves

unbranched, erect, hairless stem (leafless below)

CUT-LEAVED TOOTHWORT IN FLOWER × ⅗

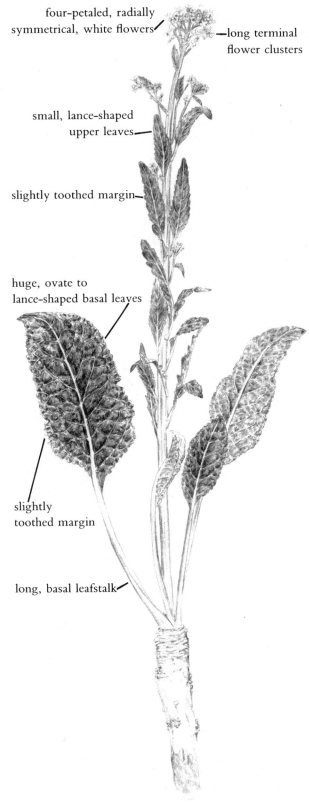

four-petaled, radially symmetrical, white flowers

long terminal flower clusters

small, lance-shaped upper leaves

slightly toothed margin

huge, ovate to lance-shaped basal leaves

slightly toothed margin

long, basal leafstalk

HORSERADISH × ¼

Horseradish does best on disturbed soil. You can find it throughout the United States, except, apparently, wherever I look for it: I've been searching high and low for horseradish since I began foraging. I once found it along a polluted highway—unusable. Next, it was in a park adjacent to a private house, and the ignorant homeowner was illegally spraying it with toxic insecticides. I've tasted fresh horseradish from a friend's garden, and it's fantastic, so I'm not giving up my search for untainted wild horseradish.

The younger leaves are wonderful in salads, and since they're so large, you can harvest plenty without harming the plant. You need only a few roots to grate and mix with vinegar to make prepared horseradish, a wonderful, pungent condiment. Simply grate the roots and cover them with vinegar. They last indefinitely in the refrigerator. Don't cook it, or you'll destroy the flavor.

Horseradish has the same medicinal constituents as *Brassica* species (page 249), although you use its roots differently. It creates a pleasant warmth in the stomach, promotes stomach secretions, and increases the flow of urine. It promotes digestion, and it's used for edema. Herbalists mix horseradish infusion with a little honey for persistent coughs. They also rub the cut root on stiff or aching joints to increase circulation and warmth. **Caution:** Prolonged contact will blister the skin, and eating it in large amounts may irritate the digestive system. Livestock have died from overconsumption.

The mustard family also supplies us with many commercial vegetables, such as cabbage, red radishes, rutabaga, broccoli, Brussels sprouts, Chinese cabbage (bok choy), commercial mustard greens, collards, turnips, and more. Other plant families that contribute to agriculture are nightshades (tomatoes, potatoes, bell peppers, and eggplant), legumes (peas and beans), goosefoot (spinach and beets), the lily family (onions and garlic), composites (lettuce, sunflowers, artichokes), and the carrot family (carrots, parsley, and celery). The tiny sample of the Earth's bounty we currently cultivate may be insufficient to meet our future needs—another reason why so many environmentally conscious people are trying to preserve our planet's biodiversity.

OTHER PLANTS OF FIELDS IN EARLY SPRING

Edible or Medicinal Plants

Bayberries, caraway roots and leaves, carrot roots and leaves, chicory leaves and roots, clover leaves, common evening primrose roots and leaves, cow parsnip roots, curly dock leaves and roots, wild onion leaves and roots, field pennycress, goatsbeard leaves and roots, Japanese knotweed shoots, orpine roots, parsnip roots, peppergrasses, sassafras twigs and roots, sheep sorrel leaves, spearmint, strawberry leaves, wild lettuce leaves, yarrow.

For Observation Only

Carrot skeletons, peach blossoms, poison ivy leaves, vines, and shrubs, thistle, sumac skeletons, yucca.

PLANTS OF THICKETS IN EARLY SPRING

Edible or Medicinal Plants

Apple blossoms, bayberry leaves, groundnut roots, hog peanut seeds, Japanese knotweed shoots, wild lettuce leaves, mugwort, nettles, sassafras roots and twigs, wild potato vine roots.

For Observation Only

Cornelian cherry flowers, hazelnut catkins, Juneberry blossoms, pecan flowers, pear blossoms, plum blossoms, poison ivy leaves, vines, and shrubs, spicebush flowers, yew flowers.

PLANTS OF WOODLANDS IN EARLY SPRING

GOUTWEED

(Aegopodium podagraria)

Goutweed is an erect, herbaceous perennial in the carrot family with smooth-stalked, three-parted, palmate-compound leaves. Sometimes the oval, pointed, toothed leaflets are partially fused. Leaves tend to enlarge, to catch more sunlight, as they evolve. Simultaneously, they tend to divide, becoming less vulnerable to strong winds and hungry predators, and allowing light to pass to the leaves below. Goutweed is in an intermediate stage of leaf division. Some leaves are completely di-

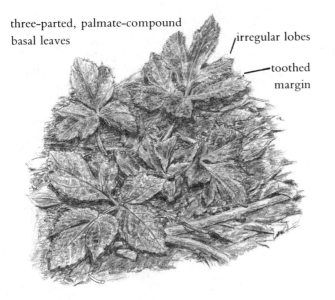

three-parted, palmate-compound basal leaves

irregular lobes

toothed margin

GOUTWEED BASAL ROSETTE × ½

vided into three parts, while others are merely lobed. In a future edition of this book (in about a million years), all the leaves should be completely divided. If not, demand a refund.

The tiny white flowers, which bloom from June to August, are arranged in compound, umbrellalike clusters. Each cluster's stem is long, and the clusters are denser than goutweed's edible look-alike, honewort (page 60). The leaves, stems, and the tiny cylindrical seeds (which appear in late summer) all have a pleasant, spicy fragrance.

Arising from rhizomes, goutweed grows in dense stands. It keeps other plants off its turf, either by shading or crowding them out, or by secreting herbicides. Look for it in partially sunny forested areas with rich, moist soil. You find either lots of it or none. I rarely see it in the countryside, probably because it began growing in urban areas in the Northeast after arriving from Europe.

Goutweed is a favorite among traditional Asians. They use the whole plant, except for the roots, as a vegetable, and they pickle the stems. I prefer the tender, young leaves, which taste like a combination of parsley and celery, similar to honewort. You can steam or sauté it as you would other vegetables, include it in soups, use it raw in salads, or dry it for future use. It also makes a great parsleylike seasoning. It's easy to collect in quantity, and you can find and use it from early spring to

late fall. The dead flower stalks often overwinter, so you may even be able to locate the following season's stands in the snow.

Goutweed is named after a disease: In the murky world of irrational common names, and the herbal medicine realm of mixed fact and fiction, it looks as if we finally have a clearly named plant that treats a disease. Nothing could be farther from the truth. Goutweed has no effect on gout, but goats love it so much, the plant was originally called goatweed. Since language changes over time, this eventually became goutweed. Then people thought it could treat gout, and the name stuck. The generic name, *Aegopodium*, means "goat-foot" in Greek.

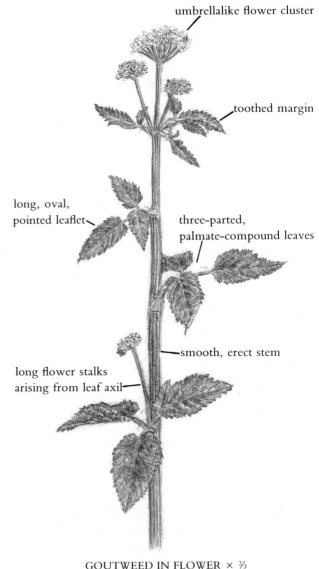

umbrellalike flower cluster

toothed margin

long, oval, pointed leaflet

three-parted, palmate-compound leaves

smooth, erect stem

long flower stalks arising from leaf axil

GOUTWEED IN FLOWER × ⅔

SOLOMON'S SEAL

(*Polygonatum* species)

Solomon's seal pokes its head out of the ground in early spring. The shoot looks like a pencil with the undeveloped leaves wrapped tightly around the stem's top. The leaves unfurl before mid-spring. The gracefully arching, single-stemmed herbaceous plant grows from 1 to 3 feet long. Its alternate, pointed, smooth-edged, oval leaves are 4 to 5 inches long, with parallel veins.

By mid-spring, you'll find clusters of one to ten paired, greenish-white, bell-shaped, stalked flowers dangling from the leaf axils. Summer transforms them into inedible blue-black berries.

This native herb, a member of the lily family, has a perennial, fleshy white rhizome more than an inch thick. A set of scars, resembling impressions of a signet ring, indicate where the previous years' stalks emerged. There is one leaf scar for each year of the rhizome's life. These "seal" impressions partially account for its common name. The generic name comes from *poly* and *gonatum*, which means "many-jointed"—referring to the rhizome's joints.

The two most common species are Solomon's seal (*Polygonatum biflorum*) and great Solomon's seal (*Polygonatum commutatum*). The former has twin flowers on each leaf axil, and it grows 1 to 3 feet. It's sometimes cultivated in gardens. The latter is larger and coarser, over 4 feet tall, with two to ten flowers in each cluster.

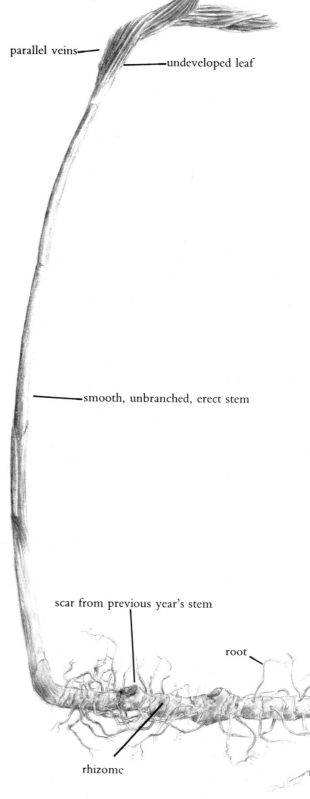

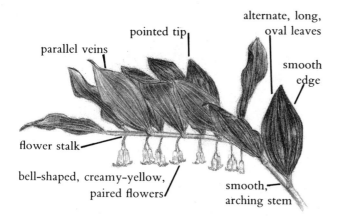

SOLOMON'S SEAL IN FLOWER × ¼

GREAT SOLOMON'S SEAL SHOOT AND RHIZOME × ⅓

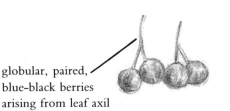

globular, paired, blue-black berries arising from leaf axil

SOLOMON'S SEAL BERRIES × 1

A similar-looking relative, false Solomon's seal (*Smilacina racemosa*), has smaller shoots. Its arched stem, 1 to 2 feet long, is somewhat zigzag. White flowers with stamen longer than the petals, then red berries (which may later turn whitish, with red speckles), grow on a long terminal cluster at the plant's tip. False Solomon's seal's yellow-beige rhizome is about ¾ inch thick. Otherwise they look alike.

Caution: False Solomon's seal's rhizome is poisonous. You'd have to soak it overnight in lye,

then boil out the lye, to make it safe. I've never tried this, and I know of no one who has.

Also, eating many of false Solomon's seal's inferior-tasting berries may cause diarrhea. Cooking them improves the flavor and reduces the purgative effect, but you must exercise great caution. Don't forage for either plant if you're an unsupervised beginner. There are too many confusing poisonous shoots, such as jack-in-the-pulpit, woodland iris, and mayapple. Because the shoots are edible when they're the most difficult to identify, locate the mature plants in flower, and return the following year to find the shoots.

Solomon's seal species grow in moist woods, bogs, and roadsides throughout eastern North America. False Solomon's seal grows in the same habitats. Its range extends from the East Coast west to Missouri and Texas.

Picking some Solomon's seal shoots where they're very plentiful does no harm—they'll soon

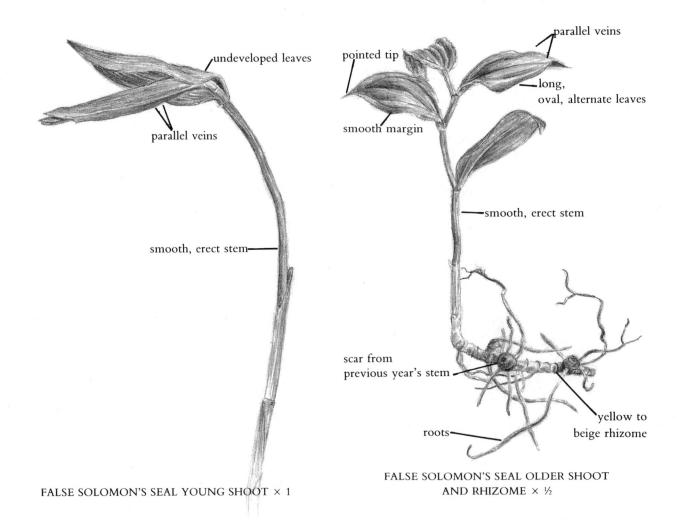

undeveloped leaves

parallel veins

smooth, erect stem

FALSE SOLOMON'S SEAL YOUNG SHOOT × 1

parallel veins

pointed tip

long, oval, alternate leaves

smooth margin

smooth, erect stem

scar from previous year's stem

roots

yellow to beige rhizome

FALSE SOLOMON'S SEAL OLDER SHOOT
AND RHIZOME × ½

regenerate. However, this plant is one of the few edible species that is not a "weed." Never collect where it's rare or legally protected. It's especially destructive to remove the rhizomes under such circumstances. False Solomon's seal is much more common.

Both plants' young shoots are edible in early spring. Solomon's-seal root is good all year, but it's best from fall to early spring.

Remove the shoots' acrid leaves, which are especially bad cooked. Add the chopped stems raw to salads, or steam or simmer them 10 to 15 minutes. They taste like asparagus, only sharper. They'll enhance virtually any entrée, vegetable dish, soup, stew, or casserole. But when the leaves unfurl, the shoots become too bitter and tough to use.

The root's flavor reminds me of burdock with a bite. Because of its slight acrid qualities, I find it best, like the shoot, in a soup or stew, where it lifts the milder flavors.

In the early 1980s, I used to attend inspirational herb classes in Harlem, taught by a charismatic folk herbalist, the late "Dr." John Moore. This man knew all the folklore and folk medicine associated with thousands of herbs, and he generously shared his knowledge and experience with everyone.

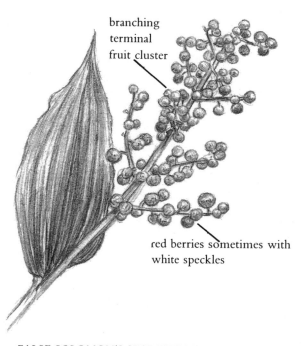

FALSE SOLOMON'S SEAL WITH BERRIES × ⅕

branching terminal fruit cluster

red berries sometimes with white speckles

As a Southern black child in the early twentieth century, he was denied a formal education. Escaping from the Arkansas cotton fields, he became a hobo, studying herbal folk traditions from Canada to Central America, traveling by boxcar. He survived the Great Depression on wild foods, and became the Surgeon General of the American Hobo Society.

A magnificent orator, "Dr." Moore treated each plant as a savior to humanity. He held Solomon's-seal root in the highest esteem, referring to it in glowing terms as part of his secret herbal formula to increase psychic abilities. This herb would allow you to "hear the trumpets of Jericho."

Solomon's seal has strong biblical associations. King Solomon of biblical Israel was renowned for his wisdom, so his seal of approval confirms the plant's usefulness. Solomon's-seal root is mucilaginous, demulcent, astringent, and expectorant. Women drink the tea, or use it as a douche for menstrual irregularities, vaginal yeast infections, and cramps. It's supposed to soothe and tone the female reproductive system, and strengthen the digestive tract.

For centuries, the decoction has been used internally to accelerate healing of wounds and broken bones. It's also good for chest and lung congestion. Externally, people apply compresses to black eyes,

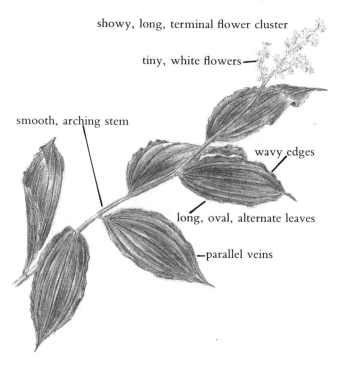

showy, long, terminal flower cluster

tiny, white flowers

smooth, arching stem

wavy edges

long, oval, alternate leaves

parallel veins

FALSE SOLOMON'S SEAL IN FLOWER × ⅓

bruises, inflammations, and benign tumors. (Be sure to get an accurate diagnosis to make sure that any tumor is really benign.)

WILD LEEK, RAMP

(Allium tricoccum)

The wild leek is a rather simple-looking herbaceous plant, although its sea-green coloring, waxy finish, and delicate texture are unusual. Ramps have long leaves with parallel veins, as do all members of the lily family. The two or three broad, long-stalked, arching, basal leaves are 4 to 12 inches long, and from less than 1 inch to nearly 2½ inches wide. Crush them to smell the strong onion flavor, absent from similar-looking poisonous lilies, such as lily of the valley. Nothing that smells like onions is poisonous.

The leaves come up in early spring and wither by late spring. Then, a domelike cluster of small, six-petaled, white flowers appears atop a bare flower stalk, ½ foot to nearly 1½ feet high. The flowers die in early summer, to be followed by three hard seedcases that look like shiny, black BBs. A few persist through the winter, confirming the identity of the next generation of leaves, and pointing to the underground locations of clusters of perennial, oblong, white bulbs—good all year.

Wild leeks thrive in partially shaded, moist, rich woodlands, where the colonies form a wondrous ground cover. They range from New England and the Great Lakes region south to northern Georgia.

Collect the greens in the spring, especially just before they wilt, when the plant doesn't need them anymore. Use the leaves and bulbs the way you'd use commercial leeks, raw in salads or cooked.

Ramps are stronger and sharper-tasting than any commercial leek or scallion, but they're terrific. I've steamed them alone for a delicious side dish, and cooked them in traditional leek and potato soups. They're incredible sautéed, perking up any dish. A wild leek and cream cheese (or tofu) dip is exquisite. Ramp bulb pickles are wonderful too. The Indians covered the bulbs with grass and

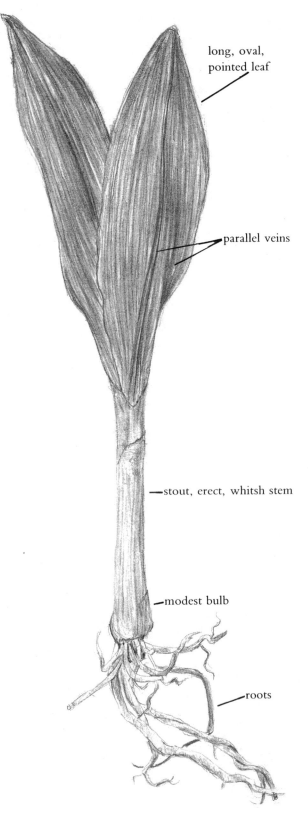

long, oval, pointed leaf

parallel veins

stout, erect, whitsh stem

modest bulb

roots

WILD LEEK SHOOT AND BULB × ⅔

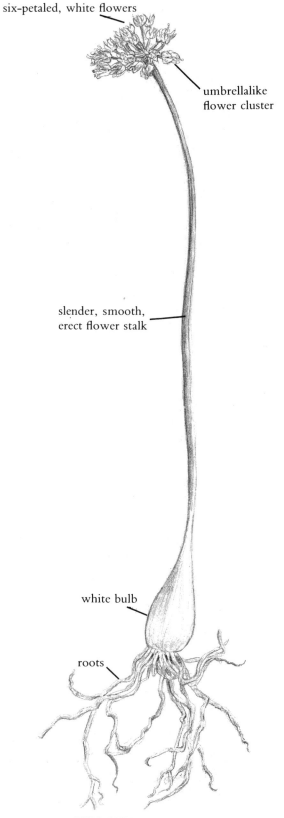

six-petaled, white flowers

umbrellalike
flower cluster

slender, smooth,
erect flower stalk

white bulb

roots

WILD LEEK IN FLOWER × ½

baked them in a bed of ashes. I imagine they came out tender and fragrant. You can also dry the leaves and bulbs. Unlike related *Allium* species, they retain their flavor.

Wild leeks have medicinal properties similar to wild onions (page 206) and garlic, but the active substances' highest concentration is in the inaccessible seeds, locked inside the hard little black capsules. Next best are the bulbs. Use a bulb infusion for colds, high blood pressure, and other problems treated by garlic.

Any wild-food aficionado will tell you that ramps are among the finest of all wild foods, so when I began hunting for edible wild plants, this savory member of the lily family was near the top of my list. Yet year after year, I'd find only paper-flavored ramps growing in field guides.

I did once find a broad, ramplike leaf with no flower in a local park, brought it home, identified it as an unknown member of the lily family, drew it, and hung the drawing on my wall.

I found my first 100 percent certain ramp in a diorama at the American Museum of Natural History in Manhattan. I stared at it for long moments, impressing the three-dimensional image into my memory. The following spring, leading a tour in a distant park, I spotted ramps from thirty yards away, and our noses soon confirmed the identification.

When I got home, I looked at the drawing on my wall and realized that ramps also grew close to home, but I had forgotten the location. I scoured the local parks the next two springs, in vain.

Three years later, leading a tour in a park close to home, I spotted one ramp. Everyone was sent running in all directions, and within five minutes, somebody had found the mother lode. Now we get unlimited quantities of one of the best spring greens every year.

Ramps were an important survival food for the pioneers, who had the greatest difficulties in the spring. They'd be vitamin deficient from living on stored food since autumn, their supplies would be running out, and they had to work hard on crops that wouldn't be ready until summer. When this delicious wild vegetable came up, they'd celebrate. In the South, they still hold ramp festivals today. Everyone spends days collecting, preparing, and

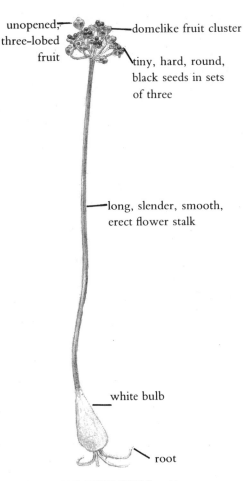

unopened, three-lobed fruit

domelike fruit cluster

tiny, hard, round, black seeds in sets of three

long, slender, smooth, erect flower stalk

white bulb

root

WILD LEEK WITH SEEDS × ½

eating ramps. You can imagine the odor permeating this event. Schools often close the next day because of bad breath.

OTHER PLANTS OF WOODLANDS IN EARLY SPRING

Edible or Medicinal Plants

Aniseroot roots and leaves, apple blossoms, birch twigs, Kentucky coffee tree pods and seeds, cow parsnip roots, curly dock leaves and roots, daylily shoots and roots, wild onion leaves and bulbs, garlic mustard leaves, groundnut roots, hog peanut seeds, horse-balm roots, jewelweed shoots, pine needles, sassafras roots and twigs, storksbill, strawberry leaves, toothwort leaves and roots, violet leaves, waterleaf leaves, wild ginger leaves and roots, wild lettuce leaves, wild potato vine roots, wintergreen leaves and berries.

For Observation Only

Blackberry canes, greenbrier vines, hazelnut catkins, hickory catkins, Juneberry blossoms, oak catkins, pear flowers, poison ivy vines, leaves, and shrubs, spicebush flowers, yew flowers.

PLANTS OF FRESHWATER WETLANDS IN EARLY SPRING

WATERLEAF, INDIAN SALAD

(*Hydrophyllum* species)

This is a native genus of flowering plants with varying leaves, 1 to 2 feet high. All species have more or less densely clustered white to blue, or purple, five-lobed, bell-shaped flowers with protruding stamens and stigmas. The leaves are mainly feather-compound, basal or alternate. Many of the twelve species have leaves with light-green splotches on them, as if stained by water, accounting for their common name. *Hydrophyllum* means

"water leaf" in Greek, also a reference to their wet habitat.

Waterleaf grows in wet places in woods and near streams, throughout much of eastern North America, and in rich, moist, shaded spots in mountains throughout the West.

Virginia waterleaf (*Hydrophyllum virginianum*) has smooth, irregularly cut leaves with five to seven lobes. It flowers from mid-spring until late summer. Large-leaf waterleaf (*Hydrophyllum macrophyllum*) looks like Virginia waterleaf, but the leaves are divided into at least seven lobes, and it's rough-hairy. Broad-leaved waterleaf (*Hydrophyl-*

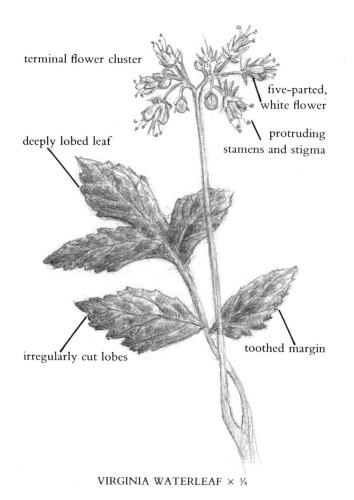

terminal flower cluster

five-parted, white flower

deeply lobed leaf

protruding stamens and stigma

irregularly cut lobes

toothed margin

VIRGINIA WATERLEAF × ¾

OTHER PLANTS OF FRESHWATER WETLANDS IN EARLY SPRING

Edible or Medicinal Plants

Apple blossoms, cattail shoots, coltstfoot flowers, cow parsnip roots, cuckooflower leaves, curly dock leaves and roots, false Solomon's seal shoots, groundnut roots, jewelweed shoots, miner's lettuce leaves, parsnip roots, Solomon's seal shoots and roots, spearmint, sweetgale, violet leaves, watercress, water mint, winter cress leaves, flower buds, and flowers.

For Observation Only

Cattail seed heads, Juneberry blossoms, plum blossoms, poison ivy vines, shrubs, and leaves, thistle, yew flowers.

PLANTS OF THE SEASHORE IN EARLY SPRING

Edible or Medicinal Plants

Bayberry, carrot roots and leaves, common evening primrose roots and leaves, curly dock leaves and roots, wild onion leaves and bulbs, field pennycress, groundnut roots, Japanese knotweed shoots, marshmallow roots, mullein leaves, parsnip roots, peppergrasses, pine needles, sheep sorrel leaves, strawberry leaves, sweet clover leaves, yarrow.

For Observation Only

Carrot skeletons, mullein skeletons, peach blossoms, poison ivy vines, leaves, and shrubs, prickly pears, sumac skeletons, yucca.

lum canadense) has maplelike leaves, with the flower stalks beneath the leaves. It flowers from late spring to early summer.

Hairless species' young leaves are good in salads when they first come up, in early spring. All varieties make a nice steamed vegetable, or soup. The flavor is like parsley, only more delicate. Collect them early in the spring, while the leaves are still tender.

PLANTS OF MOUNTAINS IN EARLY SPRING

MINER'S LETTUCE, SPANISH LETTUCE

(Montia perfoliata)

This succulent native annual consists of several unbranched stems, 3 inches to 1 foot tall, originating from a fibrous root system. Long leafstalks display wide-oval, pointed, basal leaves 1 inch long. The pairs of roundish upper leaves join right below the flower stalks, forming shallow cups around the stems. They resemble inverted parasols.

From January to July, many small, short-stalked, pink to white, five-petaled flowers top the stems. Each clawed petal is 0.1 to 0.2 inch long. The fruits are tiny, three-parted capsules, each containing one to three black seeds.

Miner's lettuce grows in shaded moist habitats, near springs, and on lower mountain slopes, often along with chickweed. It grows along the West Coast, and from the Pacific Northwest and British Columbia east to western North and South Dakota. It has also been introduced to Europe, where it's cultivated and sold as winter purslane.

Miner's lettuce is edible all year at the southern part of its range. In the north, it's best in spring and summer.

This relative of purslane (page 28) was a major food in California during the Gold Rush, and it's just as tasty and abundant today. The young upper leaves and stems are the best parts. They're great raw, with a mild flavor and tender texture. The entire plant's leaves and stems are good lightly steamed or sautéed just until the leaves wilt. Sautéed in a little butter or olive oil with a dash of salt, this plant is heavenly. You may offset the mildness with stronger-flavored greens, like sheep sorrel or wild mustard greens. You can also eat the spicy flowers, which are good for seasoning salads. Even the simmered roots are edible. They taste like water chestnuts.

Miner's lettuce is an excellent source of vitamin C. Miners used it with chickweed, as food and to prevent scurvy.

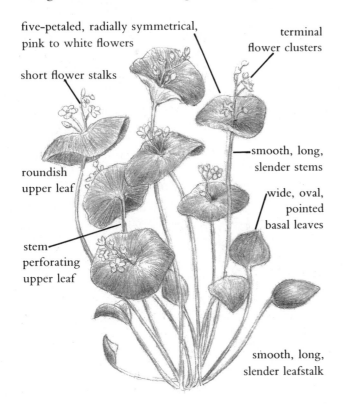

five-petaled, radially symmetrical, pink to white flowers

short flower stalks

terminal flower clusters

roundish upper leaf

smooth, long, slender stems

wide, oval, pointed basal leaves

stem perforating upper leaf

smooth, long, slender leafstalk

MINER'S LETTUCE × ½

ORPINE, STONECROP, LIVE-FOREVER

(Sedum purpureum)

Orpine is so distinctive, once you see it you'll never forget it. This stout-stemmed, herbaceous plant reaches 1 to 1½ feet tall. The light-green, oval, alternate, coarsely toothed, succulent leaves are 1 to 2⅓ inches long. They usually grow in whorls of three on the stem, although they're sometimes alternate. It's called "live-forever" because it's so fleshy, it can go for weeks without water, like a desert plant.

The small, pink-lavender five-petaled flowers, blooming from summer to early fall, grow in one

showy, umbrellalike flower head

radially symmetrical, small, pink-lavender, five-petaled flower

oval, succulent, light-green leaf

leaves often in whorls of three

smooth, succulent stem

ORPINE × ½

small, radially symmetrical, five-petaled, pink-lavender flower

ORPINE (single flower) × 3

rounded cluster on the top of the plant. Orpine's root is distinctive, with horizontal masses of fingerlike tubers.

Roseroot (*Sedum rosea*) resembles orpine, and it's also edible. It's smaller, with untoothed, spirally arranged, overlapping leaves. The yellowish-purple, or rose-pink flowers have four petals, and the flower cluster is encircled by a border of linear leaves. The large roots smell like roses, but they're unpalatable.

Orpine and roseroot leaves are best when they first come up in early spring, although they're still edible into the fall.

Roseroot and orpine don't grow together. Roseroot requires colder places—Canada, or frigid, rocky mountaintops south into New England. Or-

pine grows in more accessible places—gardens, fields, and disturbed soil, as well as mountains. It grows in the Northeast south to Maryland, and also appears scattered throughout the country wherever it can escape cultivation. Orpine and related edible relatives also grow in the West, where they prefer moist, rocky habitats.

I had thought of orpine as a pretty garden succulent until I spent one summer in the New England mountains, where orpine grows profusely. You can be sure it didn't go to waste. Everyone had orpine salad up to their ears, with no complaints. Orpine is the perfect salad green, crunchy and tender. Fortunately there were plenty of hot mustard greens growing, to complement its mild flavor. Now, when I see orpine growing in neat rows in garden plots, I think of it as that wildflower that became a garden plant by accident. In reality, orpine was first introduced into gardens from England. Orpine and roseroot leaves are also good sautéed. Their tender texture is a perfect complement to spicy, colorful foods like red peppers and tomatoes.

You can collect orpine tubers from fall to early spring—anytime the plants are not flowering. I collected the tubers one fall in wild orpine country. They were easy to dig up. Each plant came out of the ground with a large clump of fleshy, long tubers. In a few minutes I had a heavy load, with very little work and no environmental impact. You can use them raw. They're mild and crunchy, like water chestnuts. They also make incredible pickles in vinegar and spices.

Many wild foods are so concentrated, without the excess water and weight of commercial plants, that it's always a surprise to find a wild plant with more water-heavy tenderness. Of course, dig up orpine only where it's growing like a weed, and ask permission before you invade someone's garden.

Orpine has inspired more common names than most other species, including everlasting, witch's money bags, evergreen, midsummer men, common sedum, bog leaves, life-of-man, and Aaron's rod. My favorite is frog's belly. Kids love learning the reason: Rub a leaf until it separates into layers. Blow on the open end, and it will expand and look like a frog's belly.

OTHER PLANTS OF MOUNTAINS IN EARLY SPRING

Edible or Medicinal Plants

Pine needles, sassafras roots and twigs, sheep sorrel, waterleaf leaves, creeping wintergreen leaves and berries.

For Observation Only

Juneberry blossoms, spicebush flowers.

PLANTS OF DESERTS IN EARLY SPRING

For Observation Only

Prickly pear pads, yucca.

CHAPTER 7

COOKING WITH EDIBLE WILD PLANTS

WILD FOODS ARE DIFFERENT

Whether you're an experienced cook or a novice, you'll find cooking with freshly picked wild foods an exciting, creative experience. Wild plants' flavors and textures are completely different from those of processed foods, and they're a step beyond fresh garden produce. Cultivated plants have been specially developed to weigh more and can command a higher price because they contain more water than wild plants. Wild plants, on the other hand, grow to survive, so they concentrate more vitamins, minerals, fiber, and flavor.

WILD GREENS

Wild greens have more intense flavors than their commercial counterparts. I usually find I need to eat less to feel nourished, although I often eat more because they make me feel so wonderful. Each wild vegetable has so much personality that I rarely think of substituting one for another in a specific recipe. Still, their flavors and textures let us divide them into groups. Greens may be mild, bitter, tart, pungent, or salty.

The mild, rich flavor of nettles and lamb's-quarters is suited to casseroles and cream soups. They can also be steamed and served with just a little lemon juice. Mild, tender chickweed also improves soups, and its crunchy, crisp texture goes so well in salads, even the most ardent supporters of standard iceberg lettuce and tomato salads like it. Other mild-tasting, slightly tougher greens like lady's thumb and Asiatic dayflower are still good in salads, but they're even better steamed and added to other vegetables or grains.

I like bitter-tasting dandelions and watercress mixed into salads with milder greens, served along with a vinaigrette dressing to cut any bitterness. They're also good sautéed with sweet vegetables, like carrots and onions.

Tart greens like curly dock, sheep sorrel, and wood sorrel are wonderful in salads—you hardly need any dressing with them. They'll also accent any recipe requiring a tangy lift. I like using them in stuffed potatoes.

Pungent mustard greens like shepherd's purse or poor man's pepper make a nice addition to a salad with mild vegetables, or mixed with grains like rice or millet. They steam nicely too. Mustards are naturals for strong chili and curry dishes. Some species have edible seeds, great for making wild prepared mustard.

Greens in the lily and onion family have a different pungency from the wild mustards. Daylily shoots, wild leeks, and wild onions make great sauces, spreads, and dips. They complement almost any dish, and all are wonderful sautéed in olive oil with a little salt—served as a side dish with an entrée.

Some vegetables combine qualities. Purslane is mucilaginous and sour, while violets are slightly mucilaginous and somewhat pungent. Both make superb soups and stews. They're often tender and crunchy in salads as well.

EDIBLE FLOWERS

Handle edible flowers, with their delicate textures and fragrances, with care. Daylily flowers are mildly oniony and sweet, great for decorating salads or in tempura batter. Violet flowers are milder—best in salads or as garnishes, since cooking destroys their flavor. I sometimes cook wisteria, black locust, and elder flowers into fritters or pancakes. They have strong, sweet, perfumed flavors that shine through the other ingredients. You can collect them in quantity, and they're superb for wine making.

WILD FRUITS AND BERRIES

Wild fruits make incredibly tasty desserts. Because their flavors include an intricate mix of sweet, sour, flowery, fruity, bitter, and hot components, you get a more satisfying, earthier, and balanced-tasting dessert. In comparison, commercial fruits are sweeter, more watery, and less satisfying, with less flavor.

Some wild fruits, like wild apples, most wild berries, plums, and persimmons, are easy to work with. You can eat them raw, or toss them into any number of recipes. Others require more processing. Since they're not bred commercially, they may seem dry, tart, or seedy. You'll come to appreciate them through cooking experimentation, and many make excellent trail nibbles.

Sour, seedy wild fruits such as crab apples, wild grapes, currants, and highbush cranberries are wonderful cooked in fruit juice until soft, and passed through a food mill to separate the seeds. The tart puree makes an enlivening addition to other fruit recipes.

The sweet, rich wild raisins are somewhat heavy, with a modest amount of fruit and a large pit. They're wonderful cooked in prune juice and passed through a food mill. I use them to stuff pastries, the way people normally use pureed apricots or prunes.

Black cherries and chokecherries are especially interesting. They're tart and somewhat bitter, combined with the familiar cherry flavor, varying from tree to tree. Again, use a food mill to strain out the seeds.

Some fruits, like elderberries and cranberries, are much better cooked and sweetened. Elderberries add crunchiness and pungency in dessert recipes with sweet ingredients. Wintergreen berries, on the other hand, lose their flavor when you cook them.

WILD ROOTS

Wild edible roots were a big challenge for me in the beginning. Their hardiness needed the right sort of preparation. Very few are really good raw. Some are difficult to dig up, which is fine because you need less in most recipes since they have typically strong flavors.

Some roots, like dandelion, goatsbeard, cow parsnip, and wild potato vine roots, are bitter. Even cooked, they're too strong for some people. The secret is to use them in moderation, to enrich rather than embitter your stew or soup. Also, you usually have to preboil them in several changes of water to make them palatable. Repeated boiling reduces nutritional value, but it's sometimes necessary.

Some wild roots contain varying amounts of sweetness. Aniseroot's licorice flavor creates the illusion of sweetness, while wild ginger simulates commercial ginger. The mild-flavored Jerusalem artichoke and burdock contain the sugar inulin. They taste starchy and make good potato substitutes. The wild parsnip is the sweetest. Tough and coarse-looking raw, it is rendered tender and luscious by cooking.

Some wild root vegetables are unusually pungent or spicy. Most people have been brought close to tears by horseradish. Garlic mustard's and toothwort's roots are similar, while common evening primrose's tastes a little like radish and turnip. The bulbs of wild onions and wild leeks are similar to their commercial counterparts, often with even more punch.

Some roots, like hog peanuts and groundnuts, are so nutty, they need little to add to their flavor.

🍃 NUTS AND SEEDS

Wild nuts and seeds usually require some work. Yet they are all totally unique, both in preparation, and in flavor and texture:

You must shell and boil acorns to get rid of bitter tannins. The result is large, soft, protein-rich, earthy-tasting nuts—best in cooked recipes.

To eat fruity black walnuts and hickory nuts, you must first dehusk and dry them. The butternut's husk must be completely dry before you can crack it open. You'll need a heavy-duty nutcracker, vises, or hammer to crack these native nuts, but their flavors make them worth the time.

Rich and chewy gingko nuts are more like vegetables than nuts. You must separate them from their malodorous fruits and cook them before use in recipes.

Collecting seeds is usually a long, slow process. It takes about half an hour to collect about a quarter cup of mustard seeds, although they're so strong, they'll go a long way. The same holds for wild carrots and aniseroot seeds. Use them all as seasonings.

Common plantain's raw peanut-flavored seeds are mild in recipes, and you can strip them from their stems quickly. Lamb's-quarters and amaranth seeds are very plentiful, although their season is short. They're like grains, and winnowing and sorting them takes time, but you'll love their nutty crunchiness.

Seeds of wild grasses—grains—are usually too small, with too much chaff to be worthwhile. Two exceptions are wild rice and foxtail grass. For a reasonable amount of work, you get some delicious food from these plants.

ALL-NATURAL CUISINE

The recipes in this book differ from those of most wild-food books, with natural, unprocessed ingredients consistent with the high quality of the wild plants. Desserts are sweetened with fruits and fruit juices, although honey or maple syrup is sometimes called for. Everyone knows that refined sugar is unhealthful, but few people realize that it also covers up the flavors of the other ingredients— a shame when you've spent so much time studying, finding, and collecting them. If you make the additional effort to wean yourself from concentrated sugar, your taste perception will sharpen. Fruit alone will taste sweet, and the underlying flavors will come shining through.

I recommend soy milk instead of cow's milk, but the choice is up to you. We also use many whole-grain flours, including whole wheat, barley, rye, buckwheat, brown rice, and oats. They're all available in health-food stores, and other markets are starting to catch on. White flour has been stripped of its fiber, B-complex vitamins, vitamin E, and trace minerals. "Enrichment" with a few

B-complex vitamins doesn't compensate. Its main cooking advantage is that it makes breads rise well. But so does whole-wheat flour and, to a lesser extent, rye flour. They produce heavier, heartier results. The other whole-grain flours are lighter than whole wheat, but they don't rise. Use them in quick breads, to thicken sauces, for pie crusts, and other nonrising pastry recipes.

Our recipes are mostly vegetarian, with minimal use of dairy products. They're healthy and tasty. Protein comes from nuts, seeds, grains, beans, some of the vegetables, and the roots. The whole-grain breads and cakes are moist and hearty.

Some of the recipes are simple, some are elaborate. Many take time to prepare—foraging is not a quick fix. You're working with nature, not fast foods. Identifying plants, collecting and preparing wild foods is an art. It's worth it to make the most of your harvest with delicious recipes.

SPECIAL EQUIPMENT

Although our ancestors lived on wild foods as a matter of course, we have the advantage of modern kitchen equipment—examples of appropriate technology that allow us to better utilize local foods and cut down on the waste associated with the food business. A blender, electric or hand spice grinder, food mill, food processor, and food dehydrator are especially handy. You buy these devices once, and they last for years.

A hand-powered food mill is inexpensive and very helpful for separating small wild fruits from their seeds. This device is a bowl with a strainer on the bottom. You turn a crank that rotates a plate and pushes the food through the strainer. The seeds are too big to go through the holes. You put a larger bowl under the food mill to collect the fruit, and you periodically throw out the seeds.

A grain mill can be expensive, but if you plan to do much baking, it eventually pays for itself: Buying whole grain continually saves you money over buying flour or baked goods, and your breads, cakes, and pastries will be tastier and fresher.

A juicer, and a heavy-duty nutcracker are nice to have, but not indispensable. However, a food dryer is especially useful for preserving the large harvests you sometimes get. It also pays for itself, since rehydrated fruits and vegetables lower your food bills when wild foods are out of season. (See Drying Wild Food, page 275.)

CLEANING WILD FOODS

It's best to wash your fruit just before you use it. If you wash the fruit in advance and store it soggy, the additional moisture may help mold grow and multiply.

Wash large, firm fruits, such as apples, pears, and pawpaws, by rinsing them in the colander. For smaller fruits, like plums, cherries, or berries, and most leafy vegetables, it's best to set the colander in a shorter, broad bowl. The water will rise to the top of the bowl and pour out, but the colander won't overflow, so the food is completely submerged in swirling water. This will get it much cleaner than using the colander alone. Be sure the colander is taller than the bowl, so the food won't flow over the top of the colander and disappear down the drain.

Berries are sometimes mixed with debris, especially when you shake them off a bush onto a dropcloth. Try hand-sorting them on a tray before or after using the colander and bowl method.

Clean roots under running water with the coarse copper scouring pads they sell in supermarkets (not the steel-wool ones with soap). They're remarkably quick. Combine with the colander and bowl method for the excess soil.

PRESERVING WILD FOODS

Edible wild plants often grow in great abundance, and you can often harvest large quantities without harming the environment. My favorite preservation methods are freezing and drying. I freeze cooked recipes and nuts, and dry fruit and vegetables.

FREEZING WILD FOODS

Freeze your wild surpluses for the future, the same way you would commercial produce. You must precook most fruits and vegetables before you freeze them. Slice and preboil in water one minute, then drain, and immediately dunk in ice water to stop the cooking.

Parboiling destroys destructive enzymes called lysozymes. They normally obliterate protein in cells when the cells die. In their normal state, they are contained in membranes within the cells. Freezing, however, causes water to expand, and the protective packets tear as uncooked food is frozen. When frozen food is defrosted the enzymes act on it to produce an ugly, tasteless mess. Precooking destroys enzymes, preventing this from happening.

Berries are an exception. You can freeze them raw: Wash them and spin them dry in a salad spinner, or drain them on towels. Spread them in a single layer on a cookie sheet, freeze until solid, then pack them into a freezer container. Because they were frozen separately, they won't stick together, and you can defrost as many as you need. Add frozen berries to your recipes at the last minute before cooking, so they won't have time to defrost and become soggy.

Because nuts don't contain much water, you can also freeze them without processing.

DRYING WILD FOODS

The best way to dry wild foods is with a food dryer. It uses little energy, and there are a number of good models on the market. You slice the fruits and vegetables and put them on a screen. The mechanism evaporates all the food's water by blowing warm air across it, or by using convection. Since microorganisms need water to grow, the dried food won't spoil. You can store it in sealed jars for well over a year.

Remove any fruit pits, since they'd block the air from evaporating. For very sticky fruits, like pawpaws, beach plums, or cherries, use a food dryer with a blower, and put plastic wrap or other food-repellent material on the racks. You can also use this method with pureed fruit. It produces an all-natural fruit leather, similar to what they sell in stores.

Not all berries dry well. The skins of blueberries and Juneberries are watertight. You have to slice each berry (a food processor is a big help), then dry it cut side up so it won't stick to the rack, or tape plastic wrap or nonstick material to each rack to prevent sticking. Note: Never use the plastic-wrap method in your oven. If the temperature goes too high, the plastic will melt.

Slicing raspberries and blackberries does no good, since each segment is watertight. Puree them and make a leather, as described above, before you dry them. Some small fruits, such as grapes, cherries, and autumn olives, need their seeds removed. You can use a food mill to remove the seeds, and make a leather.

You can also dry sliced food on racks, over cookie sheets, in the oven. Use very low heat (an oven thermometer should read about 115° to 120°F), and keep the door ajar to release moisture. Because there's no fan, this method is less efficient.

You can air-dry small amounts of leafy vegetables in closed paper bags. This works best with culinary or medicinal herbs, like goutweed, thyme, wild mint, or mullein. Keep the bag in a warm, well-ventilated room for about two weeks.

Food is dry when it's dry to the touch, and usually crisp and brittle. To reconstitute, soak the dried food, covered with room-temperature liquid (water, fruit juice, or stock—as appropriate), and refrigerate six hours or more. You can also pour boiling liquid over the dried food, and steep, covered, away from the heat, 30 to 60 minutes.

In addition to reconstituting dry vegetables and using them in soups or stews, you can also grind the vegetables into a powder in the blender or spice grinder. Use the stronger ones, like wild leeks, wild mustard seeds, or dulse, as seasonings. Substitute some powdered vegetables for flour in baking.

UNREFINED AND WILD FLOURS

Many experienced cooks are used to using only white flour. However, there are many interesting, healthful alternatives, including whole-grain flours from the health-food store, cattail pollen, dried powdered wild vegetables, and ground nuts. Although the alternatives don't help breads rise, they're great for nonrising dough such as vegetable pie crusts, and in casseroles, stuffing, and many other recipes. For the best results, use no more than 10 to 20 percent vegetable or nut flour along with your whole-grain flours.

To substitute one type of flour for another, use the same amount of flour by weight, not by volume. One cup of all-purpose white flour weighs seven ounces, so seven ounces of any other flour makes the most precise substitute.

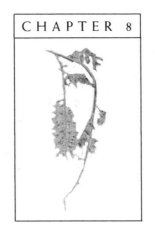

CHAPTER 8

WILD FOOD RECIPES

APPETIZERS AND SIDE DISHES

BLACK LOCUST FLOWER OATMEAL

¾ cup oatmeal
½ cup raisins
2 cups soy milk or milk
½ cup black locust flowers (or substitute
    wisteria, redbud, or finely chopped red or
    white clover flowers)

Gently simmer the oatmeal and raisins in the milk
in a covered saucepan, on low heat, for 5 minutes,
stirring often. Add the flowers, cover, and let sit
for 15 minutes.
Serves 4–6

WILD CARROT SEED MILLET

1 teaspoon wild carrot seeds
Light vegetable oil
2 cups water
Dash salt
1 cup millet

Sauté the seeds in a little light vegetable oil (sunflower or canola oil is good). When the seeds just
begin to brown, add the water and a dash of salt.
Bring to a boil and add the millet. Cover and simmer on low heat without stirring for 5 minutes,
until the millet is fluffy.
Serves 4

DANDELION SAUTÉ

This is one of the best ways to learn how to appreciate the flavor of dandelions.

3 cups chopped onions
3 tablespoons olive oil
4 cups chopped dandelion leaves
2 cups grated wild or commercial carrot
Several cloves garlic, minced
1 tablespoon wine
1 tablespoon tamari soy sauce
Black pepper to taste (optional)

Sauté the onions in the olive oil. When soft, add the dandelions, carrot, garlic, wine, and soy sauce. Cook for 10 to 20 minutes until all the flavors blend.
Serves 4–6

DENA AND GABI'S CATTAIL FLOWER PICKLES

Dena and Gabi have created one of the best ways to prepare green cattail flower heads.

Large handful green cattail flower heads (enough to stuff in one 32-ounce jar)
3 or 4 whole garlic cloves, peeled
1 teaspoon whole black peppercorns
4 to 6 wild or commercial bay leaves
Any other herbs or spices you like
¾ cup apple cider, wine, or rice vinegar
8 tablespoons tamari soy sauce
1½ cups olive oil
1¼ cups water

Boil the cattail spikes in salted water for 5 to 10 minutes. Remove from the water and shake off the excess liquid. Stuff in the jar with the garlic, black pepper, bay leaves, and herbs. Meanwhile, combine the vinegar, soy sauce, oil, and water in a saucepan. Bring to a boil, remove from the heat, and pour over the cattails. It should reach the top of the jar—if not, add a little more oil, vinegar, and water. Cover and marinate overnight in the refrigerator, stirring or shaking the jar occasionally. You can use the marinade over and over again to make more cattail pickles, or use it to flavor grains, beans, or vegetables.
Serves 6

HERBED JERUSALEM ARTICHOKES

This is a Mediterranean treatment of a native American vegetable.

2 medium onions, chopped
2 cloves garlic, minced
2 tablespoons olive oil
3 cups cubed Jerusalem artichokes
¼ teaspoon basil or oregano

¼ teaspoon thyme or rosemary
1 teaspoon sea salt
White or black papper to taste
Pinch of allspice or nutmeg
Dash of white wine (optional)

Sauté the onions and garlic in the olive oil until soft. Add the remaining ingredients, cover, and simmer over a low flame for 20 to 25 minutes. Add a tiny bit of wine or water every now and then to keep it from burning.
Serves 6

LAMB'S-QUARTERS STUFFING

2 cups grated zucchini
1 cup sliced mushrooms
1 cup grated tart apple
1 to 2 tablespoons vegetable oil
1 pound mashed tofu
2 cups lamb's-quarters leaves
¼ cup tamari soy sauce
2 to 3 tablespoons wine or sherry
2 to 3 tablespoons lemon juice
3 to 6 cloves garlic, chopped or minced, or to taste
¼ teaspoon each nutmeg, thyme, and white pepper, or to taste
A little rice and some chopped nuts (optional)

Sauté the zucchini, mushroom, and apple in the oil. When soft, cover and add the remaining ingredients. Cook 10 minutes more on low heat.
Makes 4 cups

LENTIL DELIGHT

Wild plants transform these commonplace legumes into something special.

2½ cups (1½ pounds) lentils
9 cups vegetable stock or water
4 fresh or dried wild bayberry leaves
1 tablespoon each dried epazote and cumin
½ cup bread crumbs
8 cloves garlic
½ cup dried wild leeks or onions
¼ cup olive oil

2 teaspoons sea salt, or to taste
2 teaspoons each tarragon, marjoram, and white pepper
1 teaspoon nutmeg

Cook the lentils in 6 cups of the stock with the bayberry leaves, epazote, and cumin for 20 minutes. Meanwhile, sauté the bread crumbs with the garlic and wild leeks in the oil. Mix the lentils with the sautéed vegetables and the remaining stock and herbs. Simmer for 30 minutes more.
Serves 8

NETTLE AND CARROT CASSEROLE

The incredibly rich flavor of nettle leaves is complemented by the sweetness of carrots and the tang of yogurt.

2 cups thinly sliced wild or commercial carrots
2 cups nettles
1 cup yogurt
2 eggs
⅔ cup soy milk, milk, or the reserved nettle-carrot broth (or any combination of these liquids)
2 cloves garlic, minced
¼ teaspoon dry, powdered mustard
¾ teaspoon sea salt
Dash of cayenne pepper and nutmeg

Simmer the carrots for 10 minutes in one-half inch of water. Add the nettles and simmer for another 5 minutes. Reserve the cooking water. Mince the vegetables in the blender or by hand, and mix in the remaining ingredients. Put in an oiled baking dish and bake at 275°F for 15 to 20 minutes, or until the casserole solidifies.
Serves 4

NUTTY AMARANTH

¾ cup amaranth
¾ cup millet
2 tablespoons sesame seeds or ¼ cup ground almonds
1¾ cups water

1 tablespoon tamari soy sauce
½ tablespoon vinegar

Put the amaranth, millet, and sesame seeds or almonds on separate baking dishes or cookie sheets, and place them in a 300°F oven. Roast 10 to 15 minutes or until slightly browned and fragrant, stirring often. (Amaranth takes a shorter time than millet, so keep an eye on it.) The millet should only be slightly browned. Allow to cool for a few minutes. Bring the water to a boil, add the amaranth and millet, reduce the heat, cover, and simmer about 15 minutes, or until fluffy. Add the seeds or nuts, tamari soy sauce, and vinegar.
Serves 4–6

PARK NUTS

This curried-acorn recipe also works with other wild and commercial nuts and seeds.

3 cups water
4 large garlic cloves, minced
¼ cup lemon juice
6 to 8 tablespoons curry powder
1 heaping quart of chopped white-oak acorns, leached of their tannin by repeated boiling
2 tablespoons olive oil
1 teaspoon sea salt or seasoned salt

Bring 2½ cups of the water to a boil. Put ½ cup of the water in a blender with the garlic, lemon juice, and ¼ cup of the curry powder, and blend until smooth. Add to the boiling water. Add the acorns and simmer for 5 minutes. Drain. Put in a baking dish. Mix in the rest of the curry powder, the olive oil, and the salt. Roast at 300°F for 45 to 90 minutes, or until the acorns are dry and well-roasted but not hard, stirring often.

POKEWEED AND MILKWEED: BASIC PREPARATION

This is the way to make any quantity of spring pokeweed shoots tasty and safe, and to make milkweed shoots, young leaves, unopened flower buds, and young, firm seedpods edible. It seems complicated, but it's really a very simple procedure. **Cau-**

tion: Do not experiment with these plants if you're an unsupervised beginner.

Bring a large and a medium-sized pot of water to a rolling boil. Boil the pokeweed in the medium-sized pot for 1 minute. Drain, using a colander. Add more boiling water from the large pot, which is your reservoir, and boil another minute. Drain, add more boiling water from the large pot again, and boil another 18 minutes. Drain again. Press the greens against the colander with a slotted spoon to remove excess water. Season to taste and serve. Good with vinegar, pepper, oil, lemon sauce, and bits of smoke-flavored foods.

ROCKWEED CRISPS

This seaweed bakes crunchier (and tastier) than potato chips, with more minerals than a multi-mineral pill, and without the grease.

3 cups dried rockweed
1 teaspoon of Curry Oil (page 289), or
    1 teaspoon olive oil with any seasonings you like

Put the rockweed in a large baking dish. Mix the oil in well. It's surprising how such a tiny amount of oil spreads over the surfaces of the seaweed. Resist the temptation to use more oil—it will make the seaweed taste greasy. Bake at 275°F for 20 to 25 minutes. Stir occasionally, and be careful not to let it burn. It's done when brittle, crisp, and fragrant.
Makes 3 cups

SALSIFY ROOTS IN NUT SAUCE

This is a good way to use all slightly bitter edible roots, such as salsify or dandelion.

1½ cups sliced salsify or dandelion roots
½ cup apple juice
1 cup sliced commercial or wild carrots or parsnips
1¼ cups water
3 tablespoons cashew butter

½ tablespoons arrowroot or kuzu
¼ teaspoons sea salt
1 tablespoon chopped fresh parsley
1 teaspoon each tarragon and marjoram
Dash of lemon juice (optional)

Simmer the salsify roots in the apple juice, uncovered, stirring often, for 15 to 20 minutes, or until there is almost no liquid left. Add the carrots and ½ cup of the water. Cover and simmer for another 10 minutes, or until the roots are soft. Meanwhile, blend the cashew butter, ¾ cup water, arrowroot, and salt together in the blender. Add it to the roots, slowly return to a boil, stirring constantly, until thick. Reduce the heat, cover, and simmer for another 5 minutes, stirring often. Add the herbs and optional lemon juice.

STUFFED WILD GRAPE LEAVES

Finally, a stuffed grape leaf recipe without preservatives.

30 grape leaves
2 quarts water
2 teaspoons salt

Collect grape leaves that are at least 6 inches across and cut off the stems. Add the leaves to the boiling salted water for a few minutes. Drain, and rinse the leaves under cold running water. Set aside and make the stuffing.

STUFFING
2 cups brown rice
1 cup chopped tofu
½ cup raisins
¼ cup sprouts or leftover vegetables
2 tablespoons chopped scallions or wild onions
2 tablespoons parsley or goutweed leaves
2 tablespoons sunflower seeds or chopped walnuts
1 teaspoon rosemary
1½ teaspoons chopped fresh dill or ½ teaspoon dried dill
½ teaspoon each paprika and sage
½ teaspoon salt

Pinch of cayenne pepper
1½ cups stock or water

Mix all stuffing ingredients together. Put a heaping tablespoon of stuffing on the underside of each leaf near the stem end. Fold the left and right sides of the leaf over part of the stuffing. Roll up the leaf from base to tip. Repeat this procedure with all the leaves. Put 1½ cups of stock, the garlic, and the ginger in a pressure cooker or large pot. Place a steamer rack in the pressure cooker or pot and fill the steamer with the stuffed grape leaves. Pressure-cook for 15 minutes or steam without pressure for 40 minutes. Test a grape leaf to see if it is tender. You may need to cook them longer. You may also place them in a large casserole dish, pour a small amount of water over them, and bake for 50 to 60 minutes in a 300°F oven.

SAUCE
2 cups vegetable stock
¼ cup almonds
¼ cup cashew or sesame butter (tahini)
6 cloves garlic, peeled but uncut
1½ teaspoons fresh wild or commercial ginger

2 tablespoons light miso or 1 to 2 teaspoons sea
    salt

Simmer all the sauce ingredients except the miso 10 minutes. Place everything in the blender and blend until smooth.

WATERLESS-STEAMED GREENS

This is one of the best ways to prepare chickweed, nettles, lamb's-quarters, spinach, and many other mild-tasting and commercial leafy greens. It uses very little water, and concentrates the flavor and nutrition. Use a large quantity of greens—they tend to shrink. Rinse the leaves and stems well. Sort out and discard any debris. Shake off the excess water and chop the greens. Place them in a heavy pot, with no additional water and no steamer rack. The greens cook in the water clinging to the leaves. Cover, and cook over low heat for 5 to 10 minutes, or until just wilted. Avoid overcooking. Mix in a few drops of olive or sesame oil, lemon juice, and pepper, or your choice of seasonings, and serve.

SALADS

AUTUMN GREEN SALAD

1 cup wild or commercial watercress
½ cup field-garlic leaves
½ cup chopped lady's thumb leaves and flowers
¾ cup thinly sliced Jerusalem artichokes
1 cup chickweed
½ cup chopped hickory nuts, butternuts, black
    walnuts, or other nuts

Chop the greens, mix with the remaining ingredients, and serve.
Serves 6

EARLY SPRING SEASONAL SALAD

1 cup daylily shoots
1 cup ramp or any species of wild onion leaves

or bulbs
1 cup chickweed
½ cup shepherd's purse leaves
½ cup sheep sorrel
½ cup young dandelion, chicory, sow thistle, or
    wild lettuce leaves
¼ cup curly dock leaves
¼ cup redbud or dandelion flowers (optional)
¼ cup long-leaf plantain leaves

Chop all ingredients together.
Serves 6

LATE FALL GREEN SALAD

1½ cups chickweed
1 cup wild or commercial watercress
1½ cups common mallow leaves
½ cup dandelion or sow thistle leaves

½ cup field garlic or other wild onion leaves
½ cup garlic mustard leaves
¼ cup winter cress leaves

Chop all ingredients together.
Serves 8

MID-SPRING WILD BLOSSOM SALAD

½ cup wild garlic leaves or bulbs, or other still-
  tender wild onion species
½ cup red or white clover flowers
¼ cup wild spearmint leaves
2 cups cattail shoots
2 cups violet leaves and flowers
1 cup black locust or wisteria blossoms

Chop the onions, clover, and spearmint finely.
Slice the cattails. Chop the violet leaves, and mix
everything together.
Serves 8

MILKWEED FLOWER SALAD

2 cups common milkweed flowers
½ cup roasted almonds, chopped
1 tablespoon vegetable oil
1 tablespoon lemon juice
⅛ teaspoon salt
Handful chopped scallions or wild onion leaves

Bring a pot of water to a boil. Simmer the flowers
for 1 minute. Drain, press out the excess water,
chop the flowers, and place them in a bowl with
the other ingredients. Mix well.

SEA LETTUCE SALAD

1 cup fresh sea lettuce
1 cup spinach
2 cups romaine or Boston lettuce
Handful chopped parsley
½ cup sliced radishes
1 clove garlic, minced
2 tablespoons vinegar
2 tablespoons olive oil

Sea salt to taste
Any fresh wild or commercial green herbs to
  taste
½ cup sliced olives or pickles (optional)

Wash the sea lettuce repeatedly in a tall colander
set in a bowl. Make sure all the sand is gone. Pat
or spin it dry; chop it very finely. Tear up the
lettuce and spinach, and mix all the ingredients
together.
Serves 6

SEASHORE SALAD

2 cups lamb's-quarters
1 cup sea rocket leaves
½ cup glasswort
½ cup finely chopped dulse or laver
½ cup young beach pea pods, wild bean pods, or
  snow peas
½ cup sheep sorrel
2 tablespoons field garlic seeds or minced bulbs
1 tablespoon field pennycress seeds or finely
  chopped leaves

Chop the lamb's-quarters, sea rocket, glasswort,
dulse or laver, peas or beans, and sorrel. Mix with
the remaining ingredients.
Serves 6

SUMMER GREEN SALAD

2 cups purslane
2 cups romaine lettuce
1 cup wild or commercial watercress
1 cup lamb's-quarters
1 cup wood sorrel
½ cup poor man's pepper leaves or seeds
2 cups daylily flowers (optional)

Mix all the ingredients, chop, and serve.
Serves 8

WINTER GREEN SALAD

2 cups garlic mustard leaves
2 cups field garlic (or other wild onion species)
  leaves and roots

2 cups chickweed
½ cup winter cress or creasy green leaves
2 cups mixed sprouts, or Boston or romaine
  lettuce
½ cup wintergreen berries (optional)

Chop all the greens and sprouts, and mix with the wintergreen berries.
Serves 10

SOUPS

EARLY SPRING CHICK-PEA AND VIOLET SOUP

Violets are great in any soup. I'm especially fond of them with beans.

2 cups presoaked chick-peas
5 cups water or stock
1 cup onions
1 cup sliced mushrooms
6 wild carrots or 3 to 4 large commercial carrots
5 celery stalks
3 cups chopped violets
2 to 4 tablespoons olive oil
2 wild bayberry or commercial bay leaves
¼ teaspoon white pepper
1 teaspoon salt
3 cloves garlic, minced
¼ teaspoon each savory, thyme, sage, rosemary,
  and celery seed
Dash of wine

Cook the chick-peas until soft in the water or stock. Sauté the onions, mushrooms, carrots, celery, and violets in a little olive oil. When soft, add to the soup and simmer for 25 minutes. Add the rest of the ingredients and simmer another 5 minutes.

SCANDINAVIAN FRUIT SOUP

This hot weather dessert consists of pureed fruit and chunks of fruit. It's a good way to use wild grapes and apples, which alone could be too tart.

You can vary it by including agar, resulting in a gelatinous dessert.

2 cups wild grapes
3 wild apples, cored and sliced
3 sliced bananas
2 cups water
1 cup cubed tofu
1 cup raisins
1 sliced peach
1 cup sliced watermelon
½ cup sunflower seeds
¼ cup grated coconut
Juice of 1 orange
1 teaspoon fresh orange rind
1 teaspoon crumbled dried wild or commercial
  mint
1 teaspoon cinnamon
¼ teaspoon each cloves, nutmeg, and ginger
1 egg white (optional)

Pass the wild grapes through a food mill or juicer to get rid of the pits, or briefly chop them in the blender, on low speed, and strain out the pits. Puree 2 bananas and 2 apples with the water in the blender. Place the puree and the grapes in a large pot. Add the orange juice, the sunflower seeds, the remaining apples and bananas, the strawberries, peach, mint, cinnamon, and cloves.

Bring the mixture to a boil over medium heat. Reduce the heat and simmer for 10 minutes, stirring often. Remove from the heat and chill. Just before serving, beat the egg whites until stiff with ginger and nutmeg. Place a dollop on each serving, and sprinkle with coconut.
Serves 6–8

BREADS

BLUEBERRY BREAD

2½ cup peeled mangoes, pears, or peaches
½ cup fruit juice
4 well-beaten eggs
¼ cup melted butter or oil
1 teaspoon angostura bitters (optional)
6 cups whole-grain flour
1 tablespoon cream of tartar
1½ tablespoons baking soda
1 teaspoon nutmeg
1½ teaspoons powdered ginger
2 cups blueberries that have been sprinkled with flour to prevent them from sinking into the dough

Puree the mangoes, pears, or peaches in the blender with the other liquid ingredients. Mix the remaining ingredients except the blueberries together. Mix the liquid and solid ingredients together. Fold in the blueberries. Bake in 2 bread pans in a preheated 350°F oven for 40 to 50 minutes, or until a toothpick inserted in the center comes out clean.

CLOVER CORN BREAD

3 tablespoons vinegar
½ cup oil
2 eggs
1½ cups apple juice
1 tablespoon concentrated sweetener (fruit juice concentrate, honey, barley malt, or maple syrup)
1 cup barley or whole-wheat flour
2 cups cornmeal
1 cup finely chopped dried or fresh clover flowers
1 teaspoon salt
1 teaspoon cream of tartar
2½ teaspoons baking soda

Mix the vinegar, oil, eggs, juice and sweetener. In a bowl combine the flour, cornmeal, clover, salt, cream of tartar, and baking soda. Add the liquid.

Place in an oiled baking pan. Bake at 300°F for 20 to 30 minutes or until a toothpick inserted into the center comes out clean.
Makes 1 loaf

CORN AND DAYLILY FLATBREAD

This recipe also works well using wild leeks or field garlic in place of the daylily shoots.

1 cup cooked corn
1 clove garlic
½ teaspoon salt
⅓ to ½ cup water
½ cup sliced daylily shoots
1 cup corn flour
¾ cup oat flour

Puree the corn, garlic, salt, and water in the blender. Pour into a bowl and add the daylily shoots and flours. Make a stiff dough, pat into thin patties, and sauté in a light vegetable oil on a skillet.
Makes 8 to 12 servings

PLANTAIN SEED–QUINOA PILAF

Quinoa is a light, tasty, high-protein, grainlike seed that's becoming increasingly popular in health-food circles. This dish is crunchy, light, and nutritious.

2 cups water
1 cup quinoa
¼ cup common plantain seed capsules
2 tablespoons finely chopped fresh basil or 2 teaspoons dried basil
1 clove garlic
1 teaspoon sea salt, or to taste
1 tablespoon olive oil

Bring the water to a boil and add the quinoa and plantain seeds. When the grains are fluffy turn off the heat and add the remaining ingredients. Let sit 5 minutes before serving.
Serves 4

DULSE SCONES

5 tablespoons shredded dulse
1 large onion, minced
2½ cups yogurt or buttermilk, or 1 tablespoon
    lemon juice or vinegar plus enough water to
    make 2½ cups
2⅔ cups whole-wheat flour, buckwheat flour, or
    brown-rice flour
½ teaspoon baking soda
¼ cup vegetable oil
¼ cup liquid lecithin
1 cup rolled oats
½ teaspoon thyme

Stir-fry the dulse on medium heat in 1 teaspoon of oil until brown and crisp. Put it in the blender with the onion and yogurt or buttermilk, and puree. Mix the flour and baking soda in large bowl. Mix the oil with the liquid lecithin, and stir in the oats and thyme. Stir the liquid mixtures into the dry mixture until just mixed. Preheat the oven to 375°F. Spoon the mixture onto 2 well-oiled baking sheets, leaving plenty of space between the scones. Bake for 25 to 45 minutes, or until slightly golden. Check that they don't burn on the bottom. Remove with a spatula and cool on racks.
Makes 24

IRISH SODA ELDERBERRY BREAD

⅓ cup vegetable oil or butter
⅔ cup orange juice
4 eggs
8 cups whole-wheat bread flour (or 2 pounds of
    any other whole-grain flour)
2 teaspoons salt
3¼ teaspoons baking soda
½ tablespoon cream of tartar
2 teaspoons dried crushed lemon balm or mint
    (optional)
3 cups buttermilk or yogurt (or substitute 3 cups
    soy milk less 2 tablespoons, plus 2 tablespoons
    lemon juice)

1 to 2 cups fresh or reconstituted dried
    elderberries
¼ cup caraway seeds

Preheat the oven to 375°F. Mix the butter or oil with the fruit juice. Beat in the eggs. In a separate bowl sift together the flour, salt, baking soda, cream of tartar, and mint. Alternately add the buttermilk and the flour mixture to the egg mixture until well mixed. Don't overmix. Add the elderberries and caraway. Bake in 3 8½ × 4½-inch baking dishes for 60 to 70 minutes, until a toothpick inserted in the center comes out clean. Remove from the pans and cool on racks. This bread freezes well.

PAWPAW BREAD

This is a recipe I originally created for bananas, but after I began experimenting with wild foods, it turned out to be even better with pawpaws.

3 cups pawpaw pulp (you can substitute wild
    persimmons or bananas)
1¼ cups apple juice or other fruit juice
¼ cup vegetable oil
4 eggs
1 tablespoon vanilla
1 tablespoon fresh or 1 tesapoon dried orange
    rind
1 teaspoon nutmeg
½ teaspoon powdered cardamom
6½ cups buckwheat flour or 4½ cups whole-
    wheat bread flour
2 teaspoons baking soda
1 cup chopped pecans

Mix the fruit; juice, oil, eggs, vanilla, orange rind, and spices in a blender or by hand. Mix the flour with the baking soda. Combine the wet and dry ingredients. Fold in the nuts. Put into 2 oiled 8½ × 4½-inch bread pans. bake 1 hour in a preheated 350°F oven or until a toothpick inserted in the center comes out clean.

ENTRÉES

CATTAIL SHOOTS AND CARROTS IN PEANUT SAUCE

½ cup peanut butter
¾ cup water
3 tablespoons tamari soy sauce
1 tablespoon grated wild or commercial ginger
3 to 4 cups sliced cattails and wild or commercial carrots or parsnips
Pinch of cayenne

Blend together the peanut butter, water, and tamari soy sauce. Add the ginger and reblend. Meanwhile, steam the cattails and carrots over a steamer rack for 10 to 20 minutes, or until tender. Warm the sauce and serve over the vegetables.
Serves 4–6

COCONUT–CURLY DOCK CURRY

Curly dock complements the curry perfectly.

3 cups sliced onions
2 tablespoons oil
3 stalks celery
5 wild carrots or 3 commercial carrots
5 cloves garlic
1 to 2 tablespoons grated ginger
½ cup curry powder
1 tablespoon sea salt, or to taste
9 cups coconut milk (or a mixture of coconut milk concentrate and water equal to 9 cups)
4 to 6 cups loosely chopped curly dock leaves
1 to 1½ cups chick-pea flour
Pinch of cayenne pepper (optional)

Sauté the onions in the oil. When wilted, add the celery and carrots. Cook for 5 to 10 minutes. Add the garlic, ginger, curry powder, and salt. Cook for another 5 minutes, stirring occasionally. Add most of the coconut milk (reserve 2 cups) and bring to a boil. Add the curly dock. Thoroughly mix the reserved liquid with the chick-pea flour, and stir into the boiling curry. Continue stirring for a few minutes, allowing the flour to cook and the sauce to thicken.
Serves 12–14

COMMON EVENING PRIMROSE LEAF BURGERS

These burgers were a huge success at one of my wild parties. The pungent primrose leaves are very nice with the sesame seeds and carrots.

4½ cups chopped tender, young common evening primrose leaves
4½ cups barley or whole-wheat flour
4 cups coarsely chopped, cooked, mashed wild or commercial carrots
4 cups chopped onions
3 cups cooked brown rice
3 cups water
2 cups fruit juice
2 cups roasted sesame seeds
½ cup sesame oil
3 to 5 tablespoons paprika
3 tablespoons miso
2 tablespoons fresh dill or 2 teaspoons dried dill
1 tablespoon oregano
1 teaspoon sea salt, or to taste

Mix all the ingredients together, shape into burgers, and sauté in sesame oil. You may need to add a little more barley or flour to get the right consistency for burgers that hold together. These burgers are good plain, with ketchup, or with a light lemony sauce.
Serves 12–14

COMMON EVENING PRIMROSE ROOT CHILI

Common evening primrose's turnipy-tasting roots tend to overpower other ingredients, but with this hot chili recipe, the herb has met its match.

2 tablespoons olive oil
2 onions, diced
8 garlic cloves, minced
2 teaspoons vegetable broth powder (optional)
2 cups sliced evening primrose roots
2 cups sliced wild or commercial carrots
3 Italian frying peppers
1½ cups cooked chili beans
2 cups crushed tomatoes, including their juice
2 cups pureed tomatoes (2 large tomatoes in the
    blender will do)
½ cup red wine
2 tablespoons whole-grain flour
¼ cup chili powder
6 wild or commercial bay leaves
1½ tablespoons paprika
½ tablespoon each marjoram, cayenne pepper,
    and black pepper
1 teaspoon sea salt, or to taste
½ teaspoon cumin

Heat the oil in a large pot. Add the onions, garlic, and broth powder. Sauté for 3 minutes over medium heat, stirring often. Add the carrots, evening primrose, and frying peppers. Sauté for another 10 minutes, adding more oil if necessary. Add the remaining ingredients, and bring to a boil over medium heat, stirring often. Reduce the heat and simmer for 30 to 40 minutes. Cook covered for a thinner chili, or without a lid for a thicker chili. Serve immediately, or refrigerate overnight so the flavors can blend more, reheat, and serve.
Serves 8

CURLY DOCK–CHEESE ROLLS
   IN TOMATO SAUCE

This treat is a vegetarian Sloppy Joe—messy but delicious.

3 cups ricotta cheese
6–7 scallions, sliced
3 cloves garlic, minced
1 teaspoon dried oregano
1 teaspoon dried basil
½ teaspoon pepper

½ cup chopped commercial or black walnuts
    (optional)
15 to 30 dock leaves
3 to 4 cups of your favorite tomato sauce

Mix together everything but the curly dock leaves and tomato sauce.

The curly dock leaves that are smaller than 3 inches should be chopped up and mixed into the filling. Put 1 to 2 tablespoons of filling (depending on the size of leaf) on the wide end of the leaf. Roll the leaf around the filling until you reach the tip. Place on an oiled baking sheet and pour the tomato sauce over the rolls. Bake· at 300°F for 10 to 15 minutes, or until the dock leaves are soft and shrinking.

VARIATION: Wild Stuffed Peppers. Chop up all the curly dock leaves and mix with the rest of the stuffing. Stuff into 10 green or red bell peppers and bake for 15 to 20 minutes, or until the peppers are soft.
Serves 8–10

CURRIED TOFU AND CHICKWEED

1 cup grated or finely sliced wild or commercial
    carrots
1 tablespoon vegetable oil
2 cups finely chopped chickweed
1 pound mashed tofu
3 to 5 cloves garlic, crushed
2 to 4 tablespoons curry powder, or to taste
1 tablespoon freshly grated wild or commercial
    ginger
Tamari soy sauce to taste
Dash of cayenne pepper (optional)
½ cup roasted commercial or ¼ cup roasted
    commercial walnuts and ¼ cup black walnuts

Sauté the carrots for 10 to 25 minutes in the oil. When they are soft, add the remaining ingredients except the walnuts. Turn off the heat and add the roasted walnuts.

VARIATION: Instead of oil and spices, substitute a few tablespoons of Curry Oil (page 289).
Serves 6

POTATO-PURSLANE PATTIES

Purslane always goes well with potatoes.

8 cups potatoes cut into 1 to 2 inch chunks
1½ cups water
3 cups chopped purslane stems and leaves
3 cups sautéed mushrooms
1 cup finely chopped commercial or wild onion
5 to 6 eggs
2 teaspoons rosemary
1 teaspoon black pepper

Cook the potatoes in the water until soft. Mash and add the rest of the ingredients. Mix and form patties. Bake on an oiled cookie sheet for 15 to 20 minutes in a preheated 300°F oven. Cool on racks.

SAVORY BURDOCK PATTIES

I've tried this recipe with many different types of bread crumbs, and I like it best with dried-out, slightly sweet corn crumbs.

1½ cups grated burdock root
⅔ cup bread crumbs
½ cup sliced, steamed wild or commercial carrots or parsnips
¼ cup each cooked mushrooms and chopped olives (optional)
⅓ to ½ cup water or stock (or the stock or water you cooked the burdock in)
2 eggs

½ teaspoon thyme
1 small onion, minced
¼ teaspoon sea salt
⅛ teaspoon black pepper

Steam or simmer the grated burdock root. Let it cool, then mix it with all the other ingredients. Form into small patties and bake them on an oiled cookie sheet at 300°F for 10 to 15 minutes. Turn them over and bake them another 10 to 15 minutes, or until done. Serve plain, with a sauce, or with relish.
Makes 24 patties

SUNFLOWER PATTIES

1½ cups ground sunflower seeds
½ cup grated wild or commercial carrots
½ cup finely chopped celery
2 tablespoons minced onion
1 tablespoon minced parsley
1 tablespoon minced Italian pepper
1 egg
1 clove garlic, minced
⅛ teaspoon basil
⅛ teaspoon dill
Sea salt to taste

Preheat the oven to 375°F. Mix all the ingredients into patties and place on an oiled cookie sheet. Bake for 20 to 30 minutes or until lightly browned.
Makes 8–12 patties

JAMS, DRESSINGS, SPREADS, AND SAUCES

AUTUMN OLIVE JAM

This jam is quick to make and is an all-purpose topping and spread.

16 cups autumn olive berries
1 cup agar flakes

Put the autumn olives and agar in a large saucepan over medium heat and bring to a boil, stirring often. The berries soon release enough liquid to cook in their own juice. Lower the heat, cover, and simmer for 10 to 15 minutes, or until the agar is completely dissolved, stirring often. Pass the berries through a food mill or sieve with holes just small enough to strain out the seeds. Chill and serve.
Makes 2 quarts

BLACK CHERRY JAM

You've never tasted a cherry jam like this before.

8 cups wild black cherries
2½ cups cherry or other fruit juice
½ cup black cherry concentrate (or other liquid sweetener)
1 cup agar flakes or 2 tablespoons agar powder or 2½ bars agar
2 teaspoons fresh spearmint or ¾ teaspoon dried spearmint

Bring all the ingredients to a boil over medium heat in a saucepan, stirring often. Lower the heat and simmer all the ingredients together for 10 to 15 minutes, or until the agar is completely dissolved, stirring often. Pass the mixture through a food mill or sieve with holes just small enough to strain out the seeds. Chill and serve.
Makes 6 cups

CREAMY CHICKWEED DRESSING

This dressing is low-fat, yet very rich-tasting.

2 cups water
1 to 2 cups chickweed
1 cup vegetable oil
¼ cup lemon juice
3 to 4 cloves garlic, peeled
1 large potato, baked or boiled
1 handful parsley or dill
Salt and black pepper to taste

Blend everything together in a blender until smooth.

CURRY OIL

This oil is wonderful to sauté with, to mix into grains, to season popcorn, or to use in the Rockweed Crisps recipe (page 280). It will keep refrigerated indefinitely.

½ cup olive oil
2 to 3 large cloves garlic

1 tablespoon fresh commercial or wild ginger
1 tablespoon miso (a salty soybean paste available in health-food stores and Asian markets)
1 to 3 tablespoons curry powder, to taste (or chili powder, Italian seasonings, or other herbal combinations)

Blend until smooth in a blender.
Makes ¾ cups

DRACULA'S DELIGHT— PRICKLY PEAR SAUCE

The color of the sauce holds the reason for its name. You can use this over many desserts, or pour on fruit salads and pancakes.

1 cup prickly pear pulp, seeds removed
¼ cup liquid sweetener
2 tablespoons lemon juice
1 tablespoon tapioca
1 tablespoon vegetable oil
Pinch of salt

Puree the fruit in a blender or food processor. Cook all the ingredients together over medium-high heat, stirring constantly, until it reaches a boil. Reduce the heat and simmer, stirring constantly, until it reaches a boil. Reduce the heat and simmer, stirring constantly, for 5 to 10 minutes. Serve hot or cold.

GARLIC MUSTARD HORSERADISH

This condiment is great on nori (seaweed) rolls, and anything that needs some extra pep.

1 cup chopped garlic mustard roots (or horseradish roots)
½ cup vinegar
1 teaspoon sea salt

Blend all the ingredients together in a food processor. Store in the refrigerator.
Makes 1¼ cups

GARLIC MUSTARD TOFU CREAM CHEESE

The garlic mustard adds just the right zing to the tofu's mildness.

1½ cups coarsely chopped tofu
2 tablespoons vegetable oil
1 teaspoon lemon juice
¼ teaspoon sea salt, or to taste
Dash of white pepper
1 cup garlic mustard leaves

Puree everything but the garlic mustard in the food processor. Add the garlic mustard and continue to process until finely chopped. Adjust the seasonings to taste. You can also mash the ingredients together using a fork and add finely chopped garlic mustard at the end.

POOR MAN'S MUSTARD

This mustard is similar to commercial mustard, except that it uses wild species, and that makes all the difference.

1 cup poor man's pepper seedpods (or tender, immature sea rocket seedpods, or other edible seedpods or seeds from other mustard species)
½ cup water
¼ cup apple cider vinegar
1 teaspoon tamari soy sauce
1 tablespoon white wine
1 tablespoon lemon juice
1 tablespoon water
2 teaspoons garlic
2 teaspoons turmeric
1 teaspoon tarragon (optional)
2 to 3 wild allspice berries (optional)

Mix the liquids together and bring to a boil. Put into a blender with the remaining ingredients, and blend everything together until smooth.
Makes 1¼ cups

SASSAFRAS ROOT JELLY

If you like the flavor of root beer, this is the jelly for you.

2 to 4 small, cleaned sassafras sapling roots, or 1 large root
3½ cups water
1 teaspoon powdered agar
1 tablespoon arrowroot or kuzu
¼ to ½ cup maple syrup or fruit juice concentrate

Simmer the sassafras in the water, covered, for 20 minutes. Turn off the heat and let the roots steep for another half hour. Remove the roots and measure how much tea you have left. Put 2 cups back into the pot and add the agar to it. Bring to a boil, reduce the heat, and simmer for 10 minutes, or until the agar is dissolved. Meanwhile, mix the arrowroot with another ¼ cup of the cooled sassafras tea. Stir into the simmering liquid. Add the sweetener. Cook and stir another 5 minutes, or until the mixture thickens slightly. Chill. It's ready to use when it gels.
Makes 4 cups

SUMAC HOLLANDAISE SAUCE

3 egg yolks
2 tablespoons sumac concentrate
A few sprigs of parsley
½ teaspoon nutmeg
½ teaspoon powdered mustard seed
½ teaspoon sea salt
Dash of cayenne pepper
½ cup simmering melted butter
½ cup boiling water

Place the egg yolks, sumac, herbs, spices, salt, and pepper in the blender and blend. While the motor is running, slowly add the butter and then the water. The sauce will thicken in the blender. Serve over vegetables or grains.

TOFU-GLASSWORT SPREAD

This spread is great on crackers and bread, or as a dip with chips.

1 pound mashed tofu
1 cup finely diced glasswort
1 teaspoon each basil and marjoram

2 teaspoons olive oil
1 tablespoon finely minced onion or wild leek or
  2 tablespoons chopped scallions or 2
  tablespoons chopped leaves of any other wild
  onion species
Dash of lemon juice
Pinch of salt

Mix everything together in the food processor, or
with a fork.

WILD APPLE CHUTNEY

5½ pounds apples, sliced
3 Spanish onions, chopped
6 cloves garlic, minced
Juice of a large lemon
¼ cup curry powder or to taste
¼ cup sesame seeds
2 tablespoons olive oil
2 teaspoons vegesal (natural seasoned salt) or sea
  salt

Mix all the ingredients together and bake in a covered, oiled casserole dish at 300°F for 30 minutes.
Makes 8 cups

WILD ONION SAUCE

½ cup chopped wild onion bulbs or commercial
  onions

4 cloves garlic, minced
2 tablespoons olive oil
3 tablespoons whole-wheat flour or ¼ cup barley
  flour
⅓ cup white wine
1½ cups simmering vegetable broth
2 tablespoons finely chopped fresh dill
¼ teaspoon chopped wild spicebush berries or
  commercial allspice
¼ teaspoon nutmeg
Sea salt to taste
¼ teaspoon white pepper
1½ cups chopped wild onion leaves (any species)

Sauté the onion bulbs and garlic in the oil until
they begin to brown. Add the flour and continue
sautéing, stirring often, for 5 to 10 minutes, or until
the flour becomes slightly browned and fragrant.
Slowly incorporate the wine. Bring to a boil and
gradually stir in the broth, dill, spices, salt, and
pepper, using a whisk to keep the flour from getting lumpy. Bring to a boil, reduce the heat, add
the wild onion leaves, and simmer for 5 minutes,
stirring often. For a smoother sauce, briefly place
in a blender.
Makes 4 cups

DESERTS

AUTUMN OLIVE–BANANA TOFU CREME PIE

FILLING
2 cups fruit juice
1½ cups agar flakes or 2 tablespoons agar
  powder
2 pounds tofu
3 large bananas
¼ cup oil
½ teaspoon salt
2 tablespoons vanilla

2 tablespoons lemon juice
1½ cups fruit juice concentrate or maple syrup
2 prebaked pie crusts, or your favorite cookies
  crumbled and kneaded into crusts
3 to 4 cups of warm Autumn Olive Jam (page
  288)

Boil the fruit juice and agar until the agar is dissolved. Meanwhile, put the remaining filling ingredients into a food processor or blender. Add the
juice and agar and blend. If you use a blender you

may need to leave some of the tofu out to get it to blend. You can mash the rest in with a fork afterward. Before the agar starts to thicken the mix—which will be almost immediately—pour the filling into 2 prebaked pie crusts. Pour Autumn Olive Jam on top. Place in a refrigerator or cool place to cool and harden.

❧ BLACKBERRY PEACH CRUNCH

When you harvest more wild fruit than you can eat at once, here's a good way to handle it.

FILLING
12 medium peaches, sliced
8 cups blackberries
2 cups fruit juice
6 tablespoons quick-cooking tapioca
¼ cup concentrated sweetener (optional)

TOPPING
4 cups oats
2½ cups brown rice flour
1¼ cups apple juice concentrate or honey
1 cup cornmeal
½ cup vegetable oil
2 tablespoons cinnamon
2 teaspoons almond extract
2 teaspoons sea salt

Mix the filling ingredients and place in an oiled baking dish. Mix the topping ingredients and place on top of the filling. Bake uncovered in a preheated 350°F oven for 25 to 40 minutes, or until the filling is thick and the topping is slightly brown and fragrant.

❧ CAROB-CHIP BLACK WALNUT COOKIES

That unique black walnut flavor mixed with carob, orange rind, and spices makes for incredibly pungent cookies.

⅓ cup butter or oil
½ cup maple syrup
1½ cups fruit juice
1 tablespoon vanilla
1 teaspoon angostura bitters (optional)
2 tablespoons fresh orange rind or 1 tablespoon orange extract
2 cups carob chips
1 cup black walnuts
1 cup pecans or commercial walnuts
1 teaspoon finely chopped spicebush berries, allspice, or nutmeg
1 teaspoon ground fennel seed or anise
2½ cups barley flour or 1 pound of any other flour
¼ cup carob powder
½ tablespoon baking soda

Mix together the oil, maple syrup, juice, vanilla, angostura bitters, and orange extract. Add the carob chips, nuts, and spices to the liquid mixture. Mix the flour, carob powder, and baking soda together, and stir into the liquid ingredients. Shape small cookies and bake in a preheated 350°F oven for 10 to 20 minutes, or until a toothpick inserted in the center comes out clean.

❧ CAROB-PECAN FUDGE

1½ pounds cooked fresh, canned, or rehydrated dried chestnuts
⅜ cup butter
⅜ cup liquid lecithin
½ cup carob powder
1 tablespoon vanilla
½ teaspoon angostura bitters (optional)
Your favorite liquid sweetener to taste (apple juice, barley malt, and maple syrup are good)
1 cup chopped pecans (wild or commercial)
½ cup dried, unsweetened coconut

Puree the chestnuts, butter, liquid lecithin, carob powder, vanilla and angostura bitters in a food processor. Add sweetener to taste. Add pecans and coconut, and process until chopped. If the fudge is too thin, add more carob powder or pecans. if it's too thick, add water or sweetener. Place in an oiled rectangular container. Chill. Cut into small squares.

CREAMY WILD RASPBERRY PUDDING

1 cup fruit juice
¾ cup agar flakes or 1½ tablespoons agar powder
3 tablespoons almond butter, cashew butter, or tahini
7 tablespoons arrowroot or kuzu
2½ cups apple juice or raspberry-apple juice
3½ cups of any species of wild raspberries
2 tablespoons vanilla

Simmer the fruit juice, vanilla, and agar for 10 to 15 minutes, or until the agar dissolves, over medium heat, stirring often. Blend the nut butter with the arrowroot and juice, and pour this into the agar mixture along with the raspberries. Bring to a boil, stirring constantly. When the mixture thickens, lower the heat and simmer for 5 minutes, covered, stirring often. Chill. Serve as is, or puree briefly in a food processor for a smoother texture.

CREPES WITH WILD BERRIES

Crepes are pancakes that are so thin, you can roll them around a filling. Unlike most other kinds of pancakes, they need no leavening. The liquid lecithin is not vital but greatly helps to keep the crepes from sticking. It's available in health-food stores.

1½ to 1¾ cups buckwheat flour, or 1¾ cups whole-wheat pastry flour
1 teaspoon nutmeg
1 tablespoon cinnamon
½ teaspoon cardamom
3 eggs
5 teaspoons vegetable oil
1 teaspoon liquid lecithin (optional)
3 cups soy milk or milk
2 cups wild berries (any species that doesn't require you to remove the seeds)
½ cup plain yogurt

Combine the flour, nutmeg, cinnamon, and cardamom together in a large bowl. Blend the eggs, oil, lecithin, and soy milk in a blender. Add the liquid ingredients to the dry ingredients and mix. Let the batter stand, refrigerated, for an hour or more if possible. Lightly oil a crepe pan, griddle, or frying pan with a mixture of equal amounts of oil and liquid lecithin (or just oil the surface) to create a nonstick surface. Heat over a moderate flame until hot and pour on a thin coat of batter, rotating the pan so the entire surface is coated. You may need to add a little more flour or liquid to your batter to get the right consistency. (If you're using a large griddle, you may want to make a number of crepes at once.) When the bottom of the crepe begins to brown, loosen it with a metal spatula and turn it over. When both sides are lightly browned, remove the crepe from the pan and cover it with the wild berries, yogurt, or any topping you desire. Makes 2 dozen crepes. The batter freezes well, and you can also freeze the cooked crepes.

JAPANESE KNOTWEED SURPRISE

You'll be surprised how tasty this simple recipe is.

5 cups sliced apples
2 cups sliced young Japanese knotweed shoots
1 cup apple juice

Simmer all the ingredients for 10 to 20 minutes or until soft.
Serves 8

MARILLEN KNÖDEL

This is a healthful version of traditional Hungarian apricot dumplings.

1 cup loosely packed dry wild or commercial apricots
1 pound soft cheese or pureed tofu with 1 teaspoon vegetable oil
2 eggs, lightly beaten
2 to 3 cups whole-wheat or buckwheat flour
1 teaspoon sea salt
Lukewarm soy milk or milk as needed
¼ teaspoon powdered cloves, or wild or commercial dried, powdered ginger
2 tablespoons fruit juice or honey

Steam the apricots in a steamer over a small amount of boiling water until soft. Meanwhile, mix the cheese or tofu, eggs, flour, and salt, then add just enough milk to produce a solid but soft dough. Knead until elastic. Press or roll into dumplings: Make round flat shapes large enough for an apricot to fit inside. Mix the apricots with the seasoning and sweetener, and place one apricot into each dumpling. Seal the apricot inside, and roll into a ball if desired. Let the dumplings stand on a lightly floured surface, 1 hour if possible, to dry out. Boil for 15 minutes in lightly salted water. You can serve these plain or topped with toasted chopped nuts, seeds, or butter.

MULBERRY CRUMBLE

FILLING
4 cups mulberries
1½ cups fresh orange juice
1 cup presoaked raisins or commercial currants
3 tablespoons arrowroot or kuzu
4 teaspoons concentrated sweetener (fruit juice concentrate, honey, or maple syrup)
1 tablespoon lemon juice or 1 teaspoon lemon oil
Rind of 1 orange

TOPPING
¾ cup oats
½ cup whole-grain flour
½ cup chopped nuts
¼ cup fruit juice
2 tablespoons vegetable oil or butter
2 teaspoons cinnamon

Mix together the filling ingredients in an oiled baking dish. Mix together the topping ingredients and pour over the filling. Bake in a preheated 350°F oven for 30 minutes, or until the filling is thick and bubbly.
Serves 10–12

MULBERRY PUDDING

Black cherry concentrate adds punch to the somewhat mild mulberries.

1½ cups tapioca
8½ cups milk or soy milk
14 cups mulberries
2 cups fruit juice (orange is good)
1½ cups black cherry concentrate
4 teaspoons freshly grated orange rind
2 tablespoons vanilla
2 cups cherries (sour ones are best)
½ cup lemon juice
¼ cup almond or cashew butter (optional)

Soak the tapioca in some of the milk in the refrigerator for a few hours, then add the mulberries, fruit juice, fruit concentrate, the rest of the milk, the orange rind, and the vanilla. Bring to a boil on low heat, stirring often. When the pudding starts to simmer and thicken, add the cherries. Cook 5 more minutes on low heat, stirring often. Add the lemon juice. If you want a richer pudding, preblend the nut butter with the lemon juice and stir it in after the cooking is done.
Serves 16–18

AMERICAN PERSIMMON PUDDING

This is the richest, fruitiest wild fruit pudding ever.

6 cups milk or soy milk
5 cups pitted wild American persimmons
1 cup apple juice concentrate or ¾ cup honey
4 to 5 tablespoons orange rind
2 tablespoons vanilla

Puree everything together.
Serves 12–14

TOFU-PLUM PIE

You can put this double-decker pie filling into any pie crust, although graham cracker crust suits it especially well. The creamy tofu layer is the perfect foil for the sweet-sour plum layer.

TOP LAYER
8 cups pitted plums, wild or beach
10½ tablespoons tapioca

1 tablespoon minced spicebush berries, or ½
  tablespoon allspice
1 teaspoon dried powdered mint
½ cup chopped pecans

BOTTOM LAYER
1½ pounds tofu
3 to 6 tablespoons cherry concentrate or other
  liquid sweetener
¼ cup sunflower or other light vegetable oil
¾ teaspoon sea salt
2 tablespoons lemon juice
1 tablespoon freshly grated orange rind
1 teaspoon vanilla extract
1 teaspoon each nutmeg, cinnamon, and ginger
⅛ teaspoon white pepper
1 cup dried, grated coconut
3 cups chopped pecans

Mix together all the top-layer ingredients except
the ½ cup of pecans. In a food processor, puree all
the bottom ingredients except the coconut and the
3 cups of pecans. Mix the coconut and pecans into
the bottom layer. Make enough pie crust for three
9-inch double-crusted pies. Fill each crust with the
bottom layer and some of the top layer. Sprinkle
the reserved pecans, and add more of the top layer.
Bake in a preheated oven at 425°F for 10 minutes.
Reduce to 325°F, and bake for another 25 minutes.
Makes 3 pies.

❧ WILD APPLE CAKE

This is a rather elaborate cake that is incredibly
delicious, and a meal in itself. Because it's eggless,
it's dense and chewy. Apples that are tart or sweet
are good to use in this recipe, as are meaty crabap-
ples that aren't too acidic.

3 cups cooked, mashed butternut squash
2 cups apple juice
1 cup orange juice
1 cup pineapple juice concentrate
2 tablespoons fresh orange rind
½ cup vegetable oil
1 cup soy milk or milk
2 tablespoons apple cider vinegar

3 tablespoons cinnamon
2 teaspoons nutmeg
2 teaspoons coriander
1 teaspoon ginger
1 tablespoon almond extract
4 cups grated or well-chopped wild apples
2 cups grated wild or commercial carrots
1 cup raisins
1 cup walnuts
5 cups barley flour
½ tablespoon cream of tartar
1½ tablespoons baking soda

Mix together the mashed squash, juices, concentrated juice, rind, oil, soy milk, vinegar, spices,
extract, apples, carrots, raisins, and walnuts. Mix
the flour, cream of tartar, and soda. Stir into the
liquid ingredients. Pour into oiled cake or baking
pans and bake in a preheated 300°F oven until the
cakes feel firm to the touch, and a toothpick inserted in the center comes out clean. Good served
with applesauce, or the following icing recipe.
Serves 16

ICING
24 ounces tofu or a soft cheese
6 tablespoons fruit juice or other liquid
  sweetener
¼ cup vegetable oil
1 tablespoon fresh or 1 teaspoon dried orange
  rind
2 teaspoons salt
2 teaspoons lemon juice
2 teaspoons vanilla

Mix in a food processor, with an electric mixer, or
with a fork.

❧ WILD STRAWBERRY PARFAIT

Wild strawberries are hard to collect. They're quite
small, and they're so tasty, it's hard not to eat them
as quickly as you pick them. This delicious recipe
lets you stretch this wild fruit.

1 cup agar flakes or 2 tablespoons agar powder
8 cups apple-strawberry or apple juice

10 tablespoons arrowroot or kuzu

3 tablespoons tahini or almond butter, or cashew butter

1 tablespoon vanilla extract

1½ tablespoons lemon juice, or to taste

⅔ cup strawberry jam or jam of your choice (fruit juice–sweetened jam is the most healthful)

1 to 2 cups wild strawberries

Boil the agar in 7 cups of the apple-strawberry juice for 5 to 10 minutes. Mix the arrowroot into the remaining 1 cup of cold juice, and stir into the boiling liquid. Reduce the heat and simmer for a few minutes, stirring constantly. Chill for a few hours, or until thick. Put ⅓ of the chilled mixture into a food processor and combine with the tahini or nut butter, vanilla, and lemon juice. When pureed, add the rest of the thickened mixture, along with 6 ounces of the strawberry jam and 1 cup of strawberries. Puree. In 6 to 8 clear glass cups, make 3 layers: red, white, and red. Top with wild strawberries.

Serves 6–8

BIBLIOGRAPHY AND REFERENCES

FOR FURTHER READING

Belzer, Thomas J. *Roadside Plants of South Carolina.* Missoula, Mont.: Mountain Press Publishing Co., 1984.

Blackwell, Will H. *Poisonous and Medicinal Plants.* Englewood Cliffs, N.J.: Prentice Hall, 1990.

Britton, Nathaniel, and Addison Brown. *An Illustrated Flora of the Northern United States and Canada.* New York: Dover Publications, Inc., 1913.

Elias, Thomas S., and Peter A. Dykeman. *Field Guide to North American Edible Wild Plants.* New York: Outdoor Life Books, 1982.

Elliot, Doug. *Roots: An Underground Botany.* Old Greenwich, Conn.: The Chatham Press, 1976.

Erichsen-Brown, Charlotte. *Medicinal and Other Uses of North American Plants.* New York: Dover, 1979.

Foster, Steven, and James A. Duke. *A Field Guide to Medicinal Plants.* Boston: Houghton Mifflin Co., 1990.

Gibbons, Euell. *Stalking the Healthful Herbs.* New York: Van Rees Press, 1966.

———. *Stalking the Wild Asparagus.* New York: David McKay Co. Inc., 1962.

Grieve, M. *A Modern Herbal.* New York: Dover Publications, Inc., 1931.

Gunther, Erna. *Ethnobotany of Western Washington.* Seattle: University of Washington Press, 1945.

Hall, Alan. *Wild Food Trailguide.* New York: Henry Holt, 1973.

Hardin, James W., and Jay M. Arena, M.D. *Human Poisoning from Native and Cultivated Plants.* Durham, N.C.: Duke University Press, 1969.

Hutchens, Alma R. *Indian Herbology of North America.* Ontario: Merco, 1969.

Lust, John. *The Herb Book.* New York: Bantam, 1974.

Martin, Laura C. *Wildflower Folklore.* Chester, Conn.: The Globe Pequot Press, 1984.

Mowrey, Daniel. *The Scientific Validation of Herbal Medicine.* Lehi, Utah: Cormorant Books, 1986.

Neiring, William A. *The Audubon Society Field Guide to North American Wildflowers.* New York: Alfred A. Knopf, 1979.

Newcomb, Lawrence. *Newcomb's Wildflower Guide.* Boston: Little, Brown and Co., 1977.

Peterson, Lee. *A Field Guide to Edible Wild Plants.* Boston: Houghton Mifflin Co., 1978.

Petrides, George A. *A Field Guide to Trees and Shrubs.* Boston: Houghton-Mifflin, 1958.

Phillips, Roger. *Trees of North America and Europe.* New York: Random House, 1978.

———. *Wild Food.* Boston: Little, Brown and Co., 1989.

Pijper, Dick, Jac. Constant, G. Constant, and Jansen, Kees. *The Complete Book of Fruit.* New York: Gallery Books, 1985.

Pond, Barbara. *A Sampler of Wayside Herbs.* New York: Crown, 1974.

Richardson, Joan. *Wild Edible Plants of New England.* Chester, Conn.: The Globe Pequot Press, 1981.

Runyon, Linda. *A Survival Acre.* Saranac Lake, N.Y.: The Chauncy Press, 1985.

Santillo, B.S. *Natural Healing with Herbs.* Prescott Valley, Ariz.: Hohm Press, 1984.

Saunders, Charles Francis. *Edible and Useful Wild Plants of the United States and Canada.* New York: Dover Publications, 1920.

Spencer, Edwin Rollin, Ph.D. *All About Weeds.* New York: Dover Publications, 1940.

Squier, Thomas K. *Living Off the Land.* Rutland, Vt.: Academy Books, 1989.

Stokes, Donald W. *The Natural History of Wild Shrubs and Vines.* Chester, Conn.: The Globe Pequot Press, 1989.

Symonds, George W. D. *The Shrub Identification Book.* New York: William Morrow & Co., 1963.

Taylor, Ronald J. *All About Weeds.* Missoula, Mont.: Mountain Press Publishing, 1990.

Weed, Susan S. *Healing Wise.* Woodstock, N.Y.: Ash Tree Publishing, 1989.

GENERAL INDEX

NOTE: **Boldface** numbers are for illustrations.

Abortion, 8
Acacia trees, 43
Achillea lanulosa species, 95
Achillea millefolium. See Yarrow
Acne, 36, 79
Aconitum uncinatum. See Monkshood
Acorus calamus. See Calamus
Aegopodium podagraria. See
    Goutweed
Aesculus californica. See California
    buckeye
Aesculus glabra. See Ohio buckeye
Aesculus hippocastanum. See Horse
    chestnut
Aesculus octandra. See Sweet buckeye
Aesthusa cynapium. See Fool's
    parsley
Alaria esculenta. See Edible kelp
Alcohol, toxicity of, 9
Alexanders. *See* Angelica
Alfalfa, 24, 39
Alfilaria. *See* Storksbill
Algae. *See* Seaweeds
Alisma species. *See* Water plantains
Alliaria officinalis. See Garlic mustard
Alliaria species, 255
Allium canadense. See Wild garlic
Allium cernuum. See Nodding wild
    onion
Alliums, vitamins and minerals in,
    209

Allium sativum. See Commercial
    garlic
Allium schoenoprasum. See Wild
    chive
Allium species. *See* Wild onions
Allium stellatum. See Prairie wild
    onion
Allium tricoccum. See Wild leek
Allium vineale. See Field garlic
Allspice. *See* Wild allspice
Alteratives, 8
Alternate leaves, 12–13, **12**
Althaea officinalis. See Marshmallow
Alzheimer's disease, 142
Amaranth (pigweed) (*Amaranthus*
    species), 47, 145–147, 273
  appearance of, 146
  bracts of, 146
  common seed head, **146**
  history of, 145–146, 147
  and protein, 147
  seeds of, 146, 147
  vitamins and minerals in, 147
Amaranthus species. *See* Amaranth
Amaryllidaceae. See Daffodil
Amelanchier species. *See* Juneberry
American basswood (*Tilia
    americana*), 63
  See also Linden
American chestnut (*Castanea
    dentata*), 168, **168**, 169

American elder. *See* Common
    elderberry
American hazelnut (*Corylus
    americana*), 156–157, **156**
American Journal of Clinical Nutrition,
    83
American pennyroyal (*Hedeoma
    pulegioides*), 52–53, **53**
American plum (*Prunus americana*),
    112
American yew (eastern yew) (*Taxus
    canadensis*), 144
Amianthium muscaetoxicum. See Fly
    poison
Ampelopsis brevipedunculata. See
    Porcelainberry
Amphicarpa bracteata. See Hog
    peanut
Analgesics, 7
Anemia, 41
  iron-deficiency, 238, 244
Angelica (alexanders) (*Angelica
    atropurpurea*), 61, 66–67, **67**
  medicinal uses of, 67
  poisons in, 67
  roots of, 67
Angelica archangelica. See European
    angelica root
Angelica atropurpurea. See Angelica
Angelica sinensis. See Dang gui
Angelica tree. *See* Hercules'-club

Aniseroot (sweet cicely; sweetroot) (*Osmorbiza* species), 212–214, **213**, 273
   flowers of, 214, **214**
   look-alikes of, 214
   medicinal uses of, 214
   seed head, **214**
Annuals, 18, 20
Anodynes, 7
Anorexia, 67
Antibiotics, 5, 8, 139
Antiseptics, 8
Antispasmodics, 7
Apios americana. See Groundnut
Apocynum species. *See* Dogbane
Apple (*Malus* species), 10, 16, 98, 109, 124, 150, 151–153
   blossoms of, 15, 123
   flowers of, 151, **151**
   history of, 153
   medicinal uses of, 153
   pectin in, 153
   poisons in, 153
   See also Custard apple
Appleseed, Johnny. *See* Chapman, Jonathan
Apricot (*Prunus armeniaca*), 108, **108**, 142, 158
   flowers of, **108**
   See also Vine apricot
Aralia spinosa. See Hercules'-club
Arctic *Vaccinium* species. *See* Bilberries
Arctium lappa. See Great burdock
Arctium minus. See Common burdock
Arctium species. *See* Burdock
Arizona walnut (*Juglans major*), 161
Armoracia lapathifolia. See Horseradish
Artemisia vulgaris. See Mugwort
Arthritis, 5, 8, 37, 39, 41, 105, 134, 224, 241, 245, 249
   See also Rheumatoid arthritis
Artichoke, 34
   See also Jerusalem artichoke
Artichoke family, 236–237
Asarum acuminatum species, 221
Asarum canadense species, 221
Asarum caudatum species, 221
Asarum species. *See* Wild ginger
Asclepias syriaca. See Common milkweed
Asclepias tuberosa. See Butterfly weed
Asian yew (*Taxus baccata*), 144
Asiatic dayflower (*Commelina communis*), 26–27, **26**, 271
Asimina triloba. See Pawpaw
Asparagus (sparrow grass) (*Asparagus officinalis*), 78–79, 263

male or female, 78
mature, **78**
medicinal uses of, 79
rhizomes of, 78–79
shoot of, **78**
stalks of, 78–79
vitamins and minerals in, 79
wild vs. commercial, 78, 79
   See also Beach asparagus
Asparagus officinalis. See Asparagus
Aspirin, 7, 224
Asthma, 25, 45, 72, 194, 249
Astringents, 7–8
Atriplex patula. See Orache
Autumn
   cultivated areas in, 139–145
   deserts in, 187–188
   disturbed areas in, 145
   fields in, 145–151
   freshwater wetlands in, 185–186
   lawns and meadows in, 137–139
   mountains in, 186
   seashore in, 186
   thickets in, 151–160
   woodlands in, 160–185
   See also Fall
Autumn olive (*Elaeagnus umbellata*), 16, 153–155, **154**
   berries of, 155
   flowers of, **153**
   relatives of, 155
   seeds of, 155

Banana-family plantain, 228
   See also False banana
Barbarea verna. See Scurvy grass
Barbarea vulgaris. See Winter cress
Barberrry family, 124
Barnes, Stephen, 25
Basal rosettes, **12**, 13, 17, 18, 189
Bayberries (*Myrica* species), 37, 79–80
   leaves of, 80
   medicinal uses of, 80
   minerals in, 80
B-complex vitamins, 273, 274
Beach asparagus. *See* Glasswort
Beach pea (*Lathyrus japonicus*), 91, 131–132, **131**
   flowers of, **131**
   poisons in, 131
Beach plums (*Prunus maritima*), 111–112, **111**
   flowers of, **110**
Beaked hazelnut (*Corylus cornuta*), 156, **157**
Bean. *See* Tepary bean; Wild bean
Beargrass. *See* Yucca
Bee-balm. *See* Oswego-tea
Bees, 4, 15, 50, 55, 59, 74
Berries, 16
   cleaning, 274

cooking, 272
freezing, 275
Beta carotene, 40, 50, 51, 85, 167, 188, 206
Betula lenta. See Black birch
Betula lutea. See Yellow birch
Biennials, 18, 20
Bilaterally symmetrical (irregular) flower, **13**
Bilberries (Artic *Vaccinium* species), 99, 100
Bindweed, 199, 203
Birch. *See* Black birch
Bird cherry. *See* Pin cherry
Birdfoot viola (*Viola pedata*), 231
Bitter dock. *See* Broad-leaf dock
Blackberries (*Rubus* species), 11, 16, 17, 96–98, 114–115, 126
   "cane," 96
   flowers of, 96
   leaves of, 97
   look-alikes of, 97
   medicinal uses of, 98
   minerals in, 98
   roots of, 98
   "wine"-on-the-bush, 98
Black birch (sweet birch) (*Betula lenta*), 214–216, **215**
   bark of, 214, 215
   easy-to-propagate, 215
   flowers of, **215**
   medicinal uses of, 215
   syrup from, 216
   vitamins and minerals in, 215
Black cherry (*Prunus serotina*), 121–123, **121**, 215, 272
   bark of, 10, 122, 123
   and cyanide, 122–123
   flowers of, **121**
   look-alike of, 122
   lumber of, 122
   medicinal uses of, 122, 123
Black currant (*Ribes nigrum*), 102
Black haw (*Viburnum prunifolium*), 183–185
   flowers of, **184**
   fruits of, **184**
Black locust (*Robinia pseudoacacia*), 15, 42–43, **42**, 210, 272
   flowers of, 43, 59
   pods of, **42**
   poisons in, 43
Black mulberry (*Morus nigra*), 126
Black mustard (*Brassica nigra*), 249, **250**
Black oak leaf, **175**
Black raspberry (*Rubus occidentalis*), 97, 113, **113**
   flowers of, **113**
Black walnut (*Juglans nigra*), 160–163, 164, 171, 172, 213, 273

vs. commercial walnuts, 163
cracking open shells, 162–163
flowers of, 160–161, **161**
husks of, 161, **161**, 163
medicinal uses of, 163
nuts of, 162, **162**
pickling, 163
syrup from tree, 163
vitamins and minerals in, 163
Bladder disease, 139, 205
Bladderwrack. *See* Rockweed
Blancmange. *See* Irish moss
Blasphemy vine. *See* Greenbrier
Blister rust fungus, 103
"Bloat, the," 25
Blueberries (*Vaccinium* species;
    Gaylussacia species), 98–100,
    99, 100, 119, 123, 124
  bark of, 99
  leaves of, 99, 100
  medicinal uses of, 100
  pectin in, 99
Blue elder (*Sambucus cerulea*), 103
Blue flags. *See* Iris
Blue lettuce (*Lactuca pulchella*), 247
Blue-sailors. *See* Chicory
Bog bilberries, 99
Bogs, 22
Bracken (brake; pasture fern)
    (*Pteridium aquilinum*), 31–32,
    31, 65
  fiddlehead, 31–32, **31**
  frond of, 64
  poison in, 32
Brake. *See* Bracken
Brassicaceae. See Mustard family
Brassica nigra. See Black mustard
Brassica rapa. See Field mustard
Brassica species, 258
Breast cancer, 25, 41, 86, 230
Broad-leaf dock (bitter dock)
    (*Rumex obtusifolius*), 34,
    236
Broad-leaf parsley, 212
Broad-leaved waterleaf
    (*Hydrophyllum canadense*),
    266–267
Bronchitis, 72, 249
Buckhorn. *See* English plantain
Buckthorn (*Rhammus* species), 119,
    119
Buckwheat family, 27, 237
Bulbs, 16–17, **16**
Bull thistle, immature
  flowers of, **201**
  flower stalks, **201**
Burdock (*Arctium* species), 15, 17,
    32–36, 139, 200, 236–237,
    263, 273
  burrs in, 34, 36
  leaves of, 35
  medicinal uses of, 35–36

root of, 35, 36
vitamins in, 35
Bush chinquapin (*Castanea
    sempervirens*), 169
Bush morning glory (*Ipomoea
    leptophylla*), 203
Butterflies
  monarch larvae, 50
  viceroy, 50
Butterfly weed (pleurisy root)
    (*Asclepias tuberosa*), 49, **49**
Butternut (*Juglans cinerea*), 160, 161,
    163–165, **165**, 171
  and carbohydrates, 164
  cracking open shell, 164
  flowers of, 164, **164**
  husks of, 164, 273
  as laxative, 165
  medicinal uses of, 165
  nuts of, **164**
Butternut hickory. *See* Pignut
    hickory

Cacti, 17
Cactus pear. *See* Prickly pear
Caffeine, as stimulant, 8
Cakile edentula. See Sea rocket
Cakile lanceolata. See Floridian sea
    rocket
Calamus (*Acorus calamus*), 69
Calcium, 76, 117, 176, 180, 188, 199
California blackberry (*Rubus
    ursinus*), 97
California buckeye (*Aesculus
    californica*), 169
California rose (*Rosa californica*), 158
California walnut (*Juglans
    Californica*), 161
Canada moonseed (*Menis permum
    canadense*), 165
  poisons in, 165, **167**
Canada plum (*Prunus nigra*), 112
Cancer, 6, 25, 30, 123, 141, 144,
    163, 178, 206, 221, 230, 249
  See also Breast cancer; Prostate
    cancer; Stomach cancer;
    Testicular cancer
Candida, 9
Capsella bursa-pastoris. See
    Shepherd's purse
Capsules, in herbal preparation, 8–9
Carbohydrates
  complex, 6
  refined, 5–6
Cardamine pratensis. See
    Cuckooflower
Cardiacs, 8
Cardiovascular disease, 102, 167,
    239
  See also Heart disease
Carminatives, 8
Carrion flower (*Smilax renifolia*), 56

Carrot family, 245, 259
  See also Wild carrot
Carya cordiformis. See Pignut hickory
Carya illinoensis. See Pecan
Carya ovata. See Shagbark hickory
Carya species. *See* Hickories
Carya tomentosa. See Mockernut
    hickory
Castanea dentata. See American
    chestnut
Castanea pumila. See Eastern
    chinquapin
Castanea sempervirens. See Bush
    chinquapin
Castanea species. *See* Chestnut
Catbrier. *See* Greenbrier
Catnip (*Nepeta cataria*), 51–52, **51**,
    243
Cattail (*Typha* species), 16, 67–71,
    115–116
  "cigars," 68
  flower head (overwintered), **68**
  fluff pillow of, 71
  look-alikes of, 69
  male and female, 67–68
  mats of, 71
  medicinal uses of, 71
  rhizomes of, 68, **68**, 70–71
  shoots of, **68**, 69, 70
  stalks with flower heads, **69**
  vitamins and minerals in, 70
Celtis laevigata. See Sugarberry
Celtis palida. See Desert hackberry
Celtis species. *See* Hackberry
Celtis tenuifolia. See Upland
    hackberry
Cerastium species. *See* Chickweed
Cerastium vulgatum. See Mouse-ear
    chickweed
Cercis canadensis. See Redbud
Cercis occidentalis species, 65
Chamaesyce species. *See* Spurge
Chamomile (*Matricaria chamomilla*),
    37
Chapman, Jonathan (Johnny
    Appleseed), 152
Chavez, César, 168
Checkerberry. *See* Wintergreen
Chenopodium album. See Lamb's-
    quarters
Chenopodium capitatum. See
    Strawberry blite
Cherries (*Prunus* species), 110, 112,
    119–123
  cyanide in, 122–123
  See also Cornelian cherry; Ground
    cherry
Chestnut (*Castanea* species),
    168–170, **168**
  calories in, 169
  look-alikes of, 169
  See also Water chestnuts

Chestnut oak (*Quercus primus*), 175
  with acorns, **175**
Chickweed (hen's inheritance)
  (*Stellaria* species; *Cerastium*
  species), 3, 137–139, **138,**
  256, 268, 271
  medicinal uses of, 138–139
  as nutritional powerhouse, 138
  personality of, 138
  for skin irritations, 139
  vitamins and minerals in, 138
Chicory (blue-sailors; wild succory)
  (*Cichorium intybus*), 190, 198,
  234–236, 246, 247
  flower of, **235**
  taproot and basal rosette, **235**
Childbirth, 100, 124
Childhood diseases, 52
Chives. *See* Wild chive
Chokecherry (*Prunus virginiana*),
  122, **122,** 123, 272
  medicinal uses of, 123
Cholesterol, 55, 86, 153, 167, 228
Chondrus crispus. See Irish moss
Cichorium intybus. See Chicory
Cicuta maculata. See Water hemlock
Cinnamon fern, fiddlehead of, 64
Circulatory disorders, 141
Cirsium species. *See* True thistles
Citronella. *See* Horse balm
Cleaning wild food, 274–275
Clearweed (*Pilea pumila*), 243
Cloudberry (*Rubus chamaemorus*), 97
Clovers (*Trifolium* and *Melilotus*
  species), 15, 24–26, 30
  leaves of, 26
  medicinal uses of, 25
  vitamins and minerals in, 25
Colds, 35, 64
Colic, 52
Collinsonia canadensis. See Horse
  balm
Colors, and insects, 4
Coltsfoot (son-before-father;
  coughwort) (*Tussilago
  farfara*), 71–73, **73,** 76
  flowers of, 71–72, **72**
  medicinal uses of, 72
  minerals in, 72
  toxicity of, 73
Commelina communis. See Asiatic
  dayflower
Commelina virginica. See Virginia
  dayflower
Commelin brothers, 27
Commercial garlic (*Allium sativum*),
  209
Common blackberry (*Rubus
  allegheniensis*), 96–97, **97**
  flowers of, **96**
Common blue violet (*Viola
  papilionacea*), 15, 55–56,
  229–231, 272

flowers of, 230, **230**
  in literature, 231
  medicinal uses of, 230
  in mythology, 230–231
  poisons in, 229–230
  rhizome of, **230**
  young, **230**
Common burdock (*Arctium minus*),
  32, 34
Common cattail (*Typha latifolia*), 69
Common chickweed (*Stellaria
  media*), 137
Common cinquefoil (*Potentilla
  simplex*), 92–93
Common dandelion (*Taraxacum
  officinale*), 10, 15, 20, 47,
  190–192, 197, 198, 200, 202,
  234–236, 246, 247, 249, 252,
  254, 256, 271, 272
  basal rosette and taproot, **190**
  as diuretic, 191–192
  look-alikes of, 190
  medicinal uses of, 191–192
  reproductive efficiency of, 192
  rosette flowers and seedhead, **190**
  vitamins and minerals in, 191
  worldwide growth of, 190–191
Common elderberry (American
  elder) (*Sambucus canadensis*),
  15, 103–105, **103,** 272
  collecting problems, 103
  cosmetic uses of, 105
  flowers of, 103, **104**
  magical associations of, 105
  medicinal uses of, 105
  minerals and vitamins in, 105
  poisons in, 105
Common evening primrose
  (*Oenothera biennis*), 17,
  193–195, 273
  basal rosette and taproot, **193**
  flower stalk of, **193,** 194
  medicinal uses of, 194–195
Common juniper (*Juniperus
  communis*), 211
Common mallow (*Malva neglecta*),
  76–77, **77**
Common milkweed (*Asclepias
  syriaca*), 5, 7, 17, 47–50
  flowers of, **47, 48**
  look-alike of, 49
  medicinal uses of, 50
  seedpods of, **47**
  shoot and rhizome of, **47**
Common nightshade, 199
Common parsnip (*Pastinaca sativa*),
  196–197
  basal rosette and taproot, **196**
  with flower seeds, **197**
  poisons in, 197
Common plantain (*Plantago major*),
  228, 273
  basal rosette of, **228**

Common sow thistle (*Sonchus
  oleraceus*), 201
  basal rosette and taproot, **200**
  flowers of, **200**
Common spicebush (wild allspice;
  feverbush) (*Lindera benzoin*),
  100, 180–181
  with berries, **181**
  flowers of, **181**
  medicinal uses of, 181
Common strawberry (*Fragaria
  virginiana*), 16, 75, 92–93
  in cosmetics, 93
  flowers of, **92**
  medicinal uses of, 93
  vitamins and minerals in, 93
Common sunflower (*Helianthus
  annus*), 15, 71, 93–94, **93**
  history of, 94
  minerals in, 94
  seeds, 16, 94
  sprouting, 94
  tubers in, 94
Common wax myrtle (southern
  bayberry) (*Myrica cerifera*), 79
Common wood sorrel (*Oxalis
  montana*), 30
Complex carbohydrates, 6
Composite family, 15, 259
Composite flower, **13**
Compound leaves, 12
Compress, in herbal preparation, 9
Concord grapes, 165
Conium maculatum. See Poison
  hemlock
Conservation, in foraging, 3
Cooking, 271–276
  berries, 272
  cleaning wild food, 274–275
  equipment for, 274
  and flour, 276
  flowers, 272
  fruits, 272
  greens, 271–272
  natural, 273–274
  preserving wild food, 275–276
  roots, 272–273
Corm, **16,** 17
Cornelian cherry (*Cornus mas*),
  100–101, **100**
  with fruit, **100**
  look-alikes of, 100
Cornus kousa. See Kousa dogwood
Cornus mas. See Cornelian cherry
Corylus americana. See American
  hazelnut
Corylus cornuta. See Beaked hazelnut
Corylus species. *See* Hazelnuts
"Cossack's asparagus," 69
Coughs, 10, 123, 194
Coughwort. *See* Coltsfoot
Cow cress (field peppergrass)
  (*Lepidium campestre*), 251, 253

Cow parsnip (*Heracleum maximum*), 43–45, 61, 272
  basal rosette and root, **44**
  look-alikes of, 43
  mature plant, **44**
Crab apples, 152–153, **152**, 272
Cramp bark (guelder rose) (*Viburnum opulus*), 182, 185
Cranberries (*Vaccinium* species), 224–225, 272
  medicinal uses of, 225
  vitamins and minerals in, 225
  See Highbush cranberry
Cranberry tree (*Viburnum pauciflorum*), 182
Crataegus species. *See* Hawthorns
Creasy greens. *See* Scurvy grass
Creeping wintergreen (*Gaultheria humifusa*), 223
Crystotaenia canadensis. See Honewort
Cuckooflower (lady's-smock) (*Caramine pratensis*), 257, **257**
Cuisine, all-natural, 273–274
Cultivated areas
  in autumn, 139–145
  in early spring, 232–234
  in late fall through early spring, 192–193
  in mid- to late spring, 30
  in summer, 89
Curly dock (yellow dock) (*Rumex crispus*), 34, 236–238, 243, 257, 271
  basal rosette of, **237**
  leaves of, 238
  medicinal uses of, 238
  with seeds, **236**
  vitamins and minerals in, 238
Currants (*Ribes* species), 16, 101–103, **101**, 272
  flowers of, **101**
  medicinal uses of, 102
  vitamins and minerals in, 102
Custard apple. *See* Pawpaw
Cut-leaf (evergreen) blackberry (*Rubus laciniatus*), 97
Cut-leaved toothwort (*Dentaria laciniata*), 257
  flowers of, **258**
Cystitis, 36, 38, 212, 225

Daffodil (*Amaryllidaceae*), 69
Daisies, 71
Dandelion. *See* Common dandelion
Dang gui (*Angelica sinensis*), 67
Darwin, Charles, 10, 14
Date plum. *See* Persimmon
Daucus carota. See Wild carrot
Daylily (*Hemerocallis fulva*), 14, 27, 232–234, 272
  flowers of, 233–234, **233**
  medicinal uses of, 234

negative reactions to, 233
shoot and tubers of, **232**
Death camas (*Zygadenus* species), 208
Decoction, in herbal preparation, 9
Deerberry. *See* Wintergreen
Deer tick, 4, 5
Delphinium tricorne. See Dwarf larkspur
Dementia Botanica, 2
Demulcents, 8
Dental problems, 223
Dentaria diphylla. See Toothwort
Dentaria laciniata. See Cut-leaved toothwort
Dentaria maxima. See Large Toothwort
Dermatitis, contact, 39, 58
Desert hackberry (*Celtis palida*), 170
Deserts, 17, 21, 23
  in autumn, 187–188
  in early spring, 270
  in late fall through early spring, 226
  in summer, 132–136
Devil's-walking-stick. *See* Hercules'-club
Dewberry (*Rubus flagellaris*), 97
Diabetes, 35, 67, 100, 153, 168, 192, 196, 197, 214
Diaphoretics, 8
Diarrhea, 7, 50, 93, 98, 150, 159, 188, 249, 262
Digger pine (*Pinus sabiniana*), 218
Digitalis, 7, 50, 150
Dioscorides, 211, 249
Diospyros texana. See Texas persimmon
Diospyros virginiana. See Persimmon
Disk flower, **13**
Disturbed areas, 21
  in autumn, 145
  in early spring, 234–247
  in late fall through early spring, 193–203
  in mid- to late spring, 31–42
  in summer, 90
Diuretics, 8
Dogbane (*Apocynum* species), 3, 5, 49, **49**
Dog tick, 4, 5
Dogwood, 100
  See also Kousa dogwood
Double-toothed leaf, **12**, 13
Douches, 9, 95
Drugs, 6–7
  vs. herbs, 7
Drying wild foods, 275
Duckweed (*Lemna* species), 256
Duke, Jim, 43
Dulse (*Palmaria palmata*), 82, 84–85, **84**
  vitamins and minerals in, 85

Dutch-elm disease, 170
Dwarf larkspur (spring larkspur) (*Delphinium tricorne*), 231
  poison of, **231**
Dwarf sumac (winged sumac) (*Rhus copallina*), 115
Dysentery, 171

Ear infections, 249
Early spring
  cultivated areas in, 232–234
  deserts in, 270
  disturbed areas in, 234–247
  fields in, 247–259
  freshwater wetlands in, 266–267
  lawns and meadows in, 227–231
  mountains in, 268–270
  seashore in, 267
  thickets in, 259
  woodlands in, 259–266
  See also Late fall through early spring
Eastern chinquapin (*Castanea pumila*), 169, **169**
Eastern prickly pear (*Opuntia humifusa*), 187, 188
  flowers of, **187**
Eastern yew. *See* American yew
Eclectic Materia Medica, 83
Eczema, 36, 245
Edible kelp (kelp; winged kelp; sweet kelp; wakame) (*Alaria esculenta*), 85–86, **86**
  medicinal uses of, 86
  vitamins and minerals in, 85–86
Edible Wild Plants of Northeastern North America (Peterson), 25
Elaeagnus angustifolia. See Russian olive
Elaeagnus commutata. See Silverberry
Elderberries. *See* Common elderberry
Elliot, Doug, 203
Emmenogogues, 8
Emollients, 8
Endometriosis, 247
Engelman's prickly pear (*Opuntia engelmannii*), 187
English plantain (long-leaf plantain; buckhorn) (*Plantago lanceolata*), 228
  basal rosette of, **229**
Epazote (Mexican tea; wormseed; Jerusalem oak) (*Teloxys ambrosioides*), 36–37, **36**, 46
  poisoning sysmptoms of, 36
  poisons in, 36
Epilepsy, 45, 195
Equipment, 4–5
  for cooking, 274
Erikson, Leif, 166
Erodium cicutarium. See Storksbill

European angelica root (*Angelica archangelica*), 67
European common apple (*Malus sylvestris*), 152
European linden (*Tilia europaea*), 63
European mandrake (*Mandragora officinarum*), 126
Evelyn, John, 244
Evergreen blackberry. *See* Cut-leaf blackberry
Exotic species, 18–19
Expectorants, 8
Eye problems, 168

Fall. *See* Autumn; Late fall through early spring
False banana. *See* Pawpaw
False Solomon's seal (*Smilacina racemosa*), 262–263
    with berries, **263**
    flowers of, **263**
    older shoot and rhizome of, **262**
    poisons in, 262
    young shoot of, **262**
Fat, in diet, 5–6
Feather-compound leaf, 12, **12**
Ferns, 32
    See also Cinnamon fern; Ostrich fern
Feverbush. *See* Common spicebush
Fiber, in diet, 6, 273
Fibrous roots, **16**, 17
Field garlic (*Allium vineale*), 206–207
    flowers of, **206**
    immature, **207**
Field mustard (rape; wild mustard) (*Brassica rapa*), 249–250
Field pennycress (*Thlaspi arvense*), 253–254, **253**
Field peppergrass. *See* Cow cress
Fields, 20
    in autumn, 145–151
    in early spring, 247–259
    in late fall through early spring, 203–210
    in mid- to late spring, 42–56
    in summer, 90–96
Field sorrel. *See* Sheep sorrel
Field sow thistle (*Sonchus arvensis*), 202
Filaree. *See* Storksbill
Fire cherry. *See* Pin cherry
Fish oil, 29
Flannelleaf. *See* Mullein
Floridian sea rocket (*Cakile lanceolata*), 256
Flours, unrefined and wild, 276
Flowers, 11
    cooking, 272
    identifying types of, 13–15, **13**
    parts of, 10, **13**, 14–15

Flu, 53, 64
    See also Influenza
Fly poison (*Amianthium muscaetoxicum*), 207–208, **208**
Fomentation, in herbal preparation, 9
Fool's parsley (*Aesthusa cynapium*), 129, 204–205, 245
Foraging
    conservation in, 3
    preparing for, 2–4
    safety in, 2–4
Forests, in summer, 119–128
Foxglove, 7, 50
Fox grape (*Vitis labrusca*), 165, **167**
    vine flowers of, **166**
Foxtail grass (*Setaria* species), 147–148, **148**, 273
Foxtail species (*Setaria italica*), 148
Fragaria vesca. *See* Wood strawberry
Fragaria virginiana. *See* Common strawberry
Free radicals, damage of, 6
Freezing wild foods, 275
Freshwater wetlands
    in autumn, 185–186
    in early spring, 266–267
    in late fall through early spring, 224–226
    in mid- to late spring, 66–77
    in summer, 128–131
Fruit
    cooking, 272
    identifying, 15–16
Fucus vesicylosus. *See* Rockweed
Fungal infections, 163, 165, 170, 238

Galen, 211
Gallstones, 191
Gambel's oak (*Quercus gambelii*) 175–176
Gangrene, 206
Garlic. *See* Commercial garlic; Field garlic; Wild garlic
Garlic mustard (Jack-by-the-hedge; sauce-alone) (*Alliaria officinalis*), 56, 254–255, 273
    basal rosette of, **254**, 255
    flowers of, **254**
Gaultheria humifusa. *See* Creeping wintergreen
Gaultheria procumbens. *See* Wintergreen
Gaultheria shallon. *See* Salal wintergreen
Gaylussacia species. *See* Blueberries; Huckleberries
Gentian, 192
Gerard (medieval herbalist), 200, 249
Gibbons, Euell, 43, 75, 94, 172, 203, 239

Gill-over-the-ground, 56, 255
    See also Ground ivy
Ginger. *See* Wild ginger
Gingko (maidenhair tree) (*Gingko biloba*), 139–142, **140**
    collecting, 141
    flowers of, **140**
    fruit of, 141
    as greatest botanical sensation of day, 140
    history of, 140
    medicinal uses of, 141–142
    nuts of, **140**, 141, 142, 273
    poisons in, 141
    smell of, 141
Gingko biloba. *See* Gingko
Ginseng, 8
Glandular mesquite (*Prosopis glandulosa*), 132–133
Glasswort (bleach asparagus) (*Salicornia* species), 80–81, **81**
Glecoma hederacea. *See* Ground ivy
Goatsbeard (salsify; oyster plant) (*Tragopogon* species), 197–198, 272
Goiters, 83
Goldenseal (*Hydrastis canadensis*), 114, 192
Gooseberries (*Ribes* species), 101–102, **102**
    flowers of, **102**
    medicinal uses of, 102
    vitamins and minerals in, 102
Goosefoot. *See* Lamb's-quarters
Gould, David, 169
Gout, 30, 36, 64, 260
Goutweed (*Aegopodium podograria*), 60, 259–260
    basal rosette of, **260**
    flowers of, 260, **260**
Granadilla. *See* Passionflower
Grapes (*Vitis* species), 15, 165–168, 239, 272
    as diuretic, 167
    European uses for, 167–168
    leaves of, 166–167
    medicinal uses of, 167–168
    resveratrol in, 167
    vitamins and minerals in, 167
Grass. *See* Foxtail grass
Great burdock (*Arctium lappa*), 32
    in basal rosette, **32**
    bristly burr of, **33**
    flowers of, 33, **34**
    in pre-flower stage, **33**
Great Solomon's seal (*Polygonatum commutatum*), 261
    shoot and rhizome of, **261**
Greenbrier (catbrier; blasphemy vine; stretchberry) (*Smilax* species), 56–57
    berries of, 57, **57**

shoot of, 57, **57**
vine of, 17, 57, **57**
Green laver, *See* Sea lettuce
Greens, cooking, 271–272
Ground cherry (tomatillo; husk tomato; Japanese lantern) (*Physalis* species), 16, 91–92, **91**
poisons in, 92
Ground ivy (gill-over-the-ground) (*Glecoma hederacea*), 55–56, **55**
Groundnut (hopniss)(*Apios americana*), 16, 17, 210–211, 273
flowers of, **210**
historical uses of, 210
look-alikes of, 211
and poisons, 211
tubers of, 210, **210**
vitamins in, 210
Guelder rose. *See* Cramp bark
Gum disease, 80, 118
Gymnocladus dioica. See Kentucky coffee tree

Habitats, 19–22
deserts, 21
disturbed areas, 21
fields, 20
lawns and meadows, 19–20
mountains, 21
seashore, 21
thickets, 20
wetlands, 22
woodlands, 20
Hackberry (*Celtis* species), 170–171
flowers of, **170**
medicinal uses of, 171
tree with fruit, **171**
Hairy lettuce (*Lactuca hirsuta*), 246
Hawthorns (*Crataegus* species), 7, 148–150, 151
flowers of, **149**
fruits of, **149**, 150
medicinal uses of, 150
thorn of, 150
vitamins and minerals in, 150
Hay fever, 15, 38, 244
Hazelnuts (wild filberts) (*Corylus* species), 16, 155–157
Headaches, 55, 160, 209, 223
See also Migraine headaches
Heart disease, 6, 8, 29, 64, 150, 202, 209, 228
See also Cardiovascular disease
Hedeoma pulegioides. See American pennyroyal
Hedge mustard (*Sisymbrium officinale*), 250
Helianthus annus. See Common sunflower

Helianthus maximiliani, 94
Helianthus tuberosus. See Jerusalem artichoke
Hemerocallis fulva. See Daylily
Hemlock. *See* Poison hemlock; Water hemlock
Hemorrhages, 80
Hemorrhoids, 249
Hen's inheritance. *See* Chickweed
Hepatitis, 191, 229, 244
Heracleum maximum. See Cow parsnip
Herbaceous plants, 18
Herbicides, 3
Herbs
actions of, 7–8
active ingredients of, 7
vs. drugs, 7
eating, 8
preparing and using, 8–10
smoking, 9
teas, 189
Hercules'-club (Devil's-walking-stick; angelica tree) (*Aralia spinosa*), 57–58, **58**, 103
berries of, **58**
poisons in, 57–58
shoots of, 58, **58**
Herpes simplex, 41, 52
Hibiscus palustris. See Swamp rose mallow
Hickories (*Carya* species), 171–173
cracking shells of, 172
as instruments of child abuse, 172
nuts of, 161, 164, 172, **172**, 273
vitamins and minerals in, 172
High blood pressure, 82, 87, 153, 209
Highbush blueberry (*Vaccinium corymbosum*), 99
flowers of, **98**
Highbush cranberry (*Viburnum trilobum*), 182–183, **182**, 272
bark of, 185
Himalayan yew (*Taxus wallichiana*), 144
Hind's walnut (*Juglans hindsii*), 161
Hippocrates, 35, 105, 240, 256
Hives, 71
Hobblebush (wayfaring tree; witches' hobble) (*Viburnum alnifolium*), 183
berries of, **183**
flowers of, **183**
Hog peanut (*Amphicarpa bracteata*), 91, 216–217, **216**, 273
Homeopathy, 39
Honewort (wild chervil) (*Crystotaenia canadensis*), 60–61, **61**
basal rosette of, **60**
medicinal uses of, 61

Honey locust, 61
Honey mesquite (*Prosopis juliflora*), 132–133
Hop clover (*Trifolium agrarium*), 25–26
Horse balm (stoneroot; citronella) (*Collinsonia canadensis*), 54, **54**
medicinal uses of, 54
Horse chestnut (*Aesculus hippocastanum*), 169, **170**
leaf of, **169**
poisons in, 169, 170
Horseradish (*Armoracia lapathifolia*), 250, 255, 257–259, **258**, 273
blistering effect of, 259
and digestive system, 259
medicinal uses of, 258–259
Horsetail kelp (tangle; kombu) (*Laminaria digitata*), 86–87, **86**
medicinal uses of, 87
Hoxsey formula, 25
Huckleberries (*Vaccinium* species; *Gaylussacia* species), 98–100
bark of, 99
fruit of, 99
leaves of, 99, 100
Husk tomato. *See* Ground cherry
Hydrastis canadensis. See Goldenseal
Hydrophyllum canadense. See Broad-leaved waterleaf
Hydrophyllum macrophyllum. See Large-leaf waterleaf
Hydrophyllum species. *See* Waterleaf
Hydrophyllum virginianum. See Virginia waterleaf
Hyperactive children, 106
Hypoglycemia, 35, 67, 153, 168, 192, 196, 212, 214
Hypothermia, 178

Iceland moss, 83–84
Identifying wild plants, 10–18
flowers, 13–15
fruit, 15–16
leaves, 11–13
roots, 16–17
sharpening skills, 189
shoots, 17–18
thorns, 17
universal scientific names for, 14
Immune system, 8, 29, 35, 141, 181
Impatiens biflora. See Spotted touch-me-not
Impatiens pallida. See Pale touch-me-not
Impatiens species. *See* Jewelweed
Indian fig. *See* Prickly pear
Indian salad. *See* Waterleaf
Influenza, 41
See also Flu
Infusion, in herbal preparation, 9
Inkberry. *See* Pokeweed

Insects, 4–5, 15, 150
    See also Bees; Butterflies;
        Mosquitoes; Ticks; Wasps
Insomnia, 106
Iodine deficiency, 83
Ipomoea leptophylla. See Bush
    morning glory
Ipomoea pandurata. See Wild potato
    vine
Iris (blue flags) (*Iris* species), 26, 69
    poisonous, **233**
    woodland, 262
Irish moss (sea moss; blancmange;
    jelly moss) (*Chondrus crispus*),
    83–84, **84**
    medicinal uses of, 84
    minerals in, 84
Iris species. *See* Iris
Iron-deficiency anemia, 238, 244
Ivy. *See* Ground ivy; Poison ivy

Jack-by-the-hedge. *See* Garlic
    mustard
Jack-in-the-pulpit, 262
Jacob's-staff. *See* Mullein
Japanese knotwood (*Polygonum
    cuspidatum*), 17, 167, 214,
    238–240
    flowers of, **239**
    as laxative, 239
    medicinal uses of, 239
    shoots of, **239**
Japanese lantern. *See* Ground cherry
Japanese yew (*Taxus cuspidata*), 144
Jaundice, 191
Jelly moss. *See* Irish moss
Jerusalem artichoke (sun choke)
    (*Helianthus tuberosus*), 94,
    195–196, 273
    flowers of, **195**
    tubers of, 196, **196**
    vitamins and minerals in, 196
Jerusalem oak. *See* Epazote
Jewelweed (touch-me-not)
    (*Impatiens* species), 15, 39,
    71, 73–75, **74**, 243
    flowers of, 74
    leaves of, 74
    medicinal uses of, 74–75
    seeds of, 74
    selenium in, 74
    shoot of, **73**, 74
Judas tree. *See* Redbud
Juglans californica. See California
    walnut
Juglans cinerea. See Butternut
Juglans hindsii. See Hind's walnut
Juglans major. See Arizona walnut
Juglans nigra. See Black walnut
Juglans species, 171
Juneberry (serviceberry; shadbush)
    (*Amelanchier* species), 99,
    123–124, **123**

flowers of, **123**
    name of, 124
Juniper (*Juniperus* species), 211–212,
    211
    as diuretic, 211–212
    medicinal uses of, 211–212
    poisons in, 211
Juniperus communis. See Common
    juniper
Juniperus species. *See* Juniper

Kelp. *See* Edible kelp
Kentucky coffee tree (*Gymnocladus
    dioica*), 61–62
    flowers of, **62**
    leaf of, **61**
    look-alikes of, 61
    pods of, 61–62, **62**
    poisons in, 62
    pulp of, 62
    seeds of, 62
Kidney disease, 176, 197, 200, 212
Kidney stones, 30, 64, 205
Kousa dogwood (*Cornus Kousa*),
    142, **143**
    flowers of, 142, **142**
    fruit of, 142

Lactuca canadensis. See Tall lettuce
Lactuca hirsuta. See Hairy lettuce
Lactuca pulchella. See Blue lettuce
Lactuca scariola. See Prickly lettuce
Lactuca species. *See* Wild lettuce
Lady's-smock. *See* Cuckooflower
Lady's thumb (redleg) (*Polygonum
    persicaria*), 3, 27–28, **27**, 271
Lakes, 22
Lamb's-quarters (goosefoot;
    pigweed) (*Chenopodium
    album*), 36, 37, 45–47, **45**,
    256, 271, 273
    and leaf miner, 46
    leaves of, 46, 146–147
    names of, 46
    and nitrates in soil, 46
    odorlessness of, 46
    shoots of, **45**, 46
    vitamins and minerals in, 47
Laminaria digitata. See Horsetail kelp
Laportea canadensis. See Wood nettle
Laportea species. *See* Nettles
Large cranberry (*Vaccinium
    macrocarpon*), 225, **225**
Large-leaf waterleaf (*Hydrophyllum
    macrophyllum*), 266
Large toothwort (*Dentaria maxima*),
    257
Late fall through early spring
    cultivated areas in, 192–193
    deserts in, 226
    disturbed areas in, 193–203
    fields in, 203–210
    freshwater wetlands in, 224–226

lawns and meadows in, 190–192
    mountains in, 226
    seashore in, 226
    thickets in, 210–212
    woodlands in, 212–224
Lathyrus japonicus. See Beach pea
Laver (nori) (*Porphyra* species), 85,
    85
Lawns and meadows, 19–20
    in autumn, 137–139
    in early spring, 227–231
    in late fall through early spring,
        190–192
    in mid- to late spring, 24–30
    in summer, 89
Lead, as pollutant, 3
Leaves
    identifying, 11–13
    types of, 12–13, **12**
Leek. *See* Wild leek
Legume family, 15, 43, 61, 91, 211,
    259
Lemna species. *See* Duckweed
Lepidium campestre. See Cow cress
Lepidium species. *See* Peppergrasses
Lepidium virginicum. See Poor man's
    pepper
Lettuce. *See* Miner's lettuce; Sea
    lettuce; Wild lettuce
Leukemia, 41
Life cycles, of wild plants, 18
Lilium tigrinum. See Tiger lily
Lily family, 134, 207, 208, 259, 272
Lily of the valley, 78, 264
Lime tree. *See* Linden
Linden (lime tree; American
    basswood; monkey-nut tree)
    (*Tilia* species), 62–64, **63**
    flowers of, 63
    fruits of, 62–63
    medicinal uses of, 63–64
Lindera benzoin. See Common
    spicebush
Linnaeus, 14, 27
Live-forever. *See* Orpine
Liver disorders, 25, 238
L. longicruris. See Longstemmed
    kelp
Loblolly pine (*Pinus taeda*), 219
Long-leaf plantain. *See* English
    plantain
Longstemmed kelp (*L. longicruris*),
    87
Lowbush blueberry (*Vaccinium
    angustifolium*), 99
Lyme disease, 4, 5

McIntosh tree, 152
Maidenhair tree. *See* Gingko
Mallow. *See* Common mallow
Malus species. *See* Apple
Malus sylvestris. See European
    common apple

Malva neglecta. See Common mallow

Mandragora officinarum. See European mandrake

Mandrake. *See* Mayapple

Man-of-the-earth. *See* Wild potato vine

Man-under-ground. *See* Wild potato vine

Maple-leaf viburnum (*Viburnum acerifolium*), 101, 182

Maracock. *See* Passionflower

Marsh blue violet (*Viola cucullata*), 231

Marshes, 22

Marshmallow (*Althaea officinalis*; *Malva* species), 75–77, **75**
  candy, 76
  medicinal uses of, 76
  root of, 76, **76**

Matricaria chamomilla. See Chamomile

Matricaria matricarioides. See Pineapple weed

Matteuccia struthiopteris. See Ostrich fern

Mayapple (mandrake) (*Podophyllum peltatum*), 124–126, 262
  appearance of, 124
  as cathartic, 125
  flowers of, 124, **125**
  fruit of, 125
  medicinal uses of, 126
  rhizomes of, 125–126
  toxicity of, 125–126

Maypop. *See* Passionflower

Meadows, wet, 22
  See also Lawns and meadows

Melilotus alba. See White sweet clover

Melilotus officinalis. See Yellow sweet clover

Melilotus species. *See* Clovers

Menis pernum canadense. See Canada moonseed

Menstruation, 8, 36, 53, 95, 182, 223, 240

Mentha aquatica. See Water mint

Mentha arvensis. See Wild mint

Mentha piperita. See Peppermint

Mentha spicata. See Spearment

Mesquite (*Prosopis veluntina*), 132–133, **133**

Mexican tea. *See* Epazote

Mid- to late spring
  cultivated areas in, 30
  disturbed areas in, 31–42
  fields in, 42–56
  freshwater wetlands in, 66–77
  lawns and meadows in, 24–30
  mountains in, 87
  seashore in, 78–87
  thickets in, 56–60
  woodlands in, 60–66

Mid-spring, wild plants of, 23

Migraine headaches, 249

Milkweed. *See* Common milkweed

Millet, 147, 148

Miner's lettuce (Spanish lettuce) (*Montia perfoliata*), 268, **268**

Mints (many genera), 15, 50–56, 76, 243
  commercial oils of, 50
  fragrance of, 50–51
  history of, 50
  medicinal uses of, 50

Mockernut hickory (*Carya tomentosa*), 171

Monarch butterfly larvae, 50

Monarda didyma. See Oswego-tea

Monkey-nut tree, 63, **63**
  See also Linden

Monkshood (*Aconitum uncinatum*), 231

Monolepis nuttalliana. See Povertyweed

Montia perfoliata. See Miner's lettuce

Moore, "Dr." John, 263

Morus alba. See White mulberry

Morus microphylla. See Texas mulberry

Morus nigra. See Black mulberry

Morus rubra. See Red mulberry

Morus species. *See* Mulberry

Mosquitoes, 4, 74, 89

Mountain cranberry (*Vaccinium vitis-idaea*), 225

Mountains, 21
  in autumn, 186
  in early spring, 268–270
  in late fall through early spring, 226
  in mid- to late spring, 87
  in summer, 132

Mouse-ear chickweed (*Erastium vulgatum*), 137

Moxa. *See* Mugwort

Mugwort (moxa) (*Artemisia vulgaris*), 240–241
  and dreams, 241
  flowers of, **241**
  medicinal uses of, 240–241
  young, **240**

Mulberry (*Morus* species), 11, 16, 35, 115, 126–128
  bark of, 126
  gathering, 127
  as hallucinogen, 128
  as laxative, 128
  flowers of, 126, **126**

Mullein (Jacob's-staff; flannelleaf) (*Verbascum thapsus*), 72, 247–249
  basal rosette of, **248**
  as biennial, 248
  flower stalk of, **248**
  historical uses of, 249

  medicinal uses of, 248–249
  vitamins and minerals in, 248

Multiflora rose (*Rosa multiflora*), 157
  flowers of, **157**
  fruit of, **158**

Mushrooms, 2, 3

Mustard family (*Brassicaceae*), 12, 249–259, 272, 273
  commercial vegetables from, 259
  medicinal uses of, 250
  vitamins and minerals in, 249

Myrica cerifera. See Common wax myrtle

Myrica gale. See Sweet gale

Myrica pensylvanica. See Northern bayberry

Myrica species. *See* Bayberries

Nannyberries (*Viburnum lentago*), 182, 183

Narrow-leaved cattail (*Typha angustifolia*), 69

Nasturtium officinale. See Watercress

Native species, 18–19

Nectarine (*Prunus persica*), 106, 107–108

Nepeta cataria. See Catnip

Nervines, 7

Nettles (*Urtica* species; *Laportea* species), 238, 241–245, 271
  antiseptic properties of, 245
  cleansing properties of, 245
  commercial uses for, 245
  leaves of, 243
  medicinal uses of, 244–245
  vitamins and minerals in, 244

New Mexican locust (*Robinia neomexicana*), 43

Nightshade family, 92, 259
  See also Common nightshade

Nodding wild onion (*Allium cernuum*), 207

Nori. *See* Laver

Northern bayberry (*Myrica pensylvanica*), 79
  flowers of, **79**
  fruit of, **80**

Northern wild raisins (*Viburnum cassinoides*), 182, 183, 184, 272

Nutrition, and wild foods, 5–6

Nuts, 16, 273
  freezing, 275

Nymphaca species. *See* Water lilies

Oak (*Quercus* species), 15, 173–176
  acorns in, 173–174

Oenothera biennis. See Common evening primrose

Oenothera hookeri species, 194

Ohio buckeye (*Aesculus glabra*), 169

Oil of peppermint, 52

Oil of wintergreen, 215, 223, 224

Ointment, in herbal preparation, 9
Olives. *See* Autumn olive
One-leafed piñon (single-leaf piñon)
  (*Pinus monophylla*), 218
Onion family, 272
  See also Wild onions
Opium, 247
Opposite leaves, 12–13, **12**
Opuntia engelmannii. See Engelman's
  prickly pear
Opuntia humifusa. See Eastern
  prickly pear
Opuntia phaeacantha. See Western
  prickly pear
Opuntia species. *See* Prickly pear
Orache (*Atriplex patula*), 46
Orchids, 15
Ornithogalum umbellatum. See Star-
  of-Bethlehem
Orpine (stonecrop; live-forever),
  (*Sedum purpureum*), 268–270,
  269
  common names of, 270
  flowers of, 268–269
  leaves of, 269
  tubers of, 269
Osmorhiza species. *See* Aniseroot
Ostrich fern (*Matteuccia
  struthiopteris*), 64–65, **64**
  fiddlehead of, 64, **65**
  fronds of, 64, 65
  identification of, 64
  poisons in, 64, 65
Oswego-tea (bee-balm) (*Monarda
  didyma*), 53–54
Oxalic acid, 117
Oxalis montana. See Common wood
  sorrel
Oxalis species. *See* Wood sorrel
Oxalis violacea. See Violet wood
  sorrel
Oyster plant. *See* Goatsbeard

Pagoda tree, 43
Pale touch-me-not (*Impatiens
  pallida*), 74
Palmaria palmata. See Dulse
Palmate-compound leaf, 12, **12**
Parsley. *See* Broad-leaf parsley;
  Fool's parsley
Parsnip. *See* Common parsnip;
  Water parsnip; Wild parsnip
Parthenocissus quinquefolia. See
  Virginia creeper
Passiflora incarnata. See
  Passionflower
Passionflower (maypop; passion
  fruit; vine apricot; wild
  cucumber; granadilla;
  maracock) (*Passiflora
  incarnata*), 105–106, **106**
  historical basis of name, 105

  medicinal uses of, 106
  and sleep, 106
Passion fruit. *See* Passionflower
Pastinaca sativa. See Common
  parsnip
Pasture fern. *See* Bracken
Pasture rose (wild rose) (*Rosa
  carolina*), 158
Pawpaw (custard apple; false
  banana) (*Asimina triloba*),
  176–178
  bark of, 178
  as emetic, 178
  flowers of, **176**
  fruit of, **177**
  as high-calorie wild food, 178
  poisons in, 178
  seeds of, 178
Peach (*Prunus persica*), 16, 106–109,
  107, 110, 112
  flowers of, **107**
  medicinal uses of, 108–109
  toxicity in, 109
  vitamins and minerals in, 108
Pear (*Pyrus communis*), 16, 109–110,
  110, 124
  collecting, 109
  flowers of, **109**
  tree wood of, 110
  vitamins and minerals in, 110
  See also Prickly pear
Pecan (*Carya illinoensis*), 172, **172**
  flowers of, **173**
Pennyroyal. *See* American
  pennyroyal
Peppergrasses (*Lepidium* species),
  251
Peppermint (*Mentha piperita*), 51,
  52, **52**
  medicinal uses of, 52
  oil of, 52
  vitamins in, 52
Pepperwort. *See* Toothwort
Perennials, 18, 20
Persimmon (date plum) (*Diospyros
  virginiana*), 158, 178–180, 272
  dislodging, 179–180
  flowers of in tree, **178**
  with fruit, **179**
  harvest time for, 179
  medicinal uses of, 180
  ripe and unripe, 180
  vitamins and minerals in, 180
Petasites speciosa. See Western
  coltsfoot
Peterson, Lee, 25
Peterson's Field Guide to Herbs, 43
PGEI (prostaglandin), 194
Phaseolus acutifolius. See Tepary bean
Phaseolus metcalfei. See Tepary bean
Phaseolus polystachios. See Wild bean
Phragmites species. *See* Reeds

Physalis species. *See* Ground cherry
Phytolacca americana. See Pokeweed
Pickerelweed (*Pontederia cordata*),
  185–186, **185**
  water testing before eating leaves,
  186
Pignut hickory (butternut hickory)
  (*Carya cordiformis*), 172
Pigweed. *See* Amaranth; Lamb's-
  quarters
Pilea pumila. See Clearweed
Pin cherry (bird cherry; fire cherry;
  wild cherry) (*Prunus
  pensylvanica*), 121
Pine (*Pinus* species), 217–219
  bark of, 219
  and turpentine, 219
Pineapple weed (*Matricaria
  matricarioides*), 37–38, **37**
  and allergies, 38
  medicinal uses of, 38
Piñon pine (*Pinus edulis*), 218, **218**
Pinus echinata. See Shortleaf pine
Pinus edulis. See Piñon pine
Pinus lambertiana. See Sugar pine
Pinus monophylla. See One-leafed
  piñon
Pinus ponderosa. See Ponderosa pine
Pinus resinosa. See Red pine
Pinus rigida. See Pitch pine
Pinus sabiniana. See Digger pine
Pinus species. *See* Pine
Pinus strobus. See White pine
Pinus taeda. See Loblolly pine
Pitch pine (*Pinus rigida*), 217
Pitohu genus (poisonous birds), 50
Plantago juncoides. See Seaside
  plantain
Plantago lanceolata. See English
  plantain
Plantago major. See Common
  plantain
Plantago species. *See* Plantain
Plantain (*Plantago* species), 39,
  227–229, 243
  as breath freshener, 229
  as laxative, 229
  medicinal uses of, 228–229
Plant types, 18
Plaster, in herbal preparations, 9
Pleurisy, 49
Pleurisy root. *See* Butterfly weed
Pliny the Elder, 76, 211, 229, 249
Plums, 16, 272
  See also American plum; Beach
  plums; Canada plum; Wild
  plums
Podophyllum peltatum. See Mayapple
Poison hemlock (*Conium
  maculatum*), 129, 204, **205,**
  214, 245
  flowers and seeds of, **205**

Poison ivy (*Toxicodendron radicans*), 38–39, **38**, 57, 74, 75, 115, 117, 221, 229
  leaves of, 97
  medicinal uses of, 39
  poisons in, 38–39
Poison oak (*Toxicodendron quercifolium*; *Toxicodendron diversilobum*), 38
Poisonous plants, 2, 3, 4
Poison sumac (*Rhus vernix*), 117, **118**
Pokeberries, **41**
Pokeroot, 41
Poke sallet. *See* Pokeweed
Pokeweed (poke sallet; inkberry; scoke) (*Phytolacca americana*), 17, 39–41, **40**
  medicinal uses of, 41
  poisons in, 40, 41
  shoots and taproot of, 40, **40**
Polygonatum biflorum. *See* Solomon's seal
Polygonatum commutatum. *See* Great Solomon's seal
Polygonatum cuspidatum. *See* Japanese knotwood
Polygonatum persicaria. *See* Lady's thumb
Polygonatum species. *See* Solomon's seal
Ponderosa pine (*Pinus ponderosa*), 219
Pontedera, Giulio, 186
Pontederia cordata. *See* Pickerelweed
Pontederia lanceolata species, 185
Poor man's pepper (*Lepidium virginicum*), 251, **252**, 272
  flower and seed stalk of, **252**
  immature flower stem of, **252**
Porcelainberry (*Ampelopsis brevipedunculata*), 165
  poisons in, 165, **167**
Porphyra. *See* Laver
Portulaca oleracea. *See* Purslane
Potassium, 82
Potato. *See* Wild potato vine
Potentilla simplex. *See* Common cinquefoil
Poultice, in herbal preparations, 9
Povertyweed (*Monolepis nuttalliana*), 46
Prairie wild onion (*Allium stellatum*), 207
Pregnancy, 212
Preserving wild foods, 275
Prickly ash bark, 39
Prickly lettuce (*Lactuca scariola*), 246
Prickly pear (cactus pear; Indian fig) (*Opuntia* species), 17, 187–188
  medicinal uses of, 188
  vitamins and minerals in, 188

Primrose. *See* Common evening primrose
Prosopis glandulosa. *See* Glandular mesquite
Prosopis juliflora. *See* Honey mesquite
Prosopis pubescens. *See* Screwbean
Prosopis veluntina. *See* Mesquite
Prostate cancer, 244
Prostate enlargement, 86
Prunus americana. *See* American plum
Prunus armeniaca. *See* Apricot
Prunus avium. *See* Sweet cherry
Prunus cerasus. *See* Sour cherry
Prunus maritima. *See* Beach plums
Prunus nigra. *See* Canada plum
Prunus pensylvanica. *See* Pin cherry
Prunus persica. *See* Nectarine; Peach
Prunus serotina. *See* Black cherry
Prunus species. *See* Cherries; Wild plums
Prunus virginiana. *See* Chokecherry
Psoriasis, 36
Pteridium aquilinum. *See* Bracken
Purple-flowering raspberry (*Rubus odoratus*), 114, **114**
Purple goatsbeard (salsify; oyster plant) (*Tragopogon porrifolius*), 197–198
Purslane (pursley; pussley) (*Portulaca oleracea*), 28–29, **29**, 268, 272
  history of, 28
  vitamins and minerals in, 29
Pursley. *See* Purslane
Pussley. *See* Purslane
Pyrolaceae species, 223
Pyrus communis. *See* Pear

Queen Anne's lace, 94
  See also Wild carrot
Quercus alba. *See* White oak
Quercus gambelii. *See* Gambel's oak
Quercus primus. *See* Chestnut oak
Quercus species. *See* oak

Radially symmetrical (regular) flower, **13**, 15
Radiation poisoning, 62
Raisin. *See* Northern wild raisin
Ramp. *See* Wild leek
Rape. *See* Field mustard
Raspberries (*Rubus* species), 11, 16, 112–115, 126
  "canes," 112
  leaves of, 115
  look-alike of, 114–115
  medicinal uses of, 115
  root of, 98
  vitamins and minerals in, 115
Ray flower, **13**

RDA. *See* Recommended Daily Allowances
Recommended Daily Allowances (RDA), of vitamins, 6
Redbud (Judas tree) (*Cercis canadensis*), 65–66
  flowers of, 65–66, **65**
  pods of, **66**
Red clover (*Trifolium pratense*), 24, **24**
Red mulberry (*Morus rubra*), 126, **127**, 128
Red-oak acorns, 176
Red oak group, 174
Red pine (*Pinus resinosa*), 218
Red sorrel. *See* Sheep sorrel
Reeds (*Phragmites* species), 69
Refined carbohydrates, 5–6
Reproductive-system problems, 86, 102
Respiratory disorders, 202
Rhammus species. *See* Buckthorn
Rheumatism, 30
Rheumatoid arthritis, 200
Rhizomes, 16, **16**, 17
Rhodoglossum affine, 83
Rhubarb, 34, 239
Rhus copallina. *See* Dwarf sumac
Rhus glabra. *See* Smooth sumac
Rhus species. *See* Sumac
Rhus trilobata. *See* Squawbush
Rhus typhina. *See* Staghorn sumac
Rhus vernix. *See* Poison sumac
Ribes nigrum. *See* Black currant
Ribes species. *See* Currants; Gooseberries
Rice. *See* Wild rice
Rivers, 22
Robin, John and Vespasian, 43
Robinia neomexicana. *See* New Mexican locust
Robinia pseudoacacia. *See* Black locust
Rockweed (bladderwrack; wrack) (*Fucus vesicylosus*), 82–83, **83**, 87
  medicinal uses of, 83
Rocky Mountain spotted fever, 4
Roots, 3, 189
  cooking, 272–273
  fibrous, **16**
  identifying, 16–17
Rosa californica. *See* California rose
Rosa carolina. *See* Pasture rose
Rosa eglanteria. *See* Sweetbrier rose
Rosa multiflora. *See* Multiflora rose
Rosa rugosa. *See* Wrinkled rose
Rosa species. *See* Roses
Rose hips, 157, 158–159
Roseroot (*Sedum rosea*), 269
Roses (*Rosa* species), 17, 157–160, 269

Roses (cont.)
  for cosmetics, 160
  essential oil of, 160
  flowers of, 157, 158, 160
  history of, 160
  medicinal uses of, 159, 160
  petals of, 159
  as symbol, 158
  thorns in, 157
  vitamins and minerals in, 158,
    159, 160
  See also Rose hips; Rose water
Rose water, 160
Rubefacients, 8
Rubus allegheniensis. See Common
    blackberry
Rubus chamaemorus. See Cloudberry
Rubus flagellaris. See Dewberry
Rubus laciniatus. See Cut-leaf
    blackberry
Rubus occidentalis. See Black
    raspberry
Rubus odoratus. See Purple flowering
    raspberry
Rubus parviflorus. See Thimbleberry
Rubus phoenicolasius. See Wineberry
Rubus species. See Blackberries;
    Raspberries
Rubus strigosus. See Wild red
    raspberry
Rubus ursinus. See California
    blackberry
Rumex acetosella. See Sheep sorrel
Rumex crispus. See Curly dock
Rumex hastatulus. See Seaside sorrel
Rumex obtusifolius. See Broad-leaf
    dock
Runyon, Linda, 25
Russian olive (Elaeagnus angustifolia),
    155

Saint Vitus' Dance, 130
Salal wintergreen (western
    wintergreen) (Gaultheria
    shallon), 223
Salicornia species. See Glasswort
Salix species. See Willow
Salsify, 198
  seed head of, 197
  See also Goatsbeard
Salve, as herbal preparation, 9
Sambucus canadensis. See Common
    elderberry
Sambucus cerulea. See Blue elder
Sassafras (Sassafras albidum), 126,
    219–221, 221
  flowers of, 220
  medicinal uses of, 220–221
  shoot of, 220
  tea, 220, 221
Sassafras albidum. See Sassafras
Sauce-alone. See Garlic mustard
Schav. See Sheep sorrel

Scientific names (universal), for
    identifying wild plants, 14
Scoke. See Pokeweed
Screwbean (Prosopis pubescens),
    132–133, 133
Scurvy, 6, 199, 223, 268
Scurvy grass (creasy greens)
    (Barbarea verna), 251
Sea lettuce (Green laver) (Ulva
    lactuca), 83
Sea moss. See Irish moss
Sea rocket (Cakile edentula),
    256–257, 256
Seashore, 17, 21
  in autumn, 186
  in early spring, 267
  in late fall through early spring,
    189, 226
  in mid- to late spring, 78–87
  in summer, 131–132
Seaside plantain (Plantago juncoides),
    228
Seaside sorrel (Rumex hastatulus),
    199
Seasonal cycles, of wild plants, 22
Seaweeds (algae), 81–82, 189
  and freshwater algae, 82
  poisons in, 82
  and polluted waters, 82
  sodium content in, 82
Sedum purpureum. See Orpine
Sedum rosea. See Roseroot
Seeds, 273
Selenium, 74
Serviceberry. See Juneberry
Setaria italica. See Foxtail species
Setaria species. See Foxtail grass
Shadbush. See Juneberry
Shagbark hickory (Carya ovata), 171
  tree with nuts, 171
Shamrock, 30
  See also Wood sorrel
Sheep sorrel (red sorrel; field sorrel;
    schav) (Rumex acetosella), 30,
    198–200, 271
  basal rosette of, 198, 198
  flowers of, 199, 199
  look-alikes of, 199
  medicinal uses of, 200
  name of, 199
  vitamins and minerals in, 199
Shepherd's purse (Capsella
    bursapastoris), 251–253, 253,
    272
  basal rosette and root of, 252
  for bleeding, 253
Shoots, identifying, 17–18
Shortleaf pine (yellow pine) (Pinus
    echinata), 218
Shrubs, 18
Silverberry (Elaeagnus commutata), 155
Single-leaf piñon. See One-leafed
    piñon

Sisymbrium officinale. See Hedge
    mustard
Sium suave. See Water parsnip
Skin diseases/infections/irritations,
    36, 39, 41, 73, 75, 139, 163
Skunk cabbage (symplocarpus
    foetidus), 1
Sleep problems, 106
Slender nettle (Urtica gracilis), 243
Small cranberry (Vaccinium
    oxycoccus), 224–225
Smartass. See Smartweeds
Smartweeds (smartass; water
    pepper), 27–28
Smilacina racemosa. See False
    Solomon's seal
Smilax renifolia. See Carrion flower
Smilax species. See Greenbrier
Smooth sumac (Rhus glabra), 116,
    116, 117
Socrates, 204
Sodium, 82
Soldier's woundwort. See Yarrow
Solomon's seal (Polygonatum
    biflorum), 261
Solomon's seal (Polygonatum
    species), 16, 261–264
  berries of, 262
  biblical associations of, 263
  flowers of, 261
  medicinal uses of, 263–264
  root of, 263
Son-before-father. See Coltsfoot
Sonchus arvensis. See Field sow
    thistle
Sonchus asper. See Spiny-leafed sow
    thistle
Sonchus oleraceus. See Common sow
    thistle
Sonchus species. See Sow thistle
Soporifics, 7
Sores, 8, 76, 111
Sore throats, 7, 10, 80, 102, 111,
    123, 171
Sour cherry (Prunus cerasus), 120–121
Southern bayberry. See Common
    wax myrtle
Sow thistle (Sonchus species),
    200–202
  leaves of, 202
  medicinal uses of, 202
  and nitrates, 202
Spanish bayonet (Yucca aloifolia),
    134
Spanish lettuce. See Miner's lettuce
Sparrow grass. See Asparagus
Spearmint (Mentha spicata), 51, 51,
    52
Species, native and exotic, 18–19
Spiceberries, 180–181
Spicebush. See Common spicebush
Spiderwort (Tradescantia virginiana),
    27

Spiny-leafed sow thistle (*Sonchus asper*), 202
Spotted cowbane. *See* Water hemlock
Spotted touch-me-not (*Impatiens biflora*), 74
Sprains, 76
Spraying, with herbicides, 3
Spring. *See* Early spring; Late fall through early spring; Mid-spring; Mid- to late spring
Spurge (*Chamaesyce* species), 29
Squawbush (*Rhus trilobata*), 116–117, **117**
Squier, Tom, 69
Staghorn sumac (*Rhus typhina*), 115–116
Star chickweed (*Stellaria pubera*), 137
Star-of-Bethlehem (*Ornithogalum umbellatum*), 208
Staunchweed. *See* Yarrow
Stellaria media. See Common chickweed
Stellaria pubera. See Star chickweed
Stellaria species. *See* Chickweed
Sterile ray flowers, 15
Stimulants, 8
Stinging nettle (*Urtica dioica*), 241–243, **242**
flowers of (female plant), **243**
shoot of, **242**
Stomach
damage to, 9
ulcers, 7
upset, 93
Stomach cancer, 32
Stonecrop. *See* Orpine
Stoneroot. *See* Horse balm
Storksbill (filaree; alfilaria) (*Erodium cicutarium*), 245, **245**
Strawberry. *See* Common strawberry; Wood strawberry
Strawberry blite (*Chenopodium capitatum*), 46
Stretchberry. *See* Greenbrier
Sugarberry (*Celtis laevigata*), 170
Sugar pine (*Pinus lambertiana*), 218
Sumac (*Rhus* species), 115–119
lemonade, 117–118
medicinal uses of, 118
Summer
cultivated areas in, 89
deserts in, 132–136
disturbed areas in, 90
fields in, 90–96
forests in, 119–128
freshwater wetlands in, 128–131
lawns and meadows in, 89
mountains in, 132
seashore in, 131–132
thickets in, 96–119

Sunburn, 93
Sun choke. *See* Jerusalem artichoke
Sunflower. *See* Common sunflower
Swamp rose mallow (*Hibiscus palustris*), 77
Swamps, 22
Sweet birch. *See* Black birch
Sweetbrier rose (*Rosa eglanteria*), 158
Sweet buckeye (*Aesculus octandra*), 169
Sweet cherry (*Prunus avium*), 15, 120, **120,** 121
flowers of, **120**
Sweet cicely. *See* Aniseroot
Sweet coltsfoot. *See* Western coltsfoot
Sweet gale (*Myrica gale*), 79–80
Sweet kelp. *See* Edible kelp
Sweetroot. *See* Aniseroot
Symplocarpus foetidus. See Skunk cabbage
Syphilis, 221
Syrup, herbal, 9–10

Tall lettuce (*Lactuca canadensis*), 246, **246**
basal rosette of, **246**
flowers of, **246**
Taproot, 16, 17
Taraxacum officinale. See Common dandelion
Taxus baccata. See Asian yew
Taxus brevifolia. See Western yew
Taxus canadensis. See American yew
Taxus cuspidata. See Japanese yew
Taxus species. *See* Yew
Taxus wallichiana. See Himalayan yew
Tea leaf. *See* Wintergreen
Teloxys ambrosioides. See Epazote
Tepary bean (*Phaseolus acutifolius; Phaseolus metcalfei*), 90–91
Testicular cancer, 126
Texas mulberry (*Morus microphylla*), 126
Texas persimmon (*Diospyros texana*), 179
Thickets, 20
in autumn, 151–160
in early spring, 259
in late fall through early spring, 210–212
in mid- to late spring, 56–60
in summer, 96–119
Thimbleberry (*Rubus parviflorus*), 114
Thlaspi arvense. See Field pennycress
Thorns, identifying, 17
Thrice-compound leaves, 13
Thumus serpyllum. See Wild thyme
Thyroid problems, 83
Ticks
deer, 4, 5
dog, 4, 5

Tiger lily (*Lilum tigrinum*), 232
Tilia americana. See American basswood
Tilia europaea. See European linden
Tilia species. *See* Linden
Tincture, as herbal preparation, 9
Titus, Vincent, 197
Tokushima Journal of Experimental Medicine, 83
Tomatillo. *See* Ground cherry
Tomato. *See* Husk tomato
Tonics, 8
Toothed leaves, 12, 13
Toothwort (*Dentaria* species), 257, 273
Toothwort (pepperwort) (*Dentaria diphylla*), 257
Touch-me-not. *See* Jewelweed
Toxicodendron diversilobum. See Poison oak
Toxicodendron quercifolium. See Poison oak
Toxicodendron radicans. See Poison ivy
Tradescantia occidentalis species, 27
Tradescantia pinetorum species, 27
Tradescantia virginiana. See Spiderwort
Tragopogon porrifolius. See Purple goatsbeard
Tragopogon pratensis. See Yellow goatsbeard
Tragopogon species. *See* Goatsbeard
Tree ferns, 32
Trees, 18
Trifolium agrarium. See Hop clover
Trifolium pratense. See Red clover
Trifolium repens. See White clover
Trifolium species. *See* Clovers
Triglycerides, 228
True thistles (*Cirsium* species), 200–201
basal rosette of, **201**
Tuber, 16, **16,** 17
Tuberculosis, 77, 249
Tumor, 264
Tussilago farfara. See Coltsfoot
Twice-compound leaves, 12, **12,** 13
Typha angustifolia. See Narrow-leaved cattail
Typha latifolia. See Common cattail
Typha species. *See* Cattail
Typhoid, 181

Ulcers, stomach, 7
Ulva lactuca. See Sea lettuce
Universal scientific names, for identifying wild plants, 14
Upland hackberry (*Celtis tenuifolia*), 170
Urethritis, 212
Urinary stones, 117, 192

Urinary-tract infection, 100, 167, 225
Urinary-tract problems, 61, 79, 197
Urtica dioica. See Stinging nettle
Urtica gracilis. See Slender nettle
Urtica species. *See* Nettles

Vaccinium angustifolium. See Lowbush blueberry
Vaccinium corymbosum. See Highbush blueberry
Vaccinium macrocarpon. See Large cranberry
Vaccinium oxycoccus. See Small cranberry
Vaccinium species. *See* Blueberries; Cranberries; Huckleberries
Vaccinium vitis-idaea. See Mountain cranberry
Vaginal infections, 9, 41
Verbascum thapsus. See Mullein
Viburnum acerifolium. See Maple-leaf viburnum
Viburnum alnifolium. See Hobblebush
Viburnum cassinoides. See Northern wild raisins
Viburnum lentago. See Nannyberries
Viburnum opulus. See Cramp bark
Viburnum pauciflorum. See Cranberry tree
Viburnum prunifolium. See Black haw
Viburnums (*Viburnum* species), 15, 181–185
   arrowwood group of, 183
   barks of, 184
   black haw group of, 183
   maple-leaf group of, 182–183
   medicinal uses of, 184
   wayfaring tree group of, 183
Viburnum species. *See* Viburnums
Viburnum trilobum. See Highbush cranberry
Viceroy butterfly, 50
Vine apricot. *See* Passionflower
Vines, 18
Viola cucullata. See Marsh blue violet
Viola papilionacea. See Common blue violet
Viola pedata. See Birdfoot viola
Violet. *See* Common blue violet
Violet wood sorrell (*Oxalis violacea*), 30
Virginia creeper (*Parthenocissus quinquefolia*), 165
Virginia dayflower (*Commelina virginica*), 26
Virginia waterleaf (*Hydrophyllum virginianum*), 266, **267**
Vitamin A, 206
Vitamin B₁, 4, 32
   See also B-complex vitamins

Vitamin C, 6, 40, 50, 51, 56, 65, 85, 93, 100, 102, 117, 118, 121, 134, 180, 183, 188, 199, 202, 219, 223, 239, 268
Vitamin E, 9, 75, 158, 273
Vitamin K, 93, 115, 245
Vitamin levels, minimal vs. optimal, 6
Vitis labrusca. See Fox grape
Vitis species. *See* Grapes
Vodka, in herbal preparations, 9
Vulneraries, 8

Wakame. *See* Edible kelp
Walnuts, 15
   See also Arizona walnut; Black walnut; California walnut; Hind's walnut
Washing wild plants, 3
Wasps, 15, 74
Water chestnuts, 196, 269
Watercress (*Nasturtium officinale*), 3, 19, 47, 255–256, **255**, 271
   gathering, 256
   and infection by pathogenic microorganisms, 255
   medicinal uses of, 256
   vitamins and minerals in, 256
Water hemlock (spotted cowbane) (*Cicuta maculata*), 43, 67, 128–129, **129**, 214, 245
   poisons in, 128, 129
Waterleaf (Indian salad) (*Hydrophyllum* species), 266–267
   leaves of, 266
Water lilies (*Nymphaca* species), 1
Water mint (*Mentha aquatica*), 52
Water parsnip (*Sium suave*), 129
Water pepper. *See* Smartweeds
Water plantains (*Alisma* species), 228
Wayfaring tree. *See* Hobblebush
Weeds, 19
   habitats of, 11
Western coltsfoot (sweet coltsfoot) (*Petasites speciosa*), 73
Western prickly pear (*Opuntia phaeacantha*), 187, 188
   with fruits, **188**
Western wintergreen. *See* Salal wintergreen
Western yew (*Taxus brevifolia*), 144
Wetlands, 22
   See also Freshwater wetlands
White blood cells, 41
White clover (*Trifolium repens*), 24, **24**
White mulberry (*Morus alba*), 126–127
   importation of, 127
White oak (*Quercus alba*), **174**, 175
   acorns of, **174**, **175**, 176

bark of, 176
group, 174–176
vitamins and minerals in, 176
White pine (*Pinus strobus*), 217, **217**
White pines, 102–103
White sweet clover (*Melilotus alba*), **25**, 26
Whole-grain flours, 274
Whole-wheat flour, 274
Wild allspice. *See* Common spicebush
Wild apples, 272
   See also Apple
Wild bean (wild kidney bean) (*Phaseolus polystachios*), 90–91, **90**
   pods of, 91
   poisons in, 91
Wild carrot (Queen Anne's lace) (*Daucus carota*), 129, 191, 203–206, 273
   basal rosette and taproot of, 203, 204, **204**, 205
   flower head of, 94
   medicinal uses of, 205–206
   tea, 205–206
Wild chervil. *See* Honewort
Wild chive (*Allium schoenoprasum*), 207
Wild cucumber. *See* Passionflower
Wild filberts. *See* Hazelnuts
Wild foods, value of, 5–6
Wild garlic (*Allium canadense*), 207
   flowers of, **208**
   medicinal uses of, 209, 265
Wild ginger (Indian ginger) (*Asarum* species), 221–222, 273
   flowers and rhizome of, **221**
   medicinal uses of, 222
Wild grapes. *See* Grapes
Wild kidney bean. *See* Wild bean
Wild leek (ramp) (*Allium tricoccum*), 264–266, 272, 273
   leaves of, 264
   medicinal uses of, 265
   with seeds, **265**, **266**
   shoot and bulb of, **264**
Wild lettuce (*Lactuca* species), 190, 200, 234, 235, 246–247
   and commercial lettuce, 246
   medicinal uses of, 247
Wild mint (*Mentha arvensis*), 53, **53**
Wild mustard. *See* Field mustard
Wild onions (*Allium* species), 206–209, 265, 272, 273
   collecting, 209
   look-alikes of, 206, 207–208
   medicinal uses of, 209
   vitamins and minerals in, 209
Wild parsnip, 129, 273
Wild plums (*Prunus* species), 100, 110–112, 119
   medicinal uses of, 111

pitted, 110
vitamins and minerals in, 110–111
Wild potato vine (man-of-the-earth; man-under-ground) (*Ipomoea pandurata*), 202–203
flowers of, **202**, 203
leaves of, 202
roots of, 272
Wild raisin. *See* Northern wild raisin
Wild red raspberry (*Rubus strigosus*), 113, **114**
Wild rice (*Zizania aquatica*), 129–130, **130**, 273
poisons in, 130
vitamins and minerals in, 130
Wild rose, 14
See also Pasture rose
Wild succory. *See* Chicory
Wild thyme (*Thumus serpyllum*), 54–55, **55**
honey, 55
medicinal uses of, 55
oil of, 55
and pregnancy, 55
vitamins in, 55
Willow (*Salix* species) bark tea, 7
Wineberry (*Rubus phoenicolasius*), 98, 113, **113**
flowers of, **112**
Winged kelp. *See* Edible kelp
Winged sumac. *See* Dwarf sumac
Winter annuals, 18
Winter cress (*Barbarea vulgaris*), 250–251
basal rosette of, **250**
flowers of, 251, **251**
look-alikes of, 251
Wintergreen (checkerberry; tea leaf; deerberry) (*Gaultheria procumbens*), 222–224

berries, **222**, 223, 272
flowers of, **223**
and Lenape Indian legend, 224
medicinal uses of, 223–224
oil of, 215, 223, 224
origin of, 224
poisons in, 224
tea, 223
Wister (and wisteria), 59
Wisteria (*Wisteria* species), 15, 58–60, 210, 272
flowers of, 59
pod of, 59, **59**, 60
poisons in, 59–60
vine of, **59**, 61
Wisteria species. *See* Wisteria
Witches'-broom, 170
Witches' hobble. *See* Hobblebush
Woodlands, 20
in autumn, 160–185
in early spring, 259–266
in late fall through early spring, 212–224
in mid- to late spring, 60–66
Wood nettle (*Laportea canadensis*), 243, **243**
Wood sorrel (shamrock) (*Oxalis* species), 15, 29–30, **30**
distinctiveness of, 30
history of, 30
medicinal uses of, 30
taste-alike of, 29
vitamins and minerals of, 30
Wood strawberry (*Fragaria vesca*), 92
Worms, 181
Wormseed. *See* Epazote
Wrack. *See* Rockweed
Wrinkled rose (*Rosa rugosa*), 157–158
flowers of, **158**
fruit of, **159**

Yarrow (staunchweed; soldier's woundwort) (*Achillea millefolium*), 15, 94–95, **95**
as composite, 94
flower heads of, 94
legendary history of, 95
medicinal uses of, 95
toxicity of, 95
Yeast infections, 229
Yellow birch (*Betula lutea*), 215
Yellow dock. *See* Curly dock
Yellow goatsbeard (*Tragopogon pratensis*), 197
Yellow pine. *See* Shortleaf pine
Yellow sweet clover (*Melilotus officinalis*), **25**, 26
Yew (*Taxus* species), 143–144
flowers of, **143**
fruit of, 144, **144**
medicinal uses of, 144
and Native Americans, 144
needles of, 143, 144, 217
poisons in, 143–144, 217
seeds of, 144
taxol in, 144
Yucca (beargrass) (*Yucca filamentosa*), 134–135, **134**
and fires, starting, 134
flower stalk of, **135**
poisons in, 134
roots of, 134
Yucca aloifolia. See Spanish bayonet
Yucca filamentosa. See Yucca

Zizania aquatica. See Wild rice
Zygadenus species. *See* Death camas

INDEX OF WILD FOOD RECIPES

Acorns. *See* White-oak acorns
Amaranth, nutty, 279
Appetizers and side dishes, 277–281
   black locust flower oatmeal, 277
   dandelion sauté, 277–278
   Dena and Gabi's cattail flower pickles, 278
   herbed Jerusalem artichokes, 278
   lamb's-quarters stuffing, 278
   lentil delight, 278–279
   nettle and carrot casserole, 279
   nutty amaranth, 279
   park nuts, 279
   pokeweed and milkweed, 279–280
   rockweed crisps, 280
   salsify roots in nut sauce, 280
   stuffed wild grape leaves, 280–281
   waterless-steamed greens, 281
   wild carrot seed millet, 277
Apples
   in Scandinavian fruit soup, 283
   wild cake, 295
   wild chutney, 291
Apricots, in Marillen knödel, 293–294
Autumn green salad, 281
Autumn olive–banana tofu creme pie, 291–292
Autumn olive jam, 288

Bananas, with autumn olive jam and tofu in creme pie, 291–292
Berries. *See* Wild berries
Blackberry peach crunch, 292
Black cherry
   concentrate, in mulberry pudding, 294
   jam, 289
Black locust
   blossoms, in mid-spring wild salad, 282
   flower oatmeal, 277
Black walnut carob-chip cookies, 292
Blueberry bread, 284
Bread crumbs, in savory burdock patties, 288
Breads, 284–285
   blueberry, 284
   clover corn, 284
   corn and daylily flat, 284
   dulse scones, 285
   Irish soda elderberry, 285
   pawpaw, 285
   plantain seed-quinoa pilaf, 284
Burdock, savory patties, 288
Burgers, common evening primrose leaf, 286

Cake, wild apple, 295
Carob-chip black walnut cookies, 292

Carob-pecan fudge, 292
Carrots
   with cattail shoots in peanut sauce, 286
   and nettle casserole, 279
Casserole, nettle and carrot, 279
Cattail flowers, Dena and Gabi's pickles from, 278
Cattail shoots and carrots in peanut sauce, 286
Cheese rolls, with curly dock, in tomato sauce, 287
Chestnuts, in carob-pecan fudge, 292
Chick-peas, with violets in early spring soup, 283
Chickweed
   creamy dressing, 289
   and tofu, curried, 287
Chili, common evening primrose root, 286–287
Clover corn bread, 284
Coconut–curly dock curry, 286
Common evening primrose
   leaf burgers, 286
   root chili, 286–287
Cookies, carob-chip black walnut, 292
Corn
   clover bread, 284
   and daylily flat bread, 284
Crab apples, in wild apple cake, 295

Cream cheese. *See* Tofu cream
  cheese
Creamy chickweed dressing, 289
Creamy wild raspberry pudding,
  293
Crepes with wild berries, 293
Curly dock
  cheese rolls in tomato sauce, 287
  with coconut and curry, 286
Curried tofu and chickweed, 287
Curry
  with coconut and curly dock, 286
  oil, 289
  with tofu and chickweed, 287
  with white-oak acorns, 279

Dandelion sauté, 277–278
Daylily and corn flat bread, 284
Dena and Gabi's cattail flower
  pickles, 278
Desserts, 291–296
  autumn olive–banana tofu creme
    pie, 291–292
  blackberry peach crunch, 292
  carob-chip black walnut cookies,
    292
  carob-pecan fudge, 292
  creamy wild raspberry pudding,
    293
  crepes with wild berries, 293
  Japanese knotweed surpise, 293
  Marillen knödel, 293–294
  mulberry crumble, 294
  mulberry pudding, 294
  persimmon American pudding,
    294
  tofu-plum pie, 294–295
  wild apple cake, 295
  wild strawberry parfait, 295–
    296
Dracula's delight—prickly pear
  sauce, 289
Dressings. *See* Jams, dressings,
  spreads, and sauces
Dulse scones, 285
Dumplings. *See* Marillen knödel

Early spring chick-pea and violet
  soup, 283
Early spring seasonal salad, 281
Elderberry bread, Irish soda, 285
Entrées, 286–288
  cattail shoots and carrots in
    peanut sauce, 286
  coconut–curly dock curry, 286
  common evening primrose leaf
    burgers, 286
  common evening primrose root
    chili, 286–287
  curly dock–cheese rolls in tomato
    sauce, 287
  curried tofu and chickweed, 287
  potato-purslane patties, 288

savory burdock patties, 288
sunflower patties, 288

Filling
  for autumn olive–banana tofu
    creme pie, 291–292
  for blackberry peach crunch, 292
  for mulberry crumble, 294
Fudge, carob-pecan, 292

Garlic mustard horseradish, 289
Garlic mustard tofu cream cheese,
  290
Glasswort spread, with tofu,
  290–291
Grapes
  in Scandinavian fruit soup, 283
  stuffed leaves of wild, 280–281
Greens
  autumn, in salad, 281
  late fall, in salad, 281–282
  summer, in salad, 282
  waterless-steamed, 281
  winter, in salad, 282–283

Herbed Jerusalem artichokes, 278
Hollandaise sauce, sumac, 290
Horseradish, garlic mustard, 289

Icing, for wild apple cake, 295
Irish soda elderberry bread, 285

Jams, dressings, spreads, and
  sauces, 288–291
  autumn olive jam, 288
  black cherry jam, 289
  creamy chickweed dressing, 289
  curry oil, 289
  Dracula's delight—prickly pear
    sauce, 289
  garlic mustard horseradish, 289
  garlic mustard–tofu cream
    cheese, 290
  poor man's mustard, 290
  sassafras root jelly, 290
  sumac hollandaise sauce, 290
  tofu-glasswort spread, 290–291
  wild apple chutney, 291
  wild onion sauce, 291
Japanese knotweed surprise, 293
Jelly, sassafras root, 290
Jerusalem artichokes, herbed, 278

Knotweed. *See* Japanese knotweed

Lamb's-quarters
  in seashore salad, 282
  stuffing, 278
Late fall green salad, 281–282
Layers (top and bottom), of tofu-
  plum pie, 294–295
Lecithin (liquid), in crepes with
  wild berries, 293

Lentil delight, 278–279
Lettuce. *See* Sea lettuce salad

Marillen knödel, 293–294
Mid-spring wild blossom salad, 282
Milkweed
  flower salad, 282
  and pokeweed, 279–280
Millet
  in nutty amaranth, 279
  with wild carrot seeds, 277
Mulberry crumble, 294
Mulberry pudding, 294
Mustard, poor man's, 290
  See also Garlic mustard

Nettle and carrot casserole, 279
Nuts, park, 279
Nut sauce, salsify roots in, 280
Nutty amaranth, 279

Oatmeal, with black locust flowers,
  277
Oil, curry, 289
Olives, autumn jam, 288
Onions. *See* Wild onions

Pancakes. *See* Crepes
Park nuts, 279
Patties
  purslane and potato, 288
  savory burdock, 288
  sunflower, 288
Pawpaw bread, 285
Peach, blackberry crunch, 292
Peanut sauce, with cattail shoots
  and carrots, 286
Pear. *See* Prickly pear
Pecan and carob fudge, 292
Peppers, wild stuffed, 287
Persimmon American pudding, 294
Pickles, Dena and Gabi's cattail
  flower, 278
Pie, autumn olive–banana tofu
  creme, 291–292
Pilaf, plantain seed and quinoa, 284
Plantain seed–quinoa pilaf, 284
Plum and tofu pie, 294–295
Pokeweed, and milkweed, 279–280
Poor man's mustard, 290
Potato-purslane patties, 288
Prickly pear sauce—Dracula's
  delight, 289
Primrose. *See* Common evening
  primrose
Pudding
  creamy wild raspberry, 293
  mulberry, 294
  persimmon American, 294
Purslane and potato patties, 288

Quinoa and plantain seed pilaf,
  284

Raspberry. *See* Wild raspberry
Rockweed crisps, 280

Salads, 281–283
    autumn green, 281
    early spring seasonal, 281
    late fall green, 281–282
    mid-spring wild blossom, 282
    milkweed flower, 282
    sea lettuce, 282
    seashore, 282
    summer green, 282
    winter green, 282–283
Salsify roots in nut sauce, 280
Sassafras root jelly, 290
Sauce
    nut, with salsify roots, 280
    peanut, with cattail shoots and
        carrots, 286
    prickly pear—Dracula's delight,
        289
    for stuffed wild grape leaves,
        281
    sumac Hollandaise, 290
    tomato, with curly dock and
        cheese, 287
    wild onion, 291
    See also Jams, dressings, spreads,
        and sauces

Savory burdock patties, 288
Scandinavian fruit soup, 283
Scones, dulce, 285
Sea lettuce salad, 282
Sea rocket, in seashore salad, 282
Seashore salad, 282
Side dishes. *See* Appetizers and side
    dishes
Soups, 283
    early spring chick-pea and violet,
        283
    Scandinavian fruit, 283
Spreads. *See* Jams, dressings,
    spreads, and sauces
Strawberries. *See* Wild strawbery
    parfait
Stuffed wild grape leaves, 280–281
Stuffing, lamb's-quarters, 278
Sumac Hollandaise sauce, 290
Summer green salad, 282
Sunflower patties, 288

Tofu
    with autumn olive jam and
        bananas in creme pie,
        291–292
    and chickweed, curried, 287
    cream cheese, with garlic
        mustard, 290

and glasswort spread, 290–291
    and plum pie, 294–295
Tomato sauce, curly dock–cheese
    rolls in, 287
Topping
    for blackberry peach crunch, 292
    for mulberry crumble, 294

Violets, with chick-peas in early
    spring soup, 283

Waterless-steamed greens, 281
White-oak acorns, with curry, 279
Wild apples. *See* Apples
Wild berries, crepes with, 293
Wild blossom salad, mid-spring, 282
Wild carrot seed millet, 277
Wild grapes. *See* Grapes
Wild onion sauce, 291
Wild raspberry creamy pudding,
    293
Wild strawberry parfait, 295–296
Wild stuffed peppers, 287
Winter green salad, 282–283
Wisteria blossoms, in mid-spring
    wild salad, 282

Yogurt, in nettle and carrot
    casserole, 279

ABOUT THE AUTHORS

"Wildman" Steve Brill was born in 1949. He's a self-taught naturalist with a B.A. in psychology from The George Washington University. He first learned to identify and use wild foods as a hobby in the late 1970s—an offshoot of his interest in gourmet vegetarian cooking.

He's best known for having been arrested by undercover NYC park rangers for eating a dandelion while leading an educational tour of Central Park in 1986. The ensuing flood of national publicity caused the city to drop the charges and hire him to lead foraging tours in the New York City parks.

He's currently working free-lance again, leading public wild food and ecology tours in the surprisingly rich parks in and around New York City. He also leads and provides hands-on tours and indoor presentations for schoolchildren and other groups. He does environmental radio broadcasts on WBAI Radio, and TV shows on Queens Public Access TV and other local TV stations. His botanical illustrations and lifelike models of wild mushrooms have been on exhibition throughout the Northeast.

Evelyn Dean is an artist, writer, naturalist, and gourmet chef. She attended the High School of Art and Design, and has a B.A. in art and literature from Queens College. She began collaborating with "Wildman" Steve Brill in 1985. She's led public wild food and ecology tours throughout the Northeast and run a children's environmental program at the Prospect Park Environmental Center. She also produces and directs environmental TV shows with "Wildman" Steve Brill and other members of the Queens Green Party.